ENERGY METABOLISM IN FARM ANIMALS

CURRENT TOPICS IN VETERINARY MEDICINE AND ANIMAL SCIENCE

Control of Reproduction in the Cow, edited by J.M. Sreenan

Patterns of Growth and Development in Cattle, edited by H. de Boer and J. Martin

Respiratory Diseases in Cattle, edited by W.B. Martin

Calving Problems and Early Viability of the Calf, edited by B. Hoffmann, I.L. Mason and J. Schmidt

The Future of Beef Production in the European Community, edited by J.C. Bowman and P. Susmel

Diseases of Cattle in the Tropics: Economic and Zoonotic Relevance, edited by M. Ristic and I. McIntyre

Control of Reproductive Functions in Domestic Animals, edited by W. Jöchle and D.R. Lamond

The Laying Hen and its Environment, edited by R. Moss

Epidemiology and Control of Nematodiasis in Cattle, edited by P. Nansen, R.J. Jørgensen and E.J.L. Soulsby

The Problem of Dark-Cutting in Beef, edited by D.E. Hood and P.V. Tarrant

The Welfare of Pigs, edited by W. Sybesma

The Mucosal Immune System, edited by F.J. Bourne

Laboratory Diagnosis in Neonatal Calf and Pig Diarrhoea, edited by P.W. de Leeuw and P.A.M. Guinée

Advances in the Control of Theileriosis, edited by A.D. Irvin, M.P. Cunningham and A.S. Young

Fourth International Symposium on Bovine Leukosis, edited by O.C. Straub

Muscle Hypertrophy of Genetic Origin and its Use to Improve Beef Production, edited by J.W.B. King and F. Ménissier

Aujeszky's Disease, edited by G. Wittman and S.A. Hall

Transport of Animals Intended for Breeding, Production and Slaughter, edited by R. Moss

Welfare and Husbandry of Calves, edited by J.P. Signoret

Factors Influencing Fertility in the Postpartum Cow, edited by H. Karg and E. Schallenberger

Beef Production from Different Dairy Breeds and Dairy Beef Crosses, edited by G.J. More O'Ferrall

The Elisa: Enzyme-Linked Immunosorbent Assay in Veterinary Research and Diagnosis, edited by R.C. Wardley and J.R. Crowther

Indicators Relevant to Farm Animal Welfare, edited by D. Smidt

Farm Animal Housing and Welfare, edited by S.H. Baxter, M.R. Baxter and J.A.D. MacCormack

Stunning of Animals for Slaughter, edited by G. Eikelenboom

Manipulation of Growth in Farm Animals, edited by J.F. Roche and D. O'Callaghan

Latent Herpes Virus Infections in Veterinary Medicine, edited by G. Wittmann, R.M. Gaskell and H.-J. Rziha

Grassland Beef Production, edited by W. Holmes

Recent Advances in Virus Diagnosis, edited by M.S. McNulty and J.B. McFerran

The Male in Farm Animal Reproduction, edited by M. Courot

Endocrine Causes of Seasonal and Lactational Anestrus in Farm Animals, edited by F. Ellendorff and F. Elsaesser

Brucella Melitensis, edited by J.M. Verger and M. Plommet

Diagnosis of Mycotoxicoses, edited by J.L. Richard and J.R. Thurston

Embryonic Mortality in Farm Animals, edited by J.M. Sreenan and M.G. Diskin

Social Space for Domestic Animals, edited by R. Zayan

The Present State of Leptospirosis Diagnosis and Control, edited by W.A. Ellis and T.W.A. Little

Acute virus infections of poultry, edited by J.B. McFerran and M.S. McNulty

Evaluation and Control of Meat Quality in Pigs, edited by P.V. Tarrant, G. Eikelenboom and G. Monin

Follicular Growth and Ovulation Rate in Farm Animals, edited by J.F. Roche and D. O'Callaghan

Cattle Housing Systems, Lameness and Behaviour, edited by H.K. Wierenga and D.J. Peterse

Physiological and Pharmacological Aspects of the Reticulo-Rumen, edited by L.A.A. Ooms, A.D. Degryse and A.S.J.P.A.M. van Miert

Biology of Stress in Farm Animals: An Integrative Approach, edited by P.R. Wiepkema and P.W.M. van Adrichem

Helminth Zoonoses, edited by S. Geerts, V. Kumar and J. Brandt

Energy Metabolism in Farm Animals, edited by M.W.A. Verstegen and A.M. Henken

ENERGY METABOLISM IN FARM ANIMALS

Effects of housing, stress and disease

Edited by

M.W.A. Verstegen and A.M. Henken
Department of Animal Husbandry, Agricultural University Wageningen,
Wageningen, The Netherlands

Springer-Science+Business Media, B.V.

1987

Distributors

for the United States and Canada: Kluwer Academic Publishers, P.O. Box 358, Accord Station, Hingham, MA 02018-0358, USA
for the UK and Ireland: Kluwer Academic Publishers, MTP Press Limited, Falcon House, Queen Square, Lancaster LA1 1RN, UK
for all other countries: Kluwer Academic Publishers Group, Distribution Center, P.O. Box 322, 3300 AH Dordrecht, The Netherlands

Library of Congress Cataloging in Publication Data

Library of Congress Cataloging-in-Publication Data

Energy metabolism in farm animals.

    (Current topics in veterinary medicine and
animal science)
    1. Veterinary physiology. 2. Energy metabolism.
3. Livestock--Ecology. 4. Livestock--Diseases.
I. Verstegen, M. W. A. II. Henken, A. M. III. Series.
SF768.E54 1987   636.089'2   87-22079

ISBN 978-94-010-8010-1   ISBN 978-94-009-3363-7 (eBook)
DOI 10.1007/978-94-009-3363-7

Copyright

CONTENTS

	Page
Preface	IX
List of contributors	XI

CHAPTER I. INTRODUCTION

Energy metabolism of farm animals.
A.J.H. van Es and H.A. Boekholt....................................3

The Wageningen respiration unit for animal production research:
a description of the equipment and its possibilities.
M.W.A. Verstegen, W. van der Hel, H.A. Brandsma,
A.M. Henken and A.M. Bransen....................................21

CHAPTER II. HOUSING-SYSTEMS AND ENERGY METABOLISM

Adaptation to, and energy costs of, tethering in pregnant sows.
G.M. Cronin and J.L. Barnett....................................51

Metabolic rate of piglets between sucklings.
W. van der Hel and M.W.A. Verstegen............................63

Influence of some environmental, animal and feeding factors
on energy metabolism in growing pigs.
M.W.A. Verstegen, A.M. Henken and W. van der Hel...............70

The effects of housing conditions on energy utilization of
poultry.
E.H. Ketelaars....................................87

CHAPTER III.   CLIMATIC CONDITIONS AND ENERGY
METABOLISM

Surface temperatures as parameters.
R. Geers, W. van der Hel and V. Goedseels........................105

The influence of climatic environment on sows.
B. Kemp and M.W.A. Verstegen...................................115

Thermal requirements of growing pigs from birth to slaughter.
M.W.A. Verstegen, A.M. Henken, W. van der Hel and
H.A. Brandsma..133

A formula to describe the relation between heat production at
thermoneutral as well as below thermoneutral temperatures simultaneously.
G.F.V. van der Peet, M.W.A. Verstegen and W.J. Koops...........150

Effect of environmental temperature and air velocity two days
preslaughtering on heat production, weight loss and meat
quality in non-fed pigs.
E. Lambooy, W. van der Hel, B. Hulsegge and H.A. Brandsma.....164

Effects of climatic conditions on energy metabolism and
performance of calves.
M. Vermorel..180

Climatic conditions and energy metabolism of laying hens.
M. van Kampen...199

Climatic environment and energy metabolism in broilers.
C.W. Scheele, W. van der Hel, M.W.A. Verstegen and
A.M. Henken..217

Heat tolerance of one-day old chickens with special reference
to conditions during airtransport.
A.M. Henken, W. van der Hel, A. Hoogerbrugge and
C.W. Scheele...261

CHAPTER IV.   HEALTH AND ASPECTS OF ENERGY METABOLISM

Energy metabolism and immune function.
J.M.F. Verhagen...288

Parasite worry and restlessness caused by sarcoptic mange
in swine.
M.W.A. Verstegen, J. Guerrero, A.M. Henken,
W. van der Hel and J.H. Boon.......................................304

Respiratory diseases in pigs: incidence, economic losses and
prevention in the Netherlands.
M.J.M. Tielen...321

Mastitis in dairy cows with special reference to direct and
indirect effects of climatological factors.
F.J. Grommers...337

The effect of gastrointestinal nematodes on metabolism in
calves.
A. Kloosterman and A.M. Henken.....................................352

Energy and nitrogen metabolism of growing calves continuously
infected with Dictyocaulus viviparus.
J.H. Boon and M.W.A. Verstegen.....................................372

Respiratory diseases in calves.
P. Franken, C. Holzhauer and L.A. van Wuijckhuise-Sjouke.........388

The effect of a subclinical Haemonchus infection on the
metabolism of sheep (a pilot study).
P.W.M. van Adrichem, M.J.N. Los, J.E. Vogt and Y. Wetzlar†......400

Coccidiosis: a problem in broilers.
A.C. Voeten...410

VIII

CHAPTER V. VARIATION IN ENERGY METABOLISM CHARACTERISTICS DUE TO FEEDING LEVEL AND DIFFERENCES BETWEEN BREEDS/STRAINS

Effect of feeding level on maintenance requirements of growing pigs.
C.P.C. Wenk and M. Kronauer.....................................425

Genetic variation of energy metabolism in poultry.
P. Luiting...440

Genetic variation of energy metabolism in mice.
E.J. van Steenbergen..467

Effects of body weight, feeding level and temperature on energy metabolism and growth in fish.
L.T.N. Heinsbroek...478

PREFACE

Animal production systems have changed dramatically over the last two decades. Knowledge of energy metabolism and environmental physiology has increased as appears from many textbooks on these disciplines. The contents of the symposia on energy metabolism of farm animals show this and they have initially focussed on feed evaluation and later on comparative aspects of energy metabolism. They show part of the progress being made.

Application of knowledge of energy metabolism for animals has a long history since Lavoisier. In addition to this, studies about the environmental requirements of animals have shown that we are still far from accurate assessment of these requirements in terms of nutrients and energy. In model studies on energy metabolism researchers have recognized the interaction between the environment and the energy requirements of animals. Estimation of energy requirements has been done in physiological, physical and behavioural studies. The impact of conditions as encountered by animals in various production systems has been approached from different viewpoints related to these different disciplines. In addition, various kinds of infections (bacterial, parasitic: subclinical, clinical) have been evaluated only recently with regard to their effect on protein and/or energy metabolism and thus on production.

People working in the field of feed evaluation have defined how chemical and physical properties of nutrition influence energy to be derived for maintenance and production. The physiologists have defined and studied the environment with regard to the energy required by animals for body temperature regulation. However these studies have been made at standardized conditions which are not often encountered as such by animals in practice. Physiologists have also evaluated and described processes related to growth and production. They have quantified the necessary components in terms of energy and protein. Empirically these findings have been applied to farm conditions. However it is clear that there exists a considerable gap between performance of animals kept at more or less laboratory conditions and those producing on a commercial farm. Also genetic changes in the animal population by selection and in future also by genetic engineering, are important in this respect. Genetic differences may partly explain differences between animals in nutrient and energy requirements.

The present book emphasizes on the impact of environmental conditions (housing, climatic, parasitic) on energy metabolism and its consequences for energy requirement and performances. The studies presented are aimed at measuring such effects as they differ from studies on feed evaluation or comparative aspects of energy metabolism. The results of this kind of studies need to be appraised economically and also for feeding strategies and/or housing and management systems. The changes in requirements have to be known to evaluate modern housing systems to alter these changes in order to optimize animal production. In order to study these aspects large experimental facilities are needed since practical conditions have to be simulated. Moreover often groups of animals have to be studied together. This book focusses on this kind of interdisciplinary studies. The reader is provided with experiences from studies which will normally not appear together in one book. Mostly results of such studies are reported in a monodisciplinary oriëntated book. The relevance of the interdisciplinary approach of energy metabolism may be self-evident. The book is meant for researchers in this field and for students especially interested in energy metabolism with special reference to animal production.

In addition this book is published to mark the occasion that about 20 years ago these kinds of studies were made possible with the development of large indirect calorimeters at the Departments of Animal Production of the Agricultural University at Wageningen (The Netherlands).

M.W.A. Verstegen   Wageningen, September 1987

A.M. Henken

LIST OF CONTRIBUTORS

P.W.M. van Adrichem   Department of Animal Physiology
Agricultural University
WAGENINGEN (The Netherlands)

J.L. Barnett   Animal Research Institute
Department of Agriculture and
Rural Affairs
Werribee
VICTORIA 3030 (Australia)

H.A. Boekholt   Department of Animal Physiology
Agricultural University
WAGENINGEN (The Netherlands)

J.H. Boon   Department of Fish Culture and Fisheries
Agricultural University
P.O. Box 338
WAGENINGEN (The Netherlands)

H.A. Brandsma   Department of Animal Husbandry
Agricultural University
P.O. Box 338
WAGENINGEN (The Netherlands)

A.M. Bransen   Technical and Physical Engineering
Research Service
Ministry of Agriculture and Fisheries
P.O. Box 356
WAGENINGEN (The Netherlands)

G.M. Cronin

Animal Research Institute
Department of Agriculture and
Rural Affairs
Werribee
VICTORIA 3030 (Australia)

A.J.H. van Es

Department of Animal Physiology
Agricultural University
WAGENINGEN (The Netherlands)

P. Franken

Animal Health Service
Gelderland
P.O. Box 10
6880 BD VELP (The Netherlands)

R. Geers

Laboratory for Agricultural Building
Research, University Leuven
De Croylaan 42
B-3030 HEVERLEE (Belgium)

V. Goedseels

Laboratory for Agricultural Building
Research, University Leuven
De Croylaan 42
B-3030 HEVERLEE (Belgium)

F.J. Grommers

Department of Animal Husbandry
Faculty of Veterinary Medicine
University of Utrecht
P.O. Box 80 156
UTRECHT (The Netherlands)

J. Guerrero

MSD AGVET   Technical Services
P.O. Box 2000
Rahway
NEW JERSEY 07065-0912 (USA)

L.T.N. Heinsbroek
Department of Fish Culture and Fisheries
Agricultural University
P.O. Box 338
WAGENINGEN (The Netherlands)

W. van der Hel
Department of Animal Husbandry
Agricultural University
P.O. Box 338
WAGENINGEN (The Netherlands)

A.M. Henken
Department of Animal Husbandry
Agricultural University
P.O. Box 338
WAGENINGEN (The Netherlands)

C. Holzhauer
Animal Health Service
Gelderland
P.O. Box 10
6880 BD VELP (The Netherlands)

A. Hoogerbrugge
Department of Animal Husbandry
Agricultural University
P.O. Box 338
WAGENINGEN (The Netherlands)

B. Hulsegge
Research Institute for Animal
Production "Schoonoord"
P.O. Box 501
ZEIST (The Netherlands)

M. van Kampen
Department of Veterinary Physiology
Faculty of Veterinary Medicine
University of Utrecht
P.O. Box 80 178
UTRECHT (The Netherlands)

B. Kemp

Department of Animal Nutrition
Agricultural University
WAGENINGEN (The Netherlands)

E.H. Ketelaars

Department of Animal Breeding
Agricultural University
P.O. Box 338
WAGENINGEN (The Netherlands)

A. Kloosterman

Department of Animal Husbandry
Agricultural University
P.O. Box 338
WAGENINGEN (The Netherlands)

W.J. Koops

Department of Animal Breeding
Agricultural University
P.O. Box 338
WAGENINGEN (The Netherlands)

M. Kronauer

Institut für Nütztierwissenschaften
Gruppe Ernährung
ETH ZÜRICH (Switzerland)

E. Lambooy

Research Institute for Animal
Production "Schoonoord"
P.O. Box 501
ZEIST (The Netherlands)

M.J.N. Los

Department of Animal Physiology
Agricultural University
WAGENINGEN (The Netherlands)

P. Luiting

Department of Animal Breeding
Agricultural University
P.O. Box 338
WAGENINGEN (The Netherlands)

G.F.V. van der Peet

Institute for Livestock Feeding and
Nutrition Research (IVVO)
LELYSTAD (The Netherlands)

C.W. Scheele

Spelderholt Centre for Poultry
Research and Extension
BEEKBERGEN (The Netherlands)

E.J. van Steenbergen

Department of Animal Breeding
Agricultural University
P.O. Box 338
WAGENINGEN (The Netherlands)

M.J.M. Tielen

Department of Animal Husbandry
Faculty of Veterinary Medicine
University of Utrecht
P.O. Box 80 156
UTRECHT (The Netherlands)
and

Animal Health Service
Noord-Brabant
Molenwijkseweg 48
5282 SC BOXTEL (The Netherlands)

J.M.F. Verhagen

Department of Animal Husbandry
Agricultural University
P.O. Box 338
WAGENINGEN (The Netherlands)

M. Vermorel

Laboratoire d'Etude du Métabolisme
Energétique
I.N.R.A. Theix
63122 CEYRAT (France)

M.W.A. Verstegen

Departments of Animal Nutrition
and Animal Husbandry
P.O. Box 338
WAGENINGEN (The Netherlands)

A.C. Voeten

Animal Health Service
Noord-Brabant
Molenwijkseweg 48
5282 SC BOXTEL (The Netherlands)

J.E. Vogt

Department of Animal Physiology
Agricultural University
WAGENINGEN (The Netherlands)

C.P.C. Wenk

Institut für Nutztierwissenschaften
Gruppe Ernährung
ETH ZÜRICH (Switzerland)

Y. Wetzlar†

Former staff member of the
Department of Parasitology
Central Veterinary Institute
LELYSTAD (The Netherlands)

L.A. van Wuijckhuise-Sjouke

Animal Health Service
Gelderland
P.O. Box 10
6880 BD VELP (The Netherlands)

CHAPTER I.   INTRODUCTION

ENERGY METABOLISM OF FARM ANIMALS

A.J.H. VAN ES AND H.A. BOEKHOLT

ABSTRACT

Energy plays a major part in enabling a farm animal to produce the desired products. A considerable part of the total feed is needed for its maintenance.

Utilization of feed energy involves energy losses with faeces, urine, combustible gases and heat. A survey is given on how and to what extent feed energy is converted into metabolizable energy (ME) and how efficiently the ME is utilized for maintenance, work and synthesis - growth, milk, eggs wool -. Estimates are presented of the amount of ME needed for maintenance and its between- and within-animal variation. An equation is derived for calculation of heat production.

Finally techniques used for obtaining information on energy metabolism - feed conversion, comparative slaughter, complete energy balance using indirect or direct calorimetry - are discussed. Special attention is paid to their advantages, disadvantages and precision. Also a method to measure the physical activity of animals is described and how energy costs of activity can be estimated.

INTRODUCTION

In animal husbandry the main aim is to produce products of the desired kind and quality at lowest costs of feed, housing, animal care, etc. Thus the feed should be composed in such a way that it supplies

M. W. A. Verstegen and A. M. Henken (eds.), Energy Metabolism in Farm Animals.
ISBN 0–89838–974–7, © 1987, Martinus Nijhoff Publishers, Dordrecht.

the specific nutrients in such a form and quantity that enough of these can be absorbed. Furthermore the feed should contain so much energy as is just needed to make the production possible. That means that energy needed for maintenance in its widest sense should be as low as possible: the animals should not need additional energy because of disease, parasites, necessity to cool or warm themselves, other stresses or unnecessary physical activities. That would also reduce costs of animal care. However it might increase those of housing, indeed, sometimes costs of better stall insulation have to be compared during cold periods with those of higher feed costs for maintenance and during hot periods cost of cooling with those of economic losses due to low feed intake and resulting low production.

As quantitatively the needs for feed energy exceed by far those for specific nutrients, feedstuffs are in first instance compared according to their energy value. Of total energy metabolism maintenance metabolism is often a large part as is shown in Table 1.

Table 1. Maintenance metabolism as a percentage of total energy metabolism.

man	100	ruminating growing sheep	> 60
laying hens	> 50	ruminating growing goats	> 60
growing chickens	> 30	ruminating growing cattle	> 60
pigs	> 30	cows, producing 12 kg milk/day	50
veal calfs	> 30	cows, producing 30 kg milk/day	25

Thus it will be clear that in energetic feed evaluation due attention has to be paid to the feed's energy value for maintenance as well as that for production, values which may differ considerably.

The energy values of feeds for maintenance and various kinds of production will be discussed in the next section together with the concept of energy balance, which gives information on whether the animal is in energy equilibrium or not.

Heat is a very important byproduct of energy metabolism, it permits homeotherms to maintain a body temperature considerably above the temperature of the environment. An equation to predict an animal's heat production will be presented.

Finally methods for measuring energy balance and heat production will be described and their suitability for providing valuable data for practical animal husbandry will be discussed. Special attention will be paid to their precision and pitfalls.

ENERGY VALUE OF FEEDS AND ENERGY BALANCE

General

Life of animals depends to a large extent on sufficient supply of energy. Energy is needed for maintaining the organism in a good state as well as for production, i.e. for collection and ingestion of feed, for absorption of nutrients, for regulation and detoxification, for utilization of the absorbed nutrients for maintenance and synthesis and for physical activities as work, defence, flight, etc. Homeotherm animals, moreover, may use energy to maintain their body temperature at the desired level if the environmental temperature is too low.

Whereas most plants obtain their energy from part of the sun's radiation, animals derive their energy from degradation of organic compounds. By complete combustion in a bomb calorimeter these compounds release their energy as heat-carbohydrates about 17 kJ/g, proteins about 24 kJ/g and fats about 39 kJ/g. In the animal body this degradation, of course, concerns only the digested part, otherwise the process is different because in the body it is not a direct combustion but a stepwise biochemical degradation. For proteins the degradation is incomplete as the N is released as urea or uric acid and excreted with the urine; the so-called physiological combustion value of proteins is about 18 kJ/g. Plants as well as animals may store some of the organic compounds after some conversion for later use.

Energy losses and energy balance

Utilization of ingested food energy by animals involves several kinds of losses. Not all of the food can be digested and absorbed, the remainder is excreted in the faeces. Urea, uric acid and other detoxification compounds have to be excreted with the urine. In animals with symbiosis with microbes in the forestomachs and/or large intestine energy in gaseous form (CH_4, H_2) is lost. Also losses, as heat, occur when the absorbed nutrients are used for production of ATP needed for maintenance, physical work and synthetic purposes. Furthermore when this ATP is used for maintenance and work or for the conversion of absorbed nutrients into tissue, milk, eggs and wool, part of its energy becomes heat. Thus animals with a high production or work level have a higher heat production than non or less producing ones. In animals living in symbiosis with microbes a small additional source of heat occurs because heat of fermentation is produced, resulting from the own maintenance and production metabolism of these microbes.

As a result of the losses of energy in faeces, urine and combustible gases the gross energy of the feed (GE) is not a good measure for the energy available for the metabolism of the animal. Metabolizable energy (ME, i.e. GE corrected for these losses) is thus a better measure for the energy available for the animal for maintenance, work and production of tissues, milk or eggs. All ME used for maintenance becomes heat. Part of the remaining ME (energy for production) is also converted into heat, its size depending on the composition of the ME and the purpose for which the ME is used.

It seldom occurs that ME intake equals ME needed for maintenance, work and external production. If the intake is too low some tissue reserves are mobilized and used as a nutrient, for short periods glycogen of the liver, for longer periods mainly body fat. If intake is high then the surplus of energy is first used for replenishing the depleted glycogen stores in liver and muscles and after that for the production of body fat. Thus we get the next equations:

$$ME = GE - (E_{faeces} + E_{urine} + E_{CH_4, H_2}) \qquad kJ/day$$

$$EB = ME - H - E_{work} - E_{eggs, milk} \qquad kJ/day$$

in which EB = energy balance and H = heat production.

A synonym for EB is RE, retained energy: both EB and RE are the amounts of energy daily retained in or lost from the animal's body, mainly as fat and protein, sometimes a small amount of glycogen. The sign and size of the energy balance inform us to what extent the animal's energy intake satisfies its energy needs. It will be clear from the preceeding lines that the heat production (H) arises from various sources. For example for a lactating cow some heat comes from microbial fermentation, but the main part comes from production, i.e. conversions of absorbed nutrients and synthesis of milk.

Metabolizable energy (ME)

Feedstuffs differ considerably in ME content for two reasons, first due to the variation in their content of starch, sugars, proteins, fats and plant cellwall constituents - hemicelluloses, celluloses and lignins - and second due to the variation in digestibility of these components. The first four components can be digested with the animal's own enzymes. The plant cellwall constituents can only be digested in symbiosis with microbes, although only to a degree depending mainly on their interlinking with each other especially with lignin. The process of fermentation in the rumen results in some energy loss (CH_4, H_2; fermentation heat). In poultry such a symbiosis hardly occurs. In pigs it takes place in the large intestine, thus after the digestion of starch, sugars, proteins and fats in the small intestine, and at body weights of 50 and 150 kg this fermentation amounts to not more than 10 and 20% of the total feed, respectively. In ruminants after development of their forestomachs nearly all the feed except saturated fatty acids is subject to microbial fermentation. The extent of fermentation is hampered by a high degree of interlinking and lignification, a high rate of food passage (due to a high feeding level) and a low pH of rumen fluid (due to a high level of easily fermentable matter). It will be clear from this that for feeds containing much plant cellwall, the ME content is highest for ruminants with well-developed forestomachs, lower for pigs and still slightly lower for poultry. For feeds with very little plant cellwall the ME content is somewhat higher for poultry and pigs than for ruminants. At higher feeding levels ME contents decrease somewhat in the ruminant but hardly in pigs and poultry. At very young age most animals cannot yet digest large quantities of starch and fat. Because in calculating the ME content fermentation heat is not subtracted from GE, ruminant-ME contains not only energy in organic compounds but also about 10% as heat, thus in a form that is useless for other purposes than heating the body. ME values of feeds for pigs include only a very small part of heat, those for poultry none at all. Usually ME values of feeds for pigs are simply calculated by subtracting energy in faeces and urine from GE and neglecting CH_4 and H_2; in that case they are too high by 0.5 to 3% depending on content of plant cellwall constitutents and on the age of the animal.

Utilization of ME for maintenance

The degree of efficiency of the utilization of ME by animals depends on the composition of the ME and on the purpose for which it is used. For maintenance and physical activity the energy is for the greatest part needed in the form of ATP. Glucose-ME (from starch, sugars) yields most ATP per kJ, fat-ME some 5% less, protein-ME and ME absorbed as volatile fatty acids (VFA) resulting from microbial fermentation 15-20% less. For poultry and pigs the energy values for maintenance of the ME of feeds, indeed, differ according to the ME's composition to such an extent. Ruminant ME, other than the 10% fermentation heat - useless for ATP synthesis -, consists mainly of VFA and proteins because the microbes ferment most starch and sugars and ruminant diets seldom contain more than 5% fat. Therefore one might expect that ME of all kinds of ruminant diets would have the same energy value for maintenance. Energy balance studies however have shown that this energy value increases slightly when the GE of the feed contains more ME. If the ME as a percentage of GE increases with 1% the efficiency of utilization increases with about 0.4%. Finally, all the ME used for maintenance and about 70% of the ME used for work become heat.

Utilization of ME for synthesis

Deposition of fat in the body and probably also in milk and eggs in monogastrics has an efficiency of the utilization of the ME of about 80% for glucose-ME, 95% for fat-ME - high due to few conversion steps -and 65% for protein-ME. In ruminants body fat synthesis from ME was found to be less efficient than in monogastrics: when ME was 60% of GE the efficiency was about 50%. Again this conversion efficiency was not constant but changed by about +1.5% and - 1.5% per 1% more or less ME in GE, respectively. For milkfat produced from volatile fatty acids probably the same efficiency applies, however a considerable part is produced from fat, absorbed from the gut, at a much higher efficiency.

Measuring the efficiency of the utilization of ME for lactose synthesis during lactation is difficult because its energy costs have to be separated from the large costs for maintenance and for fat and protein synthe-sis. No accurate data are available but in view of its fairly simple bio-chemical synthesis, efficiency values will be about 90% for monogastrics (from glucose-ME) and about 75% for ruminants (from ME consisting of propionic acid with 10% fermentation heat).

According to our knowledge of biochemistry protein synthesis from amino acids would have an energetic efficiency of about 85%. Most energy balance and comparative slaughter trials with growing animals however gave efficiencies of 60-70%. Partly these low values may be due to un-derestimating the maintenance requirements of young and therefore phys-ically more active and more easily stressed animals. Partly the low values are due to the high rate of protein turnover during rapid growth - but not during lactation -. This means that rapid growth also results in higher maintenance costs of energy because of faster renewal of already existing protein tissues.

For combined productions of fat, protein and/or lactose most data are available for dairy cows. The efficiency of the conversion of ME in milk energy was on average 60% for diets with a ME/GE ratio of 57% and changed by +0.4 and -0.4% for ratio's which were 1% higher or lower than 57%, respectively, a change of similar size as for maintenance. The few efficiency values for milk produced by sows vary considerably, main-ly due to measurement errors as it is very difficult to determine daily milk energy accurately. Values of about 80% are most probable. Also the efficiency values for egg energy production vary much, between 60 and 85%, mainly depending on the assumed maintenance requirements. An ef-ficiency of 80% seems most probable.

During growth with advancing age the energy retained as protein as a percentage of total energy retention decreases - from 70% to 10% - and the energy retained as fat increases to about 90%. As a result of the change in deposition from protein to fat there is a tendency in growing non-ruminants towards a slight increase in efficiency from about 65% to about 75%. In calves fed only milk or milk replacer about 40% of the re-tained energy is deposited as protein and the efficiency value is nearly 70% up to a body weight of 150 kg. In growing beef cattle above 200 kg the major energy deposition is fat, often less than 20% is protein. Depo-sition of protein, however, mainly in muscle tissue, always involves a deposition of about three times as much water at the same time because protein in tissue always is accompanied by much more water. So a deposition of 1 g protein involves a weight gain of about 4 g. Deposition of fat, on the contrary, usually replaces some water of the tissues. A deposition of one gram fat may result in a liveweight change of only 0.90-0.95 g. Therefore the corresponding energy contents of one gram liveweight change are about 40 kJ for fatty tissue and only about 6 kJ for protein tissue. Production of lean tissues clearly requires considerably less energy per gram than fatty tissues.

It is not precisely known how efficiently ME is converted into energy in foetus and foetal membranes during pregnancy, again due to the high contribution of maintenance in the total energy metabolism in a pregnant animal. Moreover we do not know and we cannot measure whether the maintenance requirements per kg body weight of pregnant and non-pregnant animals are equal or not. Assuming that they were equal in cows resulted in 4 institutes in very low efficiencies of the conversion of ME for pregnancy, only 10-25%. The values found in the very few experiments with pigs have a great error.

It will be clear that all ME used for production that is not retained in body or product becomes heat.

ME required for maintenance

In the beginning of this section it was mentioned that all ME used for maintenance purposes becomes heat. The question left unanswered was how much ME of average diets is needed for maintenance (ME_m, kJ/d). First we consider healthy, non-stressed mature animals fed maintenance diets where ME_m between species is related to metabolic body weight, body weight in kg raised to the 3/4 power. Within the same animal species a value of the power closer to one is sometimes said to give a stronger relationship. Its precise determination is extremely difficult because one needs mature animals of the same species fed the same diet that vary widely in weight. Within species often the value of the power was derived from measurements during and after the growth period. In these cases either the animals while growing were fed a maintenance diet, what gave them stress due to underfeeding, or they received a production diet in which case correction for energy retention and increased protein turnover introduced errors. Moreover young animals are more active and more easily stressed than mature ones what still more hampers a correct interpretation of the results of the measurements. Also often the animals were not sufficiently accustomed to the experimental circumstances. Table 2 shows some values for the maintenance requirement, expressed per kg metabolic weight.

Table 2.   Maintenance requirements per kg metabolic weight: $ME_m/kg^{3/4}$.

	kJ/day
laying hen	400
growing chicken	420-450
sow	350
growing pig	420-480
veal calf	460
cattle:	
non-lactating > 200 kg	420
lactating	460
mature sheep	350

The values hold true for healthy, non-stressed animals showing normal physical activity (poultry in small groups, other animals in separate cages or fastened) receiving a diet of average composition. For other diets these maintenance values may be slightly different depending on the composition of the ME used, as explained earlier.

Horizontal walking in cattle gives an increase of about 2 kJ per km per kg body weight; walking down- and especially uphill as well as trotting increases the requirement much more.

There is hardly any between-animal variation within the same species with regard to the digestion and conversion of the GE of the same diet into ME when the diet is fed at the maintenance feeding level except when the animals are very young. Very little between animal variation is present also in the efficiency of the utilization of the same ME for the same kind of production. The production process concerns biochemical conversions which differ hardly between animals of the same species. With respect to maintenance requirements however, the value of

$ME_m/kg^{3/4}$ has a between-animal variation of 5-10%. It may be related to temperament and spontaneous physical activity.

Within-animal variation of $ME_m/kg^{3/4}$ over short periods (two weeks) under comparable circumstances is small, not more than 2%. However, there are a few data that suggest that this variation over longer periods is larger. Seasonal changes in metabolism might be the cause of it as found in wildlife. It is very difficult to prove the existence of variation of $ME_m/kg^{3/4}$ within animals over longer time intervals. This is because of the size of the error of each measurement - some 2% - and the necessity to have the same feed, the same feeding level, the same measuring conditions, the same adaptation of the animal to these conditions and the same state of its health in all periods of measurement.

Animals that are not well adapted to the experimental conditions show a higher $ME_m/kg^{3/4}$. Sheep and young animals need long adaptation periods. These periods can be reduced markedly by keeping the animals during the adaptation period and measurement in their own metabolism cage together with their neighbour(s), or together in a group. Cattle are sooner adapted when they can see a nearby neighbour. Adapted animals lie down for a longer part of the day and show lower spontaneous physical activity, reason why it is useful to get some information on activity on activity by means of e.g. Doppler-based devices.

In poultry light has a large positive effect on activity and therefore on $ME_m/kg^{3/4}$. Thus long dark periods reduce feed costs for maintenance, unless these are so long that the animals become hungry and try to eat in the dark. Very short light periods may lead to lower intake and increased activity because of fighting when not all animals can eat at the same time. Here also the use of activity meters can be recommended.

Poikilotherm animals like fish have a maintenance metabolism which is lower by a factor of 2 or 3 for every 10°C that they are kept below 38°C. However, at the same time also their production metabolism is reduced so that it takes much more time to produce a given amount of product and this affects total maintenance costs per unit of product. In this respect we found little difference in energetic efficiency between growing carp and growing chickens.

PREDICTION OF HEAT PRODUCTION

In the preceeding section it was said that all ME used for maintenance becomes heat and that the amount of ME needed for maintenance (ME_m) of healthy, non-stressed animals depended on animal species, age, body weight and composition of ME. Also part of the ME used for production (work included), equal to $ME-ME_m$, is converted into heat; its size depends on animal species and kind of production(s) and of diet. Thus total heat production (H) is equal to:

$$H = ME_m + c_1 * (ME-ME_m)$$

All elements are in kJ/day and c_1 is a constant depending on animal species and kind of production(s) and of diet. As mentioned before ME_m can also be written as a function of metabolic weight ($W^{3/4}$, with W = body weight in kg):

$$ME_m = c_2 * W^{3/4}$$

in which c_2 depends on animal species and age and on composition of the ME as has been discussed earlier. Combination of both equations gives:

$$H = c_2 * W^{3/4} + c_1*(ME - c_2 * W^{3/4}) = c_1 * ME + c_2*(1-c_1) * W^{3/4}$$

$$= c_1* ME + c_3 * W^{3/4}$$

From this equation it can be seen that the main factors determining heat production are feeding level (ME) and body weight (W). The value c_1 and thus also c_3 can be derived from the efficiencies of the utilization for the various kinds of production and diets, discussed above, by subtracting these efficiencies from 1.

The terms heat production and heat loss are used often interchangeably. However heat production is not at all moments equal to the heat loss of the animal. Heat loss depends on transfer of heat, produced by the intermediary metabolism of the body, to the environment. If both are not equal body temperature will change. Over time intervals of a day or longer heat production equals heat loss in healthy homeotherm animals.

METHODS OF MEASURING ENERGY METABOLISM, THEIR SUITABILITY, ERRORS AND PITFALLS

General

Effects of non-optimal housing, stress and disease on energy metabolism of farm animals consist mainly of an increase in maintenance requirement for ME and sometimes of a decrease in digestibility and/or a change in the size or composition of the production. Voluntary intake is usually decreased and production is reduced. Moreover, the animal may use more reserves, change its behaviour, reduce rate of protein turnover, etc. Initial excitement after a change to a less comfortable situation or at the start of a disease may subside after a while through adaptation. Thus for a good understanding of these effects on energy metabolism for practical animal husbandry it would be most useful to measure intake, digestibility, energy balance and physical activity, preferably under circumstances similar to or at least very close to those at the farm.

Feed conversion techniques

In applied studies digestibilities and energy balance are seldom measured. Usually only the effect of changes in housing, feeds, etc. on feed conversion, kg feed needed per kg product, is measured, often during rather short time intervals.

In these experiments too little attention is sometimes paid to the quantity and quality of the feed used and to the weight of the product. This makes a correct interpretation of their results still more difficult. It happens that not even the dry matter content of the feed is measured, nor its ash content whereas energy values are simply taken from feed tables. Instead of storing large well-mixed batches of feed for the whole duration of long-term experiments, each week or month new batches of these feeds are bought or prepared what may lead to further uncertainty about e.g. ME intake. Fibrous byproducts and forages show marked variation in digestibility and thus in ME content from batch to batch. Especially when these products are used as feed measuring the digestibility of the ration with a few animals would considerably improve the estimate of actual ME intake. Especially during long-term experiments often too little attention is paid to collecting and weighing spilled feed.

The other element in feed conversion, the weight of the product - eggs, milk, increase in body weight - does not give much information on total energy production. In the case of production of eggs and milk the animal may have used or produced also reserve tissues. Usually body weight change is measured also in an attempt to correct for it but this is a weak criterion for the body's change in energy content. Because of diurnal variation in total body weight and due to variation in gut fill, it is not easy to measure an aminal's change in empty live weight precisely. Predicting the energy content of that change can only be done with a large error as it may vary from 10 to 45 MJ/kg in mature animals and from 10 to 25 MJ/kg in growing ones. Some improvement in growing animals can be obtained when additionally a N-balance is made because this informs to some extent on weight increase due to deposition of protein tissue, equal to 3 to 4 times protein retention (= 6.25 × N balance). The remaining weight gain must be due to fat deposition. In this way the energy content of the weight increase can be estimated somewhat better. Measuring a N-balance accurately, however, requires careful collection of feed refusals and excreta, accurate determination of N in feed and excreta and prevention of any N losses from the latter.

Comparative slaughter techniques

In growing animals, to obtain better insight in the composition of the live weight gain, the comparative slaughter technique is sometimes used. At the start and at the end of a trial representative animals of each treatment are slaughtered and analyzed for protein and fat and sometimes for energy. From protein and fat data energy content can also be derived by calculation. Accurate results are only obtained when the time interval between start and end is long and thus the weight change is large enough so that inevitable errors made by the analysis of the slaughtered animals have not a great effect. Moreover the number of animals should be not too small, otherwise the influence of between-animal variation on the result may be too large. This makes this type of experiment rather expensive when used for cattle and pigs. Other difficulties during these long experiments are unequal feed intake and behaviour or disease incidence of different treatment groups.

Energy balance techniques

    Measuring complete energy balances of the animals during the experi-
ment solves most difficulties encountered with the methods described
above, but of course is expensive with regard to equipment as well as to
execution. The main advantages are that a complete picture of energy
metabolism is obtained, i.e. energy input and energy output, and that
short measuring periods suffice. Usually N balances are made at the
same time with little additional work and these inform on the composition
of the retained or lost energy. When the animals are well accustomed to
the experimental routine, collection periods of 7-10 days for poultry and
pigs and of 10-14 days for ruminants give sufficient precision. In these
collection periods during 2-4 days heat production has to be measured.
In this way it is possible to compare animals under different circumstances. On the other hand this permits to follow changes in metabolism
with time by performing a series of balance measurements with the same
animals. During the one or two weeks between these measurements the
experiment asks much less work because no collections have to be made.

    Measuring an energy balance requires measurement of heat production
over periods of at least 24 h. This can be done by direct or indirect calorimetry.

    Usually heat production is measured by indirect calorimetry using re-
spiration chambers. The heat production can be calculated from respiratory gas exchange (O_2 consumption, CO_2 and CH_4 production) and urinary N excretion with the equation of Brouwer (1965):

Heat $= 16.18 * O_2 + 5.02 * CO_2 - 2.17 * CH_4 - 5.99 * N$
Heat   = heat production in kJ/day
O_2 = oxygen consumption, l/day
CO_2 = carbon dioxide production, l/day
CH_4 = methane production, l/day
N   = N in urine, g/day

It has the advantage that for ruminants anyway and for pigs in principle methane production is measured too. The equation is based on combustion of starch, protein and fat with methane and urea as endproducts
of the former two. For other diets or combustion of proteins with uric
acid as endproduct the equation predicts actual heat production very
closely but may be adapted. Especially for larger animals and groups of
animals open-circuit respiration chambers are used more often than

closed-circuit ones. The former allow more frequent entrance of the chamber and require less chemicals.

Direct calorimetry is seldom used for studies lasting a day or longer. The equipment is very expensive and entrance of the chamber during a mesurement is nearly impossible. It has been proved several times that over periods of one or more days direct and indirect calorimetry give the same estimates of heat production if measurements are done correctly. The main advantage of one type of a direct calorimeter, the gradient-layer instrument, is that it measures changes in heat loss nearly instant-aneously whereas other direct calorimeters and also respiration chambers react far more slowly. Direct calorimeters also inform on insensible heat loss, i.e. water vapour production, because this has to be measur-ed separately. With little additional effort this can also be measured in indirect calorimetry.

Accuracy of measurement of energy balance

Besides high costs measuring energy balances with indirect calori-meters has some other disadvantages, even when the animals are well-accustomed to the calorimeter and the circumstances in the calorimeter are equal or close to those in practical animal husbandry. As is the case with any measurement inevitable errors are made. The measurement of the ME intake during restricted feeding has a standard deviation of 1-2% of GE intake. Furthermore 24 h. heat production has a coefficient of var-iation of 1-2%. So the standard deviation of the energy balance (EB, RE) amounts to about 2% of GE depending somewhat on the length of the ex-creta collection period and the number of 24 h. heat production measure-ments. Errors, of course, will be higher when the animals are not well-accustomed to the calorimeters, experimental routine and/or circumstan-ces. Usually comparisons are made between two or more results. If these apply to the same animal over longer time intervals this animal may have changed its behaviour and therefore has a different maintenance re-quirement. To some extent such changes can be detected by observation and by using activity meters. If the comparisons are made between ani-mals or between groups of animals, between-animal variation in mainte-nance requirements increases the uncertainty. For groups this variation of course will be less than for single animals.

Activity meters

Several times attention was drawn to the measurement of spontaneous physical activity of the experimental animals moving more or less freely in a large cage or pen. Instruments based on the Doppler principle are used at Wageningen; these are on the market as devices to detect intruders. It is our experience that it is nearly impossible to standardize the measurements fully and that the meters are rather sensitive to small changes of their position. Their results are hardly applicable for comparing different groups consisting of few animals in similar circumstances or the same small group in different circumstances. For groups of 10 animals or more applicability is better, because between-animal variation affects the results much less. For the same group in the same calorimeter activity meters may give useful information on whether in successive days or trials activity was the same or not. By regressing withing one day the results of heat production during periods of 5-15 min. on activity an estimate of activity free heat production can be obtained. Measuring heat production and activity over short periods requires rather expensive instruments for physical gas analysis and data storage. Calculation of heat production over such short intervals should include changes in O_2 and CO_2 leaving the calorimeter as well as changes in the amount of these gases inside it. Elimination of some data before regressing is useful, e.g. the data collected during one or two hours after feeding when animals are meal-fed, after switching on or off the light in chambers with poultry or after milking or entrance of the chamber by an attendant. It excludes heat production that is due partly to other causes than spontaneous physical activity.

Final remarks

Full perfection of measuring energy balances can never be achieved but with careful planning, meticulous testing of the equipment and execution of the trials and good use of common sense reliable results can be obtained.

REFERENCES (a selection)

ARC, 1980. The nutrient requirements of ruminant livestock, Slough:

Commonw. Agric. Bureau: 73-119.

ARC, 1981. The nutrient requirements of pigs. Slough: Commonw. Agric. Bureau: 1-65.

Blaxter, K.L., 1962. The energy metabolism of ruminants, London: Hutchinson.

Brouwer, E., 1965. Report of sub-committee on constants and factors. In: Proc. 3rd Symp. on Energy Metab. EAAP Publ. 11: 441-443.

E.A.A.P. (European Association for Animal Production): Proc. Symposia on Energy Metabolism - Publications $\underline{8}$ (1958), $\underline{10}$ (1961), $\underline{11}$ (1965), $\underline{12}$ (1967), $\underline{13}$ (1970), $\underline{14}$ (1973), $\underline{19}$ (1976), $\underline{26}$ (1979), $\underline{29}$ (1982) and $\underline{32}$ (1985).

Es, A.J.H. van, 1961. Between-animal variation in the amount of energy required for the maintenance of cows. PUDOC, Centrum voor Landbouwpublicaties en landbouwdocumentatie, Wageningen.

Es, A.J.H. van, 1972. Maintenance. In: Handbuch der Tierernährung II, ed. W. Lenkeit & Breirem, Hamburg, Parey: 1-54.

Es, A.J.H. van, 1986. Energy metabolism in man and animals. In: Proc. 3rd Int. Congr. Nutr. (T.G. Taylor and N.K. Jenkins, eds.) John Libbey, London: 279-283.

Ferrel, C.L., Koong, L.J. and Nienaber, J.A., 1986. Effect of previous nutrition on body composition and maintenance energy costs of growing lambs. Brit. J. Nutr. 56: 595-605.

Kleiber, M., 1961. The fire of life. New York, Wiley.

Lange, P.G.B.de, Kempen, G.J.M. van, Klaver, J and Verstegen, M.W.A., 1980. Effect of condition of sows on energy balances during 7 days before and 7 days after parturition. J. Anim. Sci. 50: 886-891.

Schiemann, R., Nehring, K., Hoffmann, L., Jentsch, W. and Chudy, A., 1971. Energetische Futterbewertung und Energienormen, VEB Deutscher Landwirtsch. Verlag.

Wenk, C. and Es, A.J.H. van, 1976. Energy metabolism of growing chickens as related to their physical activity. In: Proc. 7th Symp. on Energy Metab. E.A.A.P. publ. 19: 189-192.

THE WAGENINGEN RESPIRATION UNIT FOR ANIMAL PRODUCTION RE-
SEARCH: A DESCRIPTION OF THE EQUIPMENT AND ITS POSSIBILITIES

M.W.A. VERSTEGEN, W. VAN DER HEL, H.A. BRANDSMA,
A.M. HENKEN AND A.M. BRANSEN

ABSTRACT

Six indirect calorimeters have been built at the departments of Animal
Production of the Wageningen Agricultural University. They are used for
measuring heat production and energy balances of animals at various environmental and nutritional conditions. Three sizes of chambers can be
distinguished:
- 2 small chambers of 0.085 m³ each
- 2 medium chambers of 1.8 m³ each
- 2 large chambers of 80 m³ each.
They allow measurements with small animals (mice, rats, chicks, piglets),
medium sized animals (laying hens, pigs, goats, calves) and large animals
(pigs, cows, etc.) housed individually or kept in groups.

Climatic conditions in all chambers can be maintained between 5 and
40°C dry bulb temperature (T_d). Relative humidity (or wet bulb temperature, T_w) can be chosen independent of temperature between 40 and
90%. Temperatures (T_d and T_w) can fluctuate within a day in a preprogrammed way. Various housing systems, e.g. floor system and group
housing, can be applied. Also air velocity can be set and maintained at
values between 0.2 and 0.9 m/s.

Heat production is determined per chamber from gaseous exchange of
CO_2 and O_2 over short periods (18 minutes) up to 48 hours continuously. Activity is measured continuously with a burglar device. Energy balances are derived from measurement of metabolizable energy intake (ME
= gross energy intake minus energy in faeces, urine and combustible

M. W. A. Verstegen and A. M. Henken (eds.), Energy Metabolism in Farm Animals.
ISBN 0–89838–974–7, © 1987, Martinus Nijhoff Publishers, Dordrecht.

gases) and heat production. Similarly nitrogen (N) balances can be determined from N in feed, faeces, urine and aerial NH_3. The data obtained are used to evaluate environmental and animal conditions with respect to production characteristics. These relations should be known to attain optimal production efficiency.

INTRODUCTION

The climatic environment has a large influence on rate of growth and other production traits of animals. Parameters measured in research on growing pigs are: live weight increase, feed conversion and slaughter quality. The data are mostly obtained by exposing the growing pigs to a constant or fluctuating condition during a rather long period, i.e. from one week up to several months. They will require a rather long time because it will take some time before effects on weight can be noticed due to errors in determining body weight of animals or their tissues. However, not only growth rate as such but also body composition may be dependent on temperature (Sörensen, 1961; Hicks, 1966; Pfeiffer, 1968; Close, 1981). Indirect or direct calorimeters have the advantage that they provide the possibility to investigate quantitatively during short periods (a few hours to various days) effects on heat exchange due to small changes in climate, e.g. temperature, relative humidity or air velocity. Indirect calorimeters have the advantage over direct calorimeters that they are easier to operate. On the other hand direct calorimeters may detect alterations in heat loss sooner than changes in heat production are detected by indirect calorimeters. Both systems may be used at similar accuracy for energy balances.

In practice climatic conditions change more or less continuously within and between days. This means that some effects may only be detected with equipment combining a system which measures continuously (minutes up to one hour) with a system allowing environmental conditions to be reset continuously. Moreover, results of research on effects of climatic conditions on metabolism of pigs can only be with confidence translated to practical conditions, if the housing system is the same, i.e. group housing (Holmes, 1966; Holmes and Mount, 1967). Also Close et al. (1971) and Verstegen (1971) have shown the importance of including groups in the investigations. Also activity should be taken into account. The objective of such research is to determine the energy requirements of the animals at various conditions. This is done by measuring heat production (indirectly via production of CO_2 and consumption of O_2) and other energy balance traits of the animals. This technique is also frequently used to determine nutritive values of feed or feed components. Tables on energy values of feed or feed components are then derived from measurements of energy balances determined in animals by means of indirect calorimetry. Energy intake is measured by determination of the calorific energy value of the nutrients ingested. Similarly energy in feed residue, in urine voided, in faeces excreted and in combustible gases produced like CH_4 are determined. From energy intake and energy in faeces, urine and combustible gases metabolizable energy (ME) is calculated. The energy balance (EB) can be derived from ME minus heat production. Moreover differences in heat production can be associated with differences in housing and environmental conditions (Close, 1981). Also effects of infectious diseases and of endo- and/or ectoparasites can be evaluated with regard to their effects on energy metabolism (Kroonen et al., 1985; Verhagen, 1987). Such effects can be transferred to alterations in maintenance requirement and/or feed efficiency (Figure 1). These changes

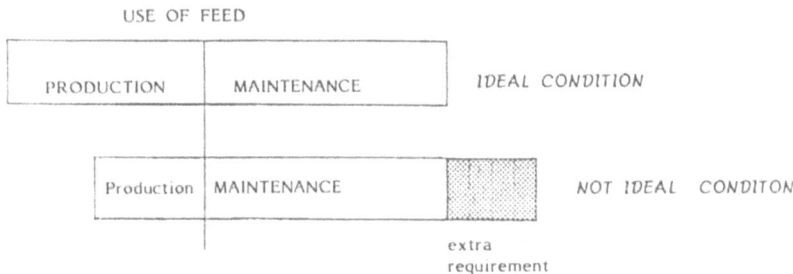

Figure 1. Effect of conditions which increase maintenance and partitioning of ingested metabolizable energy into maintenance and production.

can be used to evaluate stressfull conditions, diseases and parasites. The consequences can be assessed quantitatively with regard to its burden on the animal and with regard to the physiological and economical impacts.

In order to simulate practical conditions with special reference to housing and factorial diseases (Verhagen, 1987) calorimeters of various sizes have been developed at the departments of Animal Production of the Agricultural University in Wageningen together with the Technical and Physical Engineering Research Service.

TECHNICAL DESCRIPTION AND DIMENSIONS OF THE CALORIMETERS

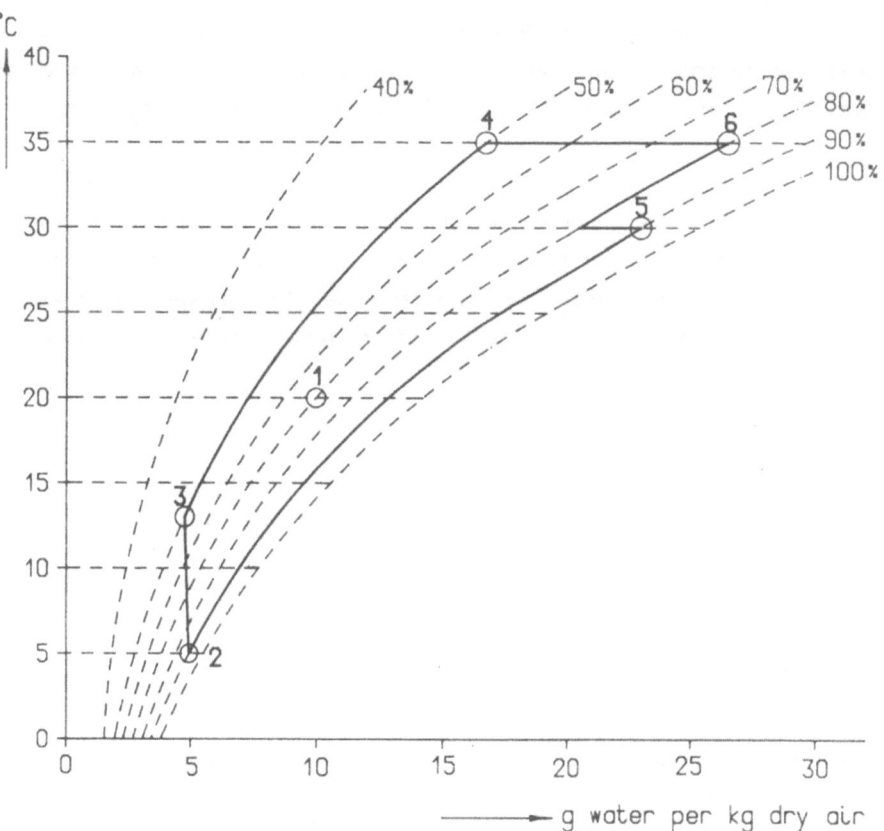

Figure 2. Climatic conditions (degrees °C) and water content (g/kg) and relative humidities to be reached and maintained in chambers 1 and 2 (for values of points 1 to 6 see Table 1).

Table 1.   Range of climatic conditions to be met by the climatic equipment (see Figure 2 for points 1 to 6) at different floor temperatures.

Point	Temperature dry bulb (°C)	Temperature wet bulb (°C)	Relative humidity (%)	Floor heating temperature (°C)
1	20	16.5	70	-*
2	5	4.0	86	25
3	13	8.1	50	30
4	35	26.1	50	-
5	30	28.6	90	-
6	35	31.9	80	30

*   - = no floor heating

Three types of calorimeters (two of each type) have been developed. The content of each of the smallest calorimeters is 0.085 m³, 1.8 m³ for each of the medium type and 80 m³ for each of the largest chambers. In the large calorimeters (chamber 1 and 2) animals can be housed under conditions which are close to practical conditions during the whole period of growing or fattening.

The area of the Mollier diagram in Figure 2 shows which conditions can be covered by the installation of chambers 1 and 2. In Table 1 these points are quantified. In all respiration chambers (1 and 2, 3 and 4 (medium), 5 and 6 (small)) the dry bulb temperature can be maintained between 5 and 40°C and the relative humidity between 50 and 90%.

The temperature (dry bulb) inside the chambers and relative humidity (wet bulb) can be maintained and altered as shown in Figure 2. In general the equipment of all chambers can alter the dry bulb temperature at least at a rate of 5°C per hour and the RH with 10% per hour, independently of one another and at maximum metabolic weight of the animals in the chambers. The heat production is measured indirectly determining respiratory gaseous exchange according to the Pettenkofer system. Measurements are done in the same way as described by Van Es (1961) and Verstegen et al. (1971).

Chambers 1 and 2

Each chamber has an inner room of (length x width x height) 6000 mm x 4000 mm x 2200 mm. To accommodate the animals the total air content, including the air conditioning unit, is 80 m³. It is large enough to contain for instance two pens for pigs of normal size. In each pen 8-10 pigs or a proportional number of other animals can be housed during several months. The chamber has external dimensions of 6200 x 5200 x 2300 mm and it is constructed from 1.5 mm and 2 mm sheet steel suitably stiffened. The chamber is insulated with 100 mm glasswool which is covered with 1.5 mm aluminium sheet.

The chamber floor consists of 35 mm non-toxic asphalt embedded in a layer of 100 mm foam glass on a concrete floor. The conductance of the floor is 0.05 W.m^{-1}.K^{-1}. The floor of chamber 1 is provided with a heat-

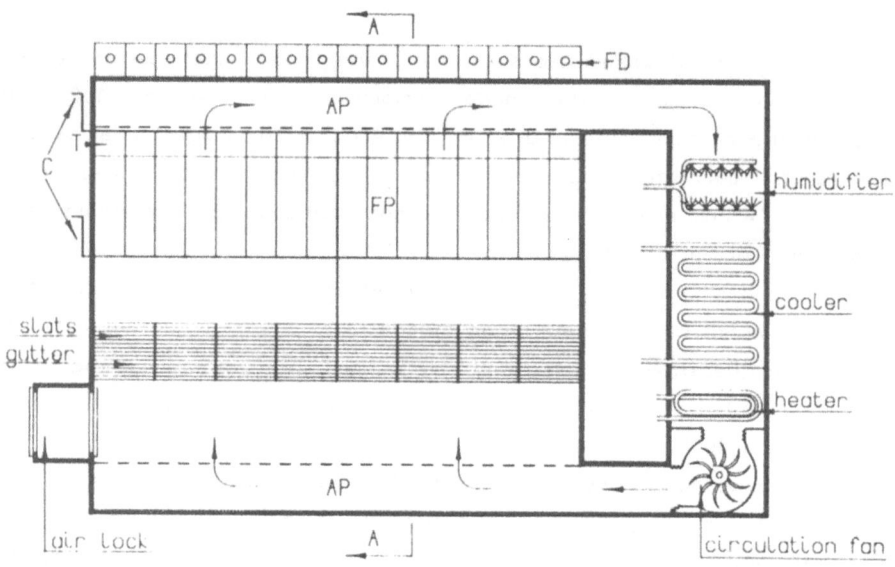

Figure 3a.   Sketch of one of the large climatic respiration chambers. AP = air plenum; FD feeding device; T = feeding trough; C = crank lever, for opening trough and/or feeding pen; FP = feeding pen; arrows indicate direction of air.

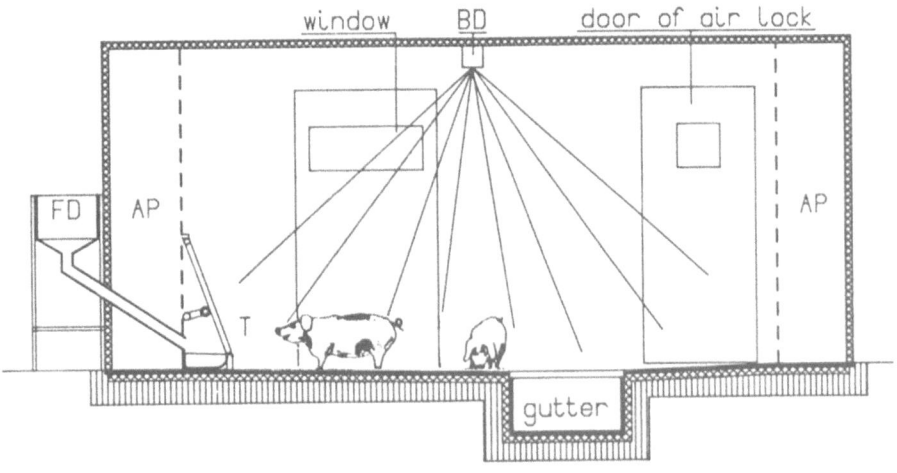

section A-A

Figure 3b. Vertical cross-section through one of the large chambers at A-A with animals. Also indicated are one of the burglar devices (BD), windows and doors.

ing sheet just underneath the asphalt on top of the foam. The capacity is such that the surface temperature can be maintained at 20°C above the chamber temperature. The maximum floor temperature to be maintained is about 30°C. The speed of change in floor temperature is about 2 to 3°C per hour. Various floor systems can be installed in the chamber, i.e. concrete, straw on concrete and slatted floors, on top of the permanent floor structure. Moreover various metabolism crates (5 per chamber for calves, sows or boars) can be placed to enable separate metabolizable energy determination per animal.

The feeding is done with 16 feeding devices outside the chamber from which tubes lead through the wall into a trough for each animal from which the lid can be opened from outside of the chamber by means of a hand crank lever (Figure 3a and 3b). Water is supplied to the animals with nipples at each feeding place and/or via a drinking bowl in each pen.

Safety switches operate an alarm. This alarm is independent of the electricity supply of the chambers and is monitored 24 hours per day. The safety pressostat acts when the pressure in the chambers rises above the desired level which is approximately 5 to 10 mm water column below the atmospheric level. The safety is also activated when dry and/ or wet bulb temperature is too high or too low. The door to the chamber is sealed off by neoprene strips. The door is tightened with lever locks. In addition to the air condition unit also separate air conditioning units can be installed to manipulate air velocity around the animals (Verhagen, 1987).

Chambers 3 and 4

These chambers have been constructed for the use with dwarf goats (Bransen and Kneepkens, 1982) but also experiments with poultry, pigs and calves were performed. The chambers have outer dimensions of 1810 × 903 × 1403 mm and inner dimensions (available for animals) are 1000 × 800 × 970 mm.

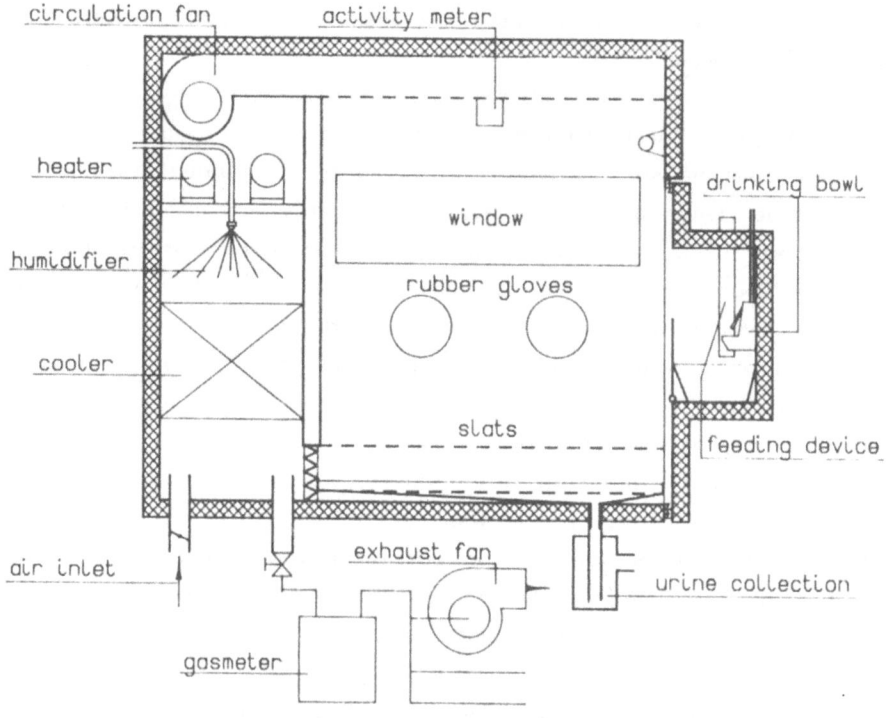

Figure 4. Lay-out of medium sized chambers (3 and 4).

Air is exhausted by means of a centrifugal fan (Figure 4). Similarly as in chambers 1 and 2 ingoing air and thus air pressure is maintained below atmospheric pressure by means of a servomotor controlled valve.

Animals can be housed in various ways. Goats can be housed in a metabolism crate. Chickens and hens can be placed with their cages in the calorimeter.

Feeding can be done from outside through a small air lock (Figure 4) and water provision can be done in various ways depending on the species or study.

Chambers 5 and 6

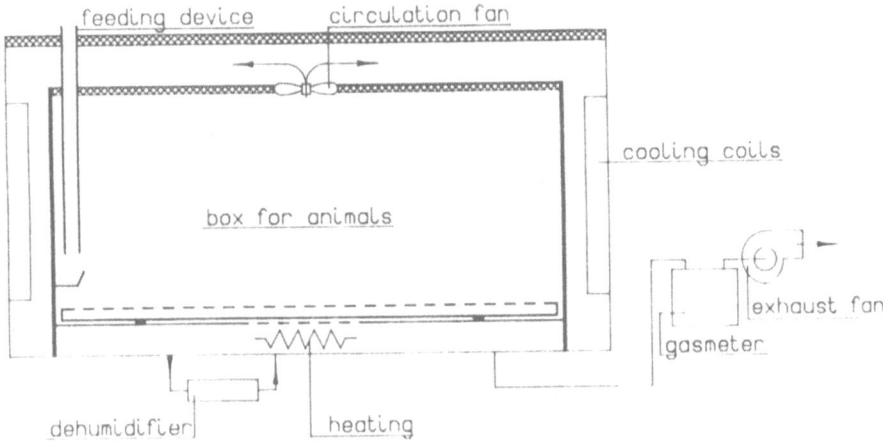

Figure 5. Lay-out of small-sized chambers (5 and 6).

These small calorimeters have been constructed for research with small or young animals, i.e. laboratory animals, chicks, etc. (Figure 5). The inner dimensions are 685 × 335 × 275 mm. Outer dimensions are 915 × 550 × 490 mm.

Air is conditioned by leading freon through a copper tubing in the wall. The regulation is on/off with a dry bulb contact thermometer. Air humidity is measured by a wet bulb contact thermometer. If dry bulb

temperature is too low and/or relative humidity is too high, then additional heating of 30 W is switched on. If wet bulb temperature is too
high a small pump recirculates about 4 l/min of chamber air through a
tube filled with silicagel or another drying substance. The dry bulb temperature can be regulated between 10 and 40°C and relative humidity
between 50 and 90%.

MODE OF OPERATION

Air conditioning and capacity of the installation

Air is exhausted from the large chambers by means of a centrifugal
fan. This air quantity can be adjusted. The quantity of air exhausted
from the chambers is replaced by outside air. The outside air is admitted
to the air conditioning circuit through motorized valves. The valve opening is regulated by the chamber pressure controller and with a servomotor. The chamber is completely air conditioned in order to maintain the
desired dry and wet bulb temperature.

In the main air circuit air is drawn from the chamber and recirculated
through the air conditioning unit with a cooling coil, electric heaters,
sprays and recirculation fan and then back to the respiration chamber.
The cooling coil in the air conditioning unit of chambers 1 and 2 is fed
with chilled glycol which is maintained at about 1°C in a separate circuit. Glycol temperature is maintained with the use of air cooled heat exchangers. Air is recirculated to provide air velocities at about 0.2 m/s
or less at the site of the animals. Before entering the chamber ingoing
air is mixed with recirculating air and drawn through the heater and
humidifier. In the chambers 1 and 2 recirculation fans are placed just
after the heater (Figure 3a). Chambers 3, 4, 5 and 6 are cooled with
freon filled coils and have each only one recirculation fan. A dry bulb
controller switches the electric heaters and cooling on/off. In chambers 1
to 4 the capacity of heating supply switched on depends on the deviation
from the desired temperature. The wet bulb temperature controls the
sprays and also the heating. Water condensated in the heat exchanger
(cooler) is collected. As a high CO_2 level is to be maintained in the
chambers for measuring purposes, the ventilation rate should be rather
low.

Before entering the air conditioning unit, part of the recirculating air, mixed with outside air, is drawn through the direct expansion coils of the dehumidification circuit and the temperature maintaining unit. After leaving the unit the air is brought back into the main circulation circuit. As the indirect calorimetry system has been selected it is required to measure CO_2, O_2 and eventually also CH_4 of the in- and outgoing air. It should be pointed out that in the part inside the air conditioning unit where overpressure might be present, i.e. just after the recirculation fan, the casing should be absolutely air tight. By keeping the chamber pressure below atmospheric errors in the determination of gaseous exchange are reduced because leakage of respiratory gas out of the chamber is prevented. No chamber air may leave the chamber or air conditioning unit in any other way then through the exhaust fan. When leakage is present only outside air is coming in. This will not cause an error because this will be registered by the gasmeter at the end.

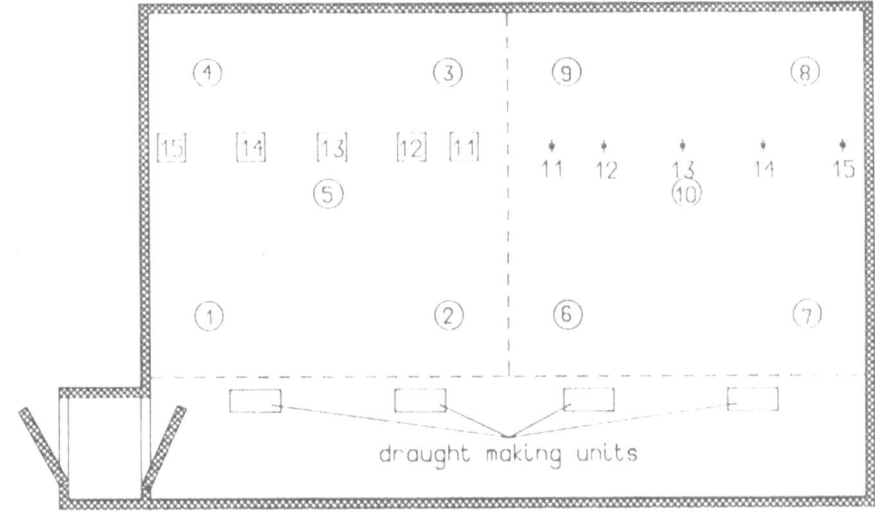

Figure 6. Location of measuring points of air velocities (Table 2 and 3). * indicates points of measuring of draught at various distances above the floor when additional air conditioner is operating (about 1m above the floor).

Table 2. Air velocities (m/s) inside chamber 1 and 2 at 5 different recirculation rates. Points of measurement are 500 mm above the floor and are indicated in Figure 6.

Place of measure-ment	Recirculation rate (1 = lowest; 5 = highest)				
	1	2	3	4	5
1	0.24	0.27	0.38	0.40	0.42
2	0.15	0.16	0.18	0.17	0.18
3	0.13	0.12	0.17	0.19	0.21
4	0.17	0.27	0.32	0.37	0.36
5	0.17	0.14	0.18	0.18	0.22
6	0.11	0.14	0.19	0.17	0.25
7	0.10	0.15	0.18	0.21	0.25
8	0.09	0.15	0.18	0.23	0.22
9	0.10	0.11	0.17	0.19	0.18
10	0.13	0.16	0.19	0.21	0.25

The volume of air recirculated is 7200 m³/h in chamber 1 and 2 (this could be increased to 9000 m³/h). In chambers 3 and 4 it is 570 m³/h and in chambers 5 and 6 about 30 m³/h. In chambers 1 and 2 various air velocities can be applied. This is done by increasing recirculation rate. In Figure 6 and Table 2 air velocities have been given at various locations in the chamber. As can be seen from Table 2 there is a gradual increase in air velocity with increased recirculation rate. However, at normal circulation rate (see rate 1, Table 2) air velocity is in general below 0.20 m/s. In chambers 3, 4, 5 and 6 it was similarly tested whether air velocity was below 0.20 m/s at the location of the animals. The conditioning unit of chamber 1 and 2 was designed to manage about 4,5 kW of heat produced by animals in addition to the heat of fans, light and transmission through walls. This is equal to the heat production of four high yielding cows together or about 20 normally growing pigs of 100 kg. In chambers 3 and 4 about 0.22 kW from animals can be cooled per chamber. Air can be humidified if needed. Normally this is not necessary since animals produce enough water vapour. The surplus vapour in the recirculating air will be collected from the heat exchangers. It has been found previously that gaseous ammonia will be trapped in condensed water (Verstegen et al., 1973). This condensed water can be sampled and analyzed for NH_3 to correct the nitrogen balance accordingly.

Measurement of the exhausted air quantity

Air is exhausted from the chamber with dry gasmeters (Schlumberger meterfabriek Dordrecht, The Netherlands). In chambers 1 and 2 two gasmeters are placed in parallel. Ventilation rate can be measured with one or two meters each with a minimum measuring rate of 0.650 m³/hour and a maximum measuring rate of 100 m³/hour. In chamber 3 and 4 air leaving the chamber is measured with a gasmeter with a maximum capacity of 6 m³/hour and a minimal rate of 0.04 m³/hour. In chambers 5 and 6 the gasmeters can measure a maximum ventilation rate of 2.5 m³/hour and a minimal rate of 0.016 m³/hour.

Sampling of air in all chambers is done in a similar way. Just before or after the gasmeter, the air is sampled in duplicate with two small membrane pumps. A sampling pump takes about 1-2 litres/minute and part of this air is collected in recipients of glass (containing about 2,5 litre each) during periods of 12, 24 or 48 hours. In the recipients pistons, covered with mercury, are gradually lowered by a synchronous motor. This sampling is in accordance with the procedure described by Van Es (1961 and 1966). At the end of the 12, 24 or 48 hour periods the sample in the recipients is analyzed on a paramagnetic Servomex OA 184 O_2 analyser (Servomex, Zoetermeer, The Netherlands) and an infrared TPA-301 CO_2 analyser (Ahrin, Rijswijk, The Netherlands). Part of the air sampled by the membrane pumps is drawn continuously through the CO_2 and O_2 analyser the contents being recorded. The two analysers are calibrated with air of a known composition from a high pressure cylinder. The composition of the air in the cylinder is analyzed with a modified Sonden apparatus similar to that described by Van Es (1958).

All data on CO_2, O_2, temperature and humidity of the air entering and leaving the chamber as well as the temperature of the gasmeter and the barometric pressure are recorded on a multichannel recorder. In addition all data measured are stored in a data storage system, checked for parity errors and send to the computer. Computation of CO_2 production and O_2 consumption from these data is done similarly as described by Verstegen et al. (1971).

Balance periods of 6 to 7 days are normally studied. In each period heat production is normally determined two times 48 hours. Measurements of gaseous exchange are done during each 18 minute period of these 48 hours. This means that analysis of chamber air is done at the start and end of each 18 minute period. Thus, determination of CO_2 production and O_2 consumption is done per period of 18 minutes. The O_2 and CO_2 content measured at a specific time is compared with the CO_2 and O_2 content 18 minutes before. In between ingoing air was also sampled and measured for CO_2 and O_2. Sometimes measurements of O_2 and CO_2 from ingoing air give different output since the previous measurements of 18 minutes ago. Then it was assumed that changes in signal magnitude have occurred gradually over time. Signals of CO_2 and O_2 output from chamber air between these two times were assumed to be effected in the same way as the signals from ingoing air. Thus a kind of smoothing procedure is applied when an output signal of ingoing air changes its magnitude. In this way large erratic fluctuations in CO_2 and O_2 measurements and thus large fluctuations in respiratory coefficients are avoided.

Measurements of activity

All three types of chambers are provided with activity meters. The principle is given in Figure 3b. This Figure shows a similar method of measurement as Wenk and Van Es (1976). Activity is measured with ul-trasound waves during every 6 minute period. The doppler effect (in microvolts) is a measure for activity of animals. Every surface change is associated with a movement of animals. The devices used are a Messl Spacegard Burglar SX15 alarm and also a Solfan microwave intrusion de-tector, model 3225. Activity is measured by placing activity meters in or above the chambers. In chambers 1 and 2 two devices are used in each chamber: one at 2 m above each pen. In chambers 3 and 4 one meter within each chamber is placed at about 1 m above the animals. For cham-bers 5 and 6 the activity meter is placed at a distance of about 0.5 m above the animals and outside the chamber. The activity meter is at-tached to the chambers thus movement of the chambers itself is not re-corded. Similarly as reported by Wenk and Van Es (1976) the correlation coefficient between activity and heat production within a day was about 0.6 to 0.9. From measurements within a day the relation between heat production and activity was calculated as:

$$H = a * c + b$$

in which
   H = heat production
   a = regression coefficient
   c = activity counts
   b = constant

The microwave burglar devices are high frequency devices. Therefore they function relatively independent of temperature and humidity. Also noise and air turbulence do not affect the measurements. The relation between the output of the activity meters and the heat production have been shown to be clearly linear (Verhagen, 1987). This also appears from the correlation coefficients as mentioned earlier.

Telemetry

To measure body temperature an automatic recording system was developed by the Technical and Physical Engineering Research Service at Wageningen. The system consists of transponders, a receiver, an interface and a microcomputer (Figure 7). The transponders can be implanted inside animals (Verhagen, 1987). Every 30 seconds a signal from a transponder is transmitted through the antenna to the receiver. The transponder is operated by a quartz clock and its identification and the measurements of temperature are transmitted in time intervals. Each transponder operates at a different frequency (about 30 MHz). The interface converts the length of the signal into milliseconds which in turn are converted to actual data by the microcomputer. Data are stored as a mean value, e.g. over 5 min or longer intervals. Transponders are calibrated before and after experiments. The weight of the transponder is about 120 grams. Dimensions of a transponder are given in Figure 7 also. The transponder is made of glass and stainless steel.

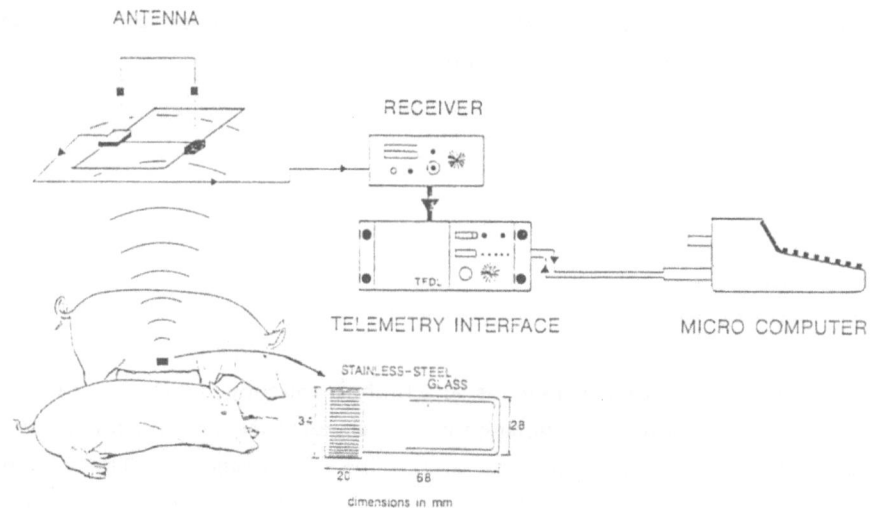

Figure 7.   System of telemetrically measurements of body temperature.

COMPUTATION OF RESPIRATORY GASEOUS EXCHANGE AND WATER VAPOUR PRODUCTION

The computation of CO_2 production, O_2 consumption, CH_4 production and H_2O production (H_2O production = water vapour production) is carried out as follows (Van Es and Verstegen, 1968):

First the measured volume of air is converted into volume at standard conditions (dry air, 0°C and 760 mm Hg). This is done with the equation:

$$V_o = V_g \times \frac{B-Pw}{760} \times \frac{273}{273 + T_g} \tag{1}$$

in which

V_o = standardized volume, dry air 0°C and 760 mm Hg (litres)

V_g = volume measured with gasmeters (litres)

B = average barometric pressure

P_w = pressure of water vapour inside the chamber

T_g = average dry bulb temperature of outgoing air as measured in the gasmeter (°C).

It is assumed that P_w in the chamber is equal to that in the gasmeters since the temperatures in the chamber and gasmeter do not differ much (only some °C). P_w is computed with the following equation derived by Brouwer (Van Es, 1961).

$$P_w = R_{ch}/100 \times (3.999 \times 0.45547 \ T_{ch} + 0.001708 \ T_{ch}^2 + 0.000468 \ T_{ch}^3) \quad (2)$$

in this formula

R_{ch} = relative humidity in the chamber

T_{ch} = dry bulb temperature in the chamber.

The volume of ingoing air at standard conditions will differ from that of outgoing air at standard conditions when the volumes of O_2 consumed and of CO_2 produced are not equal. Furthermore it is assumed that the production of N_2 by the animals (Costa et al., 1968) is so small that it has no influence on the computation of CO_2, CH_4 and O_2. The volume of ingoing air at standard conditions can be computed from

$$V_i = V_o \times (100 - C_o - O_o - M_o)/(100 - C_i - O_i) \quad (3)$$

in this formula C, O and M are % CO_2, O_2 and CH_4 respectively and the suffices i and o stand for ingoing and outgoing air. The CO_2 production, O_2 consumption and CH_4 production of the animals are calculated as:

litres CO_2 produced = $C_o \times V_o - C_i \times V_i + A_i \times (corr_c)$ (4)

litres CO_4 produced = $M_o \times V_o$ (5)

litres O_2 consumed = $O_i \times V_i - O_o \times V_o + A_i \times (corr_o)$ (6)

in which $(corr_c)$ and $(corr_o)$ represent the correction for a possible difference in CO_2 and O_2 content in the chamber at the start and at the end of the period chosen. The correction factor A_i for chamber 1 and 2 amounts to 8 litres for each 0.01 vol % difference in contents of O_2 and CO_2 at start and end respectively because the chamber content is about 80 m³. Similarly the correction A_i for chamber 3 and 4 (1.8 m³) amounts to 0.18 l for each 0.01 vol % difference in contents of O_2 and CO_2 at the start and end. For chamber 5 and 6 (0.085 m³) the factor A_i amounts to 0.0085 liter for each 0.01 vol % difference in contents of O_2 and CO_2 at the start and end of the chosen period. The concentrations of O_i, O_o, C_i and C_o content are measured with an accuracy of about 0.01%. An accuracy of 1% in CO_2 can thus only be maintained if the difference in C_i and C_o is about 1% (the variation due to this error is then 0.01/1 * 100% = 1%). However, also another source of error is involved. The chamber air changes in CO_2 content due to CO_2 production of the animals. If the chamber content is large like in chamber 1 and 2, then 0.01 vol % CO_2 of 80 m3 is about 8 liter of CO_2. Thus the error in measurement for correction of differences in chamber content is at least 8 liter of CO_2. Suppose the ventilation rate is 1000 liter and the mean CO_2 content of air leaving the chamber is about 1%. Then 10 liter of CO_2 are leaving the chamber. The error from correction is then about 80%. If the ventilation rate is 10.000 l then this error is about 8%. This means that ventilation rate should be at least as high as chamber content when 1% as measuring error for CO_2 can be made.

Heat production of mammals can be computed from data on gaseous exchange and N-excretion using the formula of Brouwer (1965):

$$H \ (kJ) = 16.18 * O_2 + 5.02 * CO_2 - 2.17 * CH_4 - 5.99 * N \qquad (7)$$

Heat production of birds is calculated with the formula of Romijn and Lokhorst (1961):

$$H = 16.20 * O_2 + 5.00 * CO_2 - 1.59 * N \qquad (8)$$

where O_2, CO_2 and CH_4 represent volumes consumed or produced (litres) and N is urinary nitrogen (g). The use of urea or uric acid N output as a measure of protein degradation is slightly inaccurate because several other nitrogeneous compounds may be present in the urine. One g of urea-N represents about 269 kJ but in uric acid it represents 485 kJ. When expressed in kcal the heat production in kJ has to be divided by 4.186. For growing pigs we may assume that the CO_2/O_2 ratio (RQ) is about 1. Moreover, CH_4 and N in equation (7) only give a very small contribution to the heat production. Only a few litres of CH_4 (Verstegen, 1971) are produced by pigs and a few grammes of N are found in the urine compared to exchange in CO_2 and O_2. Therefore we may change this formula to:

$$H = 21.20 * CO_2$$

Thus, per kJ heat about 0.02 l of CO_2 are produced. Water vapour production is computed as H_2O leaving the chamber plus water collected from the dehumidifier minus H_2O entering the chamber. Water vapour leaving the chamber (H_2O_o) with the air is:

$$H_2O_o \text{ (g)} = V_o \times 0.8036 \times P_w/(B - P_w) \tag{9}$$

in which

P_w = water vapour pressure of air leaving the chamber (formula 2).
When substituting relative humidity and temperature of ingoing air in formula (2) the water entering the chamber (H_2O_i) can be computed as:

$$H_2O_o i = V_i \times 0.8036 \times P_{wi}/(B - P_w) \tag{19}$$

in which

P_{wi} = water vapour pressure of air entering the chamber.
Subtracting the H_2O_o from H_2O_i and correcting this for the H_2O gathered in the dehumidifier will give the H_2O production inside the chamber.

ADDITIONAL PROVISIONS

Table 3. Air velocities (m/s) at various locations (see Figure 6) and various distances above the floor of chamber 1 and 2.

Measuring point	Distance above floor (mm)				
	50	100	200	400	600
11	0.63	0.67	0.77	0.70	0.53
12	0.80	0.82	0.82	0.67	0.30
13	0.70	0.76	0.74	0.63	0.45
14	0.74	0.77	0.78	0.71	0.36
15	0.51	0.61	0.56	0.54	0.43
mean	0.68	0.73	0.73	0.65	0.41

In chamber 1 and 2 the air conditioning can be set at such a scheme that any temperature fluctuation schedule can be applied within the limits of capacity. By means of a preprogrammed scheme daily fluctuation in temperature and relative humidity as occurring in practice can be applied.

In addition an extra air conditioning unit can be used inside chambers 1 and 2 (Verhagen, 1987). This unit can be used to create draught. At farm conditions a cold air stream locally reaching the animals (defined as draught) is supposed to be detrimental for animal health. The unit can be set to increase air velocity at about 300 to 500 mm above the floor to about 0.9 m/s. The temperature of this locally increased air stream can be lowered by about 4-5°C below chamber temperature. In Table 3 results of air velocity tests at various positions are given. The measuring points (11 to 15) are as indicated in Figure 6. Both pens are considered identical. Therefore means are given.

Feeding can be done from outside per individual animal regardless of the housing system used (Figure 3a). Also group feeding can be done from outside. During the respiration experiments the chambers 1 and 2 must not be entered, or, if needed, only by an air lock (Figure 3a). Therefore the animals are fed from outside through a funnel from which a tube leads to the trough. Dry feed or wet mash is put in the funnel and when a plug in the funnel is opened, the mash flows through the tube into the trough. Sometimes with some extra water the funnel can be emptied. The lid of the trough is opened from outside.

ENERGY AND NITROGEN BALANCES

The floor of chambers 1 and 2 has a fall of about 1% into a gutter (Figure 3a and 3b). At the end of each measuring period the floor can be easily cleaned with water. The amount of water used for cleansing is measured by a water meter. The mixture of water, faeces and urine is collected in the gutter. From this gutter it will flow into a tank. This tank is maintained in a cellar. After filling the tank it is lifted and weighed. The contents are sampled after thoroughly mixing. Moreover various metabolism crates (5 per chamber for calves, boars or sows) can be used to determine ME-intake per individual animal.

In chamber 3 and 4 urine is collected through a gutter underneath the animals and flows directly into a funnel outside the chambers. Faeces is collected in a vessel underneath the floor of the chambers. Feeding can be done manually through rubber gloves (Figure 4) protruding into the chambers provided that the feed is already inside. Metabolism crates or cages can be used also in these chambers, e.'g. for laying hens, when individual measurements are needed.

In chamber 5 and 6 collection of excreta can only be done after opening of the chambers.

The data on energy and N-content in faeces, urine, or the mixture from the gutter, dust catched in air filters, condens water collected from the dehumidification section and in the air leaving the chamber are used to compute energy and N-balances.

TECHNICAL TESTS OF THE SYSTEM

Climatic tests

The air conditioning unit can maintain a constant climate in the chamber with temperatures adjustable between 10 and 40°C in the summer and between 5 and 40°C in winter. In Figure 8 the rate of change in chamber conditions in chamber 1 and 2 are given for a summer (a) and winter (b) situation. Desired conditions can be reached within a short period. During the tests it was found that the temperature within an experimental period did not deviate more than 0.5°C from the desired value. Relative humidity can be adjusted between 40 to 90% and the deviation from the desired value is maximally about 5%.

Leakage test

As respiration experiments require the measurement of all CO_2 and CH_4 produced and of all O_2 consumed no air should leave the chamber except through the exhaust fan and gas measuring unit. Moreover leakage to the chamber should be reduced since also at the minimum ventilation rate the pressure should be maintained at 5-7 mm water column below atmospheric. Tests revealed that leakage was very low. It was in the chambers 1 and 2 about 0.120 to 0.150 m³/minute when the pressure was about 5-7 mm below atmospheric pressure. In chambers 3 and 4 this

42

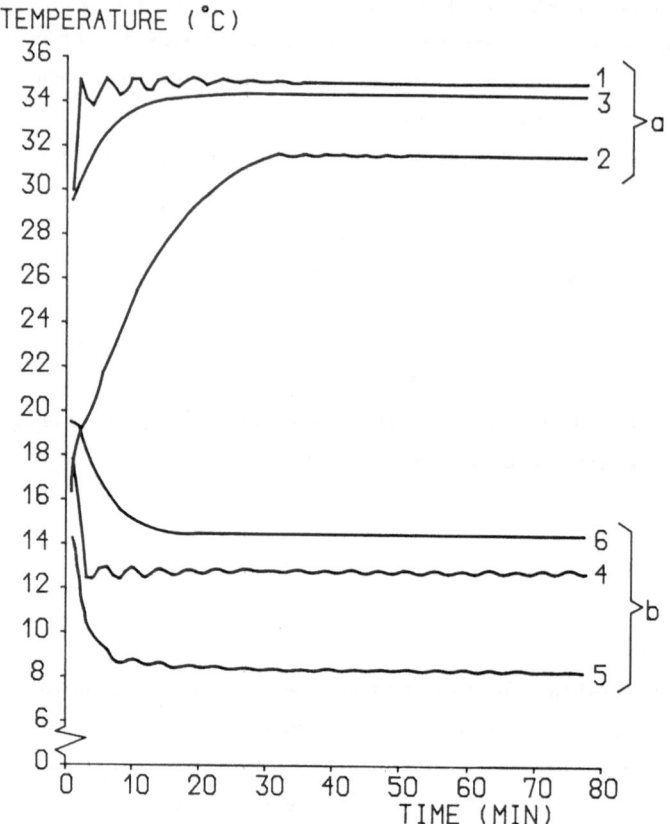

Figure 8. Rate of change in chamber conditions with change in ingoing dry and wet bulb temperature for a summer situation (a) and a winter situation (b): 1 and 4, dry bulb temperature of ingoing air; 2 and 5, wet bulb temperature of ingoing air; 3 and 6, dry bulb temperature of chamber air.

leakage was about 0.002 to 0.003 m³/minute at this underpressure and in chambers 5 and 6 about 0.001 m³/minute.

Test experiment with CO2

In order to find out whether with the equipment reproducible mea-surements of CO_2 could be made before and at the end of an experiment with animals, CO_2 was released into the chamber from high pressure cy-

Table 4. Test experiments with CO_2 in various chambers.

Chamber no.	CO_2 released from cylinder (l)	Duration (min)	Recovery (%)
1	7235	1090	100.8
1	6197	1147	100.0
1	3301	1012	101.3
1	8212	1407	100.5
1	2804	1020	101.5
2	7191	1065	99.5
2	5208	1150	100.1
2	7030	1170	101.5
2	7306	1433	100.7
2	2894	990	99.8
3	69.6	1441	102.6
3	326.5	1490	102.6
3	90.1	1501	101.7
3	120.2	1363	99.1
3	195.3	1397	102.6
4	444.1	1123	100.3
4	349.5	1290	99.1
4	54.9	1440	98.6
4	57.7	1357	97.8
4	97.6	1043	99.0
5	43.1	1280	104.1
5	48.1	1237	101.7
5	75.9	1370	107.0
6	56.1	1410	98.5
6	80.1	1340	105.1
6	80.1	1279	105.1

linders at such a rate as can be expected from animals. As the experiments are done with animals of different weights the tests were also carried out with a release of various amounts of CO_2. At the start of a test experiment CO_2 is released from the cylinder until the CO_2 content in the chamber is 0.7 to 1.0%. Then the cylinder is weighed. Together with the start of ventilation the release of CO_2 from the cylinder and the gas sampling are started. At the end of the CO_2 test, usually lasting 15-25 hours, the cylinder is weighed and the gas sample is analyzed for CO_2. The amount of CO_2 released in the chamber is measured in the same way as the CO_2 production of the animals. When the same quantity of CO_2 is recovered as was released, the volume measurement of the air leaving the chamber, the gas sampling and the CO_2 analysis are correct.

The correctness of the O_2 analysis is verified by analyzing ingoing air, having a constant composition, and gases of known contents, stored in high pressure cylinders.

The CO_2, CH_4 and O_2 analysers are calibrated by passing calibration gas of a known composition through it. The samples in the cylinder are analyzed on a Sonden as described by Van Es (1958). The other parts of the calibration curve for each analysis are also tested by analyzing air with various contents of CO_2 and O_2 on the infrared TPA CO_2 analyser and on the paramagnetic O_2 Servomex analyser and also on the Sonden. The results of CO_2 tests are given in Table 4.

As recoveries are close to 100% it was thought that no further tests with O_2 (or burning alcohol) are necessary.

As stated before differences in content before and after a respiration period must also be taken into account. It is clear from the calculation that at a low ventilation rate there is much more fluctuation from this source compared to a high ventilation rate. Apart from ventilation rate (clearance) it is important that metabolic body size is high enough to maintain:
- a high enough CO_2 percentage in the outgoing air;
- and reach a sufficient level in a short time.

The concentration is the balance between clearance by ventilation and release by production:

$$R/V = c * q$$

in which

R = release of CO_2 in l per time unit (= 0.02 l per kJ metabolic rate)

V = volume of the chamber

c = required steady concentration (C_o)

q = clearance rate (ventilation)

The errors of recovery need to be within limits of about 2% to give reliable results of gaseous exchange for shorter periods. The CO_2 and O_2
content of air entering and leaving the chamber is measured at about
0.01% accuracy. When longterm experiments are performed a recovery
which is systematically too low will influence the accuracy of comparison
made between two measurements which differ a long time.

Calibration tests of the gasmeters

Gasmeters are supposed to have an inaccuracy of less than 1%. However, as animals change in size and in metabolic rate various ventilation
rates have to be applied to maintain a similar CO_2 content in the chamber. Therefore gasmeters were tested with a mercury pump at the department of Animal Physiology (Van Es, 1966). The measurements show
that the calibration values of the gasmeters depend on ventilation rate
(Table 5). It is therefore important to calibrate a gasmeter at different
rates.

Correction for persons entering the chamber

The large calorimeters are provided with an air lock of about 1 m³
(Figure 3). These are provided to allow a person to enter the chamber
for short periods at feeding, to take measurements or to treat animals.
Entering the chamber will only give a small error. This can be seen as
follows: suppose there are animals in the chamber with an equivalent metabolic body size of 10 persons. Then the disturbance of a person will
be less than 10% during the few minutes of human presence, because the
animals are growing and their metabolic rate per unit of metabolic body
size is much higher than that of a person. In addition, the person will
replace 1.0 m³ of chamber air by ingoing air. This will compensate for
about the metabolism of the person. A person will produce about 16 l
CO_2 per hour and consume a similar amount of O_2. Suppose the chamber
has about 0.7% CO_2, then 1.0 m³ will contain about 7.0 l CO_2. Thus, if
the person stays about 25 minutes in the chamber, he will produce about

Table 5. Calibration value of gasmeters at various ventilation rates.

Chamber	Gasmeter	Flow through gasmeters (l/min)	Calibration value
1	1	494	0.9705
	1	678	0.9989
	1	778	1.0227
	1	974	1.0561
1	2	434	1.0122
	2	693	1.0430
	2	910	1.0813
	2	947	1.0965
2	1	487	0.9545
	1	680	0.9820
	1	929	1.0325
	1	1022	1.0514
2	2	446	0.9800
	2	678	0.9989
	2	833	1.0409
	2	955	1.0618
3	1	12	0.9532
	1	26	0.9534
	1	36	0.9602
	1	52	0.9579
4	1	11	0.9182
	1	26	0.9219
	1	36	0.9252
	1	46	0.9471
5	1	6.02	1.0138
	1	7.63	1.0176
	1	8.50	1.0267
6	1	5.52	0.9966
	1	8.67	0.9997
	1	9.39	1.0033

6.7 l of CO_2. Thus addition of CO_2 from the person and the replacement of 1.0 m³ chamber air via the air lock with ingoing air are similar. It is thought that no correction is needed for the person provided that he/she does not upset the animals.

REFERENCES

Brouwer, E., 1965. Report of sub committee on constants and factors. 3rd Symposium on Energy Metabolism. Troon Scotland EAAP publ. 11: 441-443.

Bransen, A.M. and Kneepkens, H.E., 1982. De geiten respiratiecellen. Koeltechniek 75: 256-258.

Close, W.H., Mount, L.E. and Start, I.B., 1971. The influence of environmental temperature and plane of nutrition on heat losses from groups of pigs. Anim. Prod. 13: 285-294.

Close, W.H., 1981. The climatic requirement of the pig. In: J.A. Clark, editor. Environmental aspects of housing for animal production. Butterworths, London: 149-161.

Costa, G.L., Ulrich, L., Kantor, F. and Holland, J.F., 1968. Production of elemental nitrogen by certain mammals including man. Nature, London, 218: 546-551.

Es, A.J.H. van, 1958. Gas analysis in open circuit respiration chambers. 1st Symposium on Energy Metabolism, Copenhagen. EAAP publ. 8: 132-137.

Es, A.J.H. van, 1961. Between-animal variation in the amount of energy required for the maintenace of cows. Publ. Landbouwk. Onderz. 67.5: 116 pp.

Es, A.J.H. van, 1966. Labour saving methods for energy balance experiments with cattle; description of equipment and methods used. Neth. J. Agric. Sci. 14: 32-46.

Es, A.J.H. van and Verstegen, M.W.A., 1968. Data processing in physiology experiments. Int. Summer School in biomathematics and data processing in animal experiments. Elsinore, Denmark: 8 pp.

Hicks, A.M., 1966. Physiological responses of growing swine to low temperatures. University of Sashatchewan, Ph.D. Thesis.

Holmes, C.W., 1966. Studies on the effect of environment on heat losses from pigs. Queens Univ. of Belfast. Ph.D. Thesis.

Holmes, C.W. and Mount, L.E., 1967. Heat loss from groups of growing pigs and under various conditions of environmental temperature and air movement. Anim. Prod. 9: 435-451.

Kroonen, J.G.E.M., Verstegen, M.W.A., Boon, J.H. and Hel, W. van der, 1986. Effect of lungworms (Dictyocaulus viviparus) on energy and nitrogen metabolism in growing calves. Brit. J. Nutr. 55: 351-360.

Pfeiffer, H., 1968. Die Quantitative und Qualitative Schlachtkörperzusammensetzung sowie der Nährstoffansatz bei Schweinen unter verschiedenen Haltungsbedingungen. Hühn. Archiv. no. 82: 1-68.

Romijn, C. and Lokhorst, W., 1961. Some aspects of energy metabolism in birds. In: 2nd Symposium on Energy Metabolism. EAAP publ. no. 10: 49-59.

Sörensen, P.H., 1961. Influence of climatic environment on pig performance. In: Nutrition of Pigs and Poultry. Proc. Univ. of Nottingham. 8th Easter School in Agric. Sci., Butterworths, London: 88-103.

Verhagen, J.M.F., 1987. Acclimation of growing pigs to climatic environment. Ph.D. Thesis. Agricultural University Wageningen: 128 pp.

Verstegen, M.W.A., 1971. Influence of environmental temperature on energy metabolism of growing pigs housed individually and in groups. Meded. Landbouwhogeschool Wageningen, 71-2: 115 pp.

Verstegen, M.W.A., Hel, W. van der, Koppe, R and Es, A.J.H. van, 1971. An indirect calorimeter for the measurement of the heat production of large groups of animals kept together. Meded. Landbouwhogeschool Wageningen 71-16: 13 pp.

Verstegen, M.W.A., Close, W.H., Start, I.B. and Mount, L.E., 1973. The effects of environmental temperature and plane of nutrition on heat loss, energy retention and deposition of protein and fat in groups of growing pigs. Brit. J. Nutr. 30: 21-35.

Wenk, C. and Es, A.J.H. van, 1976. Eine Methode zur Bestimmung des Energieaufwandes für die Körperliche Aktivität von wachsenden Küken. Schweiz. Landwirtsch. Monatshefte 54: 232-236.

CHAPTER II.   HOUSING SYSTEMS AND ENERGY METABOLISM

ADAPTATION TO, AND ENERGY COSTS OF, TETHERING IN PREGNANT SOWS

G.M. CRONIN AND J.L. BARNETT

ABSTRACT

In moving away from group-housing systems for pregnant sows during the last 25 years, tether-housing was developed to reduce construction and labour costs. However, compared to group-housing, tethering has been reported to lead to an increase in MR and blood corticosteroid levels, and to reduce the immune response and health status of sows; factors which may decrease productivity. Tethering has also been associated with the development of behaviours such as stereotypies and excessive drinking: these behaviours appear to have a stress-reducing function. The process of adaptation itself bears a production cost to the animal: initially from a sustained elevation in blood corticosteroid levels and subsequently, in association with increased levels of stereotypy performance, a higher energy requirement and MR.

Future research is therefore needed to further develop housing systems for pregnant sows that consider both the species-specific requirements of the animal (e.g. social contact, exercise) and the economic production of pig meat. A current advance in this direction is the development of the electronic sow feeding system, which incorporates the better features of both group- and tether-housing systems.

INTRODUCTION

During the last 25 years in the pig industry, there has been a trend towards intensification of husbandry conditions and a concomitant increase in capitalization. Modern pregnant-sow accommodation for example, has been designed to help reduce construction and labour costs. Indeed, feed costs now represent the major proportion (about 70%) of the total costs of keeping pregnant sows on Dutch farms.

Tether-housing has become a common form of pregnant-sow accommodation in The Netherlands as well as other European countries. Tethering involves restraining sows by means of a neck- or girth-tether (see Daelemans, 1984), and provides advantages such as minimization of floor area per sow and ease of effluent disposal compared with group-housing systems. Sows in tethers, as well as other forms of individual housing, also have the benefit of individual and controlled feeding, individual care by the stockperson and reduced injuries due to fighting compared to group-housed sows. On the negative side, tethering limits the sow's ability to exercise, to huddle when cold and increases specific health problems (Tillon and Madec, 1984). Further, the performance of social behaviours is restricted, and tethered sows may not be able to withdraw from the agression of an adjacent sow (Barnett et al., 1987a).

Restraint by tethering has been found to alter the type of behaviour, but not necessarily the quantity of activity performed, compared with group-housed sows (Svendsen and Bengtsson, 1983; Barnett et al., 1984). Stereotypic behaviours and excessive drinker-use are reported to be common behaviours performed by tethered sows (Sambraus and Schunke, 1982; Stolba et al., 1983; Cronin, 1985). By definition, stereotypic behaviours are actions that are apparently without purpose, but which are performed repetitively and continuously in an identical manner, sometimes for many hours per day. The reported stereotypies generally involved oral actions (lick, chew) which were performed against the tether chain or bars of the stall, or were sham activities. Because of the association between tether-housing of sows and the performance of stereotypic behaviours, tether-housing has at times been criticized in both the scientific and popular press with the suggestion that the welfare of sows may be reduced due to tethering.

The suitability of any housing system will depend upon the ability of sows to adapt to it. Modifications to the behaviour and physiology of sows in tethers compared to group-housing systems are presumably representative of, or associated with, the process of adaptation to tether-

housing relative to group-housing. Depending on its extent, this adaptation may impose a cost on the animal which should be measurable in production terms.

The aims of this paper are to quantify the cost of adaptation to the
animal in terms of metabolic rate and to identify some behavioural and
physiological adaptations to tether-housing.

RESPIRATION CHAMBER EXPERIMENT

Heat production (HP) is a good estimate of metabolic rate (MR). Since
MR is a major determinant of feed conversion efficiency (FCE), HP is also a useful measure of the potential profitability of different housing systems for pregnant sows. Factors that influence HP include the level of
activity and the thermal environment of the sow. While it has been found
by Svendsen and Bentsson (1983) and Barnett et al. (1984) amongst
others, that tethered sows may be as active as group-housed sows,
Cronin (1985) reported that recently-tethered sows were much less active
during the first few weeks of tethering than later on. Other experiments
comparing recently-tethered and group-housed sows have shown an increased MR in the former animals (e.g. Geuyen et al., 1984). Only one
experiment has been reported in which the HP of experienced and inexperienced tethered sows was compared. This experiment was performed
in the respiration chambers at the Agricultural University of Wageningen, and include the following treatments (see Cronin et al., 1986a for
full details):
1. tethered sows with about 6 months experience of tethering;
2. tethered sows with about 2 months experience of tethering; and
3. sows loose-housed as a group.

HP from the experienced-tethered (treatment 1) sows was 20.8 and
11.8% greater, respectively, than the treatment 2 and 3 sows. As expected, the sows were most active during the 12 hours of light (06.00 to
18.00 h). HP during this period was 35.6 and 23.9% greater for the
treatment 1 sows than the other 2 treatments, respectively (see Cronin
et al., 1986a). This increased MR of the experienced-tethered sows (as
estimated by HP), compared with the other treatments implies a physiological change had occurred due to long-term tethering. Associated with

this was a behavioural change, represented by an increased activity level, most of which was in the form of stereotypies and increased drinker use. However, increased activity alone does not account for the differences in HP as the activity-free HP of treatment 1 sows during the dark period (18.00 to 06.00 h) was about 8% greater than the treatment 2 sows and 6.5% greater than the treatment 3 sows.

There are a number of potential deficiencies associated with tether-housing, to which the sow may respond by altering her behaviour and/or physiology compared with group-housing. Such deficiencies include reduced social contact, inability to huddle when cold, restraint per se and inability to perform functions like exercise, comfort activities, exploration and other 'obligatory' behaviour sequences (Stolba, 1983). Should the performance of functionally-important behaviours remain inhibited, and should the sow fail to satisfy her requirements via some alternative strategy, then chronic stress may occur. If, as has been suggested by McBride (1980), a behavioural modification to a chronic stressor is ineffective, a change in physiology may occur, such as that reported by Cronin et al. (1986a) with an increase in MR. Further, it might be expected that a related lowering of production occurs.

In the next 2 sections, the consequences for sows of tether - compared with group-housing will be discussed in relation to physiological and behavioural change, and the implications of these changes for production.

PHYSIOLOGICAL CHANGES AS A CONSEQUENCE OF HOUSING IN NECK TETHERS

Central to the outcome of animal production is protein metabolism. While an increased MR is directly associated with a decreased FCE, the accompanying reduction in protein incorporation into tissues of growing pigs or the foetuses of pregnant sows will result in production losses, i.e. a decreased growth rate or litter size/survival. Glucocorticoids have a direct effect on nitrogen balance and at the same time are recognized as providing an indicator of a stress response (e.g. Moberg, 1985). In turn the stress response, particularly because of its detrimental effects on nitrogen balance, health and reproduction, can indicate potential losses in production. Restraint is known to be a potent stressor of the pituitary-adrenal-axis and therefore it is pertinant to examine whether pigs in tethers exhibit a chronic stress response with consequent effects on production.

Several experiments have been carried out over the last 6 years on the adaptation of sows to different housing systems, in particular comparisons have been made between pigs in neck-tethers and groups and evidence has been sought for a chronic stress response.

Pregnant pigs housed in neck-tethers can show a sustained elevation of free corticosteroid concentrations indicative of a chronic stress response, compared to group-housed pigs (2.2 and 1.3 ng/ml, respectively; Barnett et al., 1985). This elevation of corticosteroid concentrations was sufficient first to alter metabolism resulting in an increase in plasma glucose concentrations (Barnett et al., 1985), which may reflect an increased energy requirement to adapt to a tether-stall environment, and second to lower immunological reactivity to an antigen (Barnett et al., 1987b), reflecting a suppression of the immune system. These data suggest potential adverse consequences due to physiological (and psychological) processes associated with housing in neck-tethers compared to groups with the recommended space allowance of 1.4 m2/pig (Anon., 1983). However, a more recent experiment (Barnett et al., 1987a) shows that the adrenal corticosteroid responses of individually-housed pigs are not as clear as the above suggests. In that experiment the response of pregnant gilts housed in neck-tethers in the unmodified stall used in the previous experiments (vertical bars separated by 138 mm) was compared with gilts in modified stalls in which the vertical bars were covered in a steel mesh (mesh openings of 45 × 70 mm). The modification was designed to reduce the opportunity for agonistic interactions between neighbouring gilts while maintaining some social contact. The results show a reduction in the percentage of attempted agonistic interactions resulting in retaliations in the modified stalls (0%) compared to the unmodified stalls (46%) and gilts housed in a group (4%). The gilts housed in neck-tethers in the modified stalls also showed a reduction in free corticosteroid concentrations; overall mean values for gilts in modified stalls, unmodified stalls and a group measured at 8 and 12 weeks in the housing treatments were 2.7, 4.6 and 2.8 mg/ml, reflecting a reduction in the stress response attributable to the housing system.

These studies show the physiological stress response observed in housing systems with a high level of confinement (neck-tethers) may be the result of impediments to normal social behaviour (withdrawl from threat) which can be ameliorated by the design of the housing system (removal of threat). Thus restraint and/or limitations to movement per se may not be the main determinant of the stress response observed in these studies.

While the experiments of Barnett et al. have concentrated on adrenal physiology associated with housing pigs in neck-tethers, Cronin et al. (1986a) have examined the metabolic consequences of such housing. Although these studies were conducted at different times and in different countries, there is general agreement between the findings of the two approaches. The experiment of Cronin et al. (1986a) showed a higher MR of pigs in tethers compared to groups while the initial observations of Barnett et al. (1985) showed a chronic stress response and effects on nitrogen metabolism in pigs in tethers compared to groups. Further, Cronin et al. (1986a) showed experienced pigs in tethers had a higher MR than pigs in groups and inexperienced pigs in tethers while Barnett et al. (1987b) showed second parity pigs experiencing neck-tethers for the second time had higher free corticosteroid concentrations than second parity pigs in groups for the second time or second parity pigs experiencing tethers for the first time (mean concentrations were 4.7, 2.6 and 3.6 ng/ml, respectively).

One aspect that requires further examination is the effect of pregnancy. Barnett et al. (1987b) have shown that corticosteroid concentrations are greater in pregnant than non-pregnant pigs in tethers, suggesting that pregnancy per se is a stressor. While Cronin et al. (1986a) found no differences in MR between non-pregnant pigs in tethers or groups which agrees with the findings of Barnett et al. (1984) who found no differences in free corticosteroid concentrations between non-pregnant pigs housed in groups or tethers, the metabolic responses of pregnant and non-pregnant pigs in tethers have not been compared.

BEHAVIOURAL CHANGES AS A CONSEQUENCE OF HOUSING IN NECK-TETHERS

Long-term changes

It has been argued by Beilharz and Zeeb (1981) amongst others that through genetic selection domestic animals will progressively adapt to intensive husbandry systems and concomitantly their welfare will be less at risk. While information exists of behavioural and physiological comparisons of pregnant sows in tether- and group-housing, the effects of genetic selection to intensive husbandry have so far not been reported for sows. Such information may be important in determining whether sows have, or are capable of, adapting to tether-housing via selection.

In an experiment performed at the Animal Research Institute, Werribee, (Barnett et al., unpublished), the behaviour and physiology of pregnant gilts of two genotypes, in either tethers or groups was studied. The pigs were all born and reared under similar conditions in the same pig herd but were descended from two different herds. Herd A was a semi-intensive farm on which little selection had occurred during the last 30 years. Herd B was a 'modern' intensive piggery where a consistent and high level selection pressure for higher productivity was practiced.

The activity levels and the main classes of active behaviour by sows, summed over 9 weeks of observation, are shown in Table 1. Inspite of Herd B sows being more active than Herd A sows, Campbell and Taverner (1985) have found that at all levels of energy intake, growing pigs from Herd B deposited more protein and less fat than pigs from Herd A. Within genotype there were also differences in the type and level of performance of behaviours that constituted the recorded activity. Notably, the performance of investigative behaviours (e.g. sniff, lick, touch, listen) was reduced and stereotypic behaviours (e.g. sham chew) increased for Herd B sows compared with Herd A sows. There were also effects on behaviour due to the housing system: within genotypes, tethered sows were more active than group-housed sows. The performance of investigative, social and locomotive/posture changing behaviours were reduced but vigorous and aggressive-like behaviours (e.g. biting), drinking and comfortive behaviours were increased.

Table 1. Proportion of observation time that tethered- and group-housed sows of 2 genotypes were active, and the proportion of active time spent in the performance of different classes of behaviour.

| | Genotype A | | Genotype B | |
	Group	Tether	Group	Tether
Active (% of observation time)	32.0	41.2	47.6	63.2
Class of behaviour (% of active time)				
Environment directed				
investigation	58.3	49.0	43.2	28.9
aggressive-like	0.2	4.3	0.1	6.8
drinking	6.7	6.9	5.6	10.1
locomotive/posture change	7.0	2.8	6.8	2.0
conflict (e.g. escape attempt)	0.0	7.6	0.0	4.2
Social				
agonistic	1.2	0.5	0.8	0.7
non-agonistic	6.6	1.0	8.2	1.9
Self-directed				
stereotypies	17.0	22.0	32.1	36.4
comfortive	1.5	2.0	0.8	1.0
Other	1.5	3.9	2.4	8.0

Although it is not known whether the sows in Herd A and Herd B differed prior to selection, the data suggest that intense selection for high productivity under intensive husbandry may have changed sow behaviour. However, it is not possible to conclude that the Herd B sows were any better adapted to tethering than Herd A sows.

Short-term changes

Surveys performed at large piggeries in Europe by Sambraus and Schunke (1982), Stolba et al. (1983) and Cronin (1985) amongs others, have revealed that stereotypies were a major form of activity performed by tethered shows. Evidence that stereotypies and excessive drinking are related to stress reduction has been presented by Dantzer and Mormede (1983), who showed that blood corticosteroid levels in frustrated pigs were reduced with the performance of stereotypic behaviour, and Brett and Levine (1979) who found that polydipsia suppressed pituitary-adrenal activity in rats. Thus if stereotypies and excessive drinking are effective adaptive mechanisms to reduce stress, there should be no difference in production between group- and tether-housed sows. This is generally supported by the literature: of 14 published studies reviewed by Hemsworth (1982), the majority show few differences. However, there are occasional studies with conflicting results, but since behavioural data were not collected on the sows in most of these studies, it is not known whether stereotypies, etc. were common, nor at what level of performance they occurred. Within a single housing system however, there should be measurable effects on the production of sows showing higher compared with lower levels of these behaviours.

Cronin (1985) investigated the relationship between level of stereotypy performance and piglet production by tethered sows in a large commercial herd and found conflicting results depending on parity of the sows. Some possible reasons for the conflicting results of Cronin (1985) and some of the studies on tethering vs. group housing may relate to our lack of understanding of behavioural adaptation by individual animals and the consequences for production. For example, some of the variation might be accounted for by stereotypies being still 'developing' compared with 'established' (Kiley, 1977; Cronin, 1985) and by endorphins, which have been implicated in the performance of 'developing' stereotypies (Cronin et al., 1986b) and drinking (Baldwin and Parrot, 1984). These points clearly require further research.

In conclusion, the measurement of HP (to estimate MR) not only provides a valuable means for comparison of the productivity of sows in different housing systems, but also significantly contributes to our understanding of the process of adaptation to tethering by sows. At a time when the design of pig accommodation in intensive husbandry systems is influenced by concern over animal welfare, it is also important to con-sider the effect of different housing systems on sow production. In this paper we have presented information on metabolic rate (from respiration chamber studies), blood stress hormones and behaviour and their effects on sow production.

ACKNOWLEDGEMENTS

We wish to thank mr. C.G. Winfield for his constructive criticism of this manuscript and acknowledge the financial support of the Pig Re-search Council of Australia.

REFERENCES

Anonymous, 1983. Model code of practice for the welfare of animals. 1. The Pig. Australian Bureau of Animal Welfare, Canberra.

Baldwin, B.A. and Parrott, R.F., 1984. Effects of naloxone on feeding and drinking in pigs. In: Proc. Int. Congr. Applied Ethology in Farm Animals, 1-4 August 1984, Kiel, FRG, pp. 382-385.

Barnett, J.L., Cronin G.M., Winfield, C.G. and Dewar, A.M., 1984. The welfare of adult pigs: The effects of five housing treatments on behaviour, plasma corticosteroids and injuries. Appl. Anim. Behav. Sci., 12: 209-232.

Barnett, J.L., Hemsworth, P.H. and Winfield, C.G., 1987a. The effects of design of individual stalls on the social behaviour and physiological responses related to the welfare of pregnant pigs. Appl. Anim. Be-hav. Sci. (in press).

Barnett, J.L., Hemsworth, P.H., Winfield, C.G. and Fahy, V.A., 1987b. The effects of pregnancy and parity number on behavioural and physiological responses related to the welfare status of individual and group housed pigs. Appl. Anim. Behav. Sci. (in press).

Barnett, J.L., Winfield, C.G., Cronin, G.M., Hemsworth, P.H. and Dewar, A.M., 1985. The effects of individual and group housing on behavioural and physiological responses related to the welfare of pregnant pigs. Appl. Anim. Behav. Sci., 14: 149-161.

Beilharz, R.G. and Zeeb, K., 1981. Applied ethology and animal welfare. Appl. Anim. Ethol., 7: 3-10.

Brett, L.P. and Levine, S., 1979. Schedule-induced polydipsia suppresses pituitary-adrenal activity in rats. J. Comp. Physiol. Psychol., 93: 946-956.

Campbell, R.G. and Taverner, M.R., 1985. Effect of strain and sex on protein and energy metabolism in growing pigs. Proc. 10th Symp. on Energy Metabolism of Farm Animals, Airlie, Virginia, E.A.A.P. publ. no. 32: 78-81.

Cronin, G.M., 1985. The development and significance of abnormal stereotyped behaviours in tethered sows. Ph.D. Thesis, Agricultural University of Wageningen.

Cronin, G.M., Tartwijk, J.M.F.M.,van, Hel, W. van der and Verstegen, M.W.A., 1986a. The influence of degree of adaptation to tether-housing by sows in relation to behaviour and energy metabolism. Anim. Prod., 42: 257-268.

Cronin, G.M., Wiepkema, P.R. and Ree, J.M. van, 1986b. Endorphins implicated in stereotypies of tethered sows. Experientia, 42: 198-199.

Daelemans., J. 1984. Confinement of sows related to productivity. Ann. Rech. Vét., 15: 149-158.

Dantzer, R. and Mormède, P., 1983. De-arousal properties of stereotyped behaviour: evidence from pituitary-adrenal correlates in pigs. Appl. Anim. Ethol., 10: 233-244.

Geuyen, T.P.A., Verhagen, J.M.F. and Verstegen, M.W.A., 1984. Effect of housing and temperature on metabolic rate of pregnant sows. Anim. Prod., 38: 477-485.

Hemsworth, P.H., 1982. Social environment and reproduction. In: D.J.A. Cole and G.R. Coxcroft (Editors), Control of Pig Reproduction, Butterworth Scientific, London, pp. 585-601.

Kiley, M., 1977. Stereotypies and their causation. Appl. Anim. Ethol., 3: 290-291.

McBride, G., 1980. Adaptation and welfare at the man-animal interface. In: M. Wodzicka-Tomaszewska, T.N. Edey and J.J Lynch (Editors), Behaviour in relation to Reproduction, Management and Welfare of Farm Animals. Reviews in Rural Science No. IV, Univ. of New England, pp. 195-198.

Moberg, G.P., 1985 (Editor). Biological response to stress: Key to assessment of animal well-being? In: Animal Stress. American Physiological Society, Bethesda, Maryland, pp. 27-49.

Sambraus, H.H. and Schunke, B., 1982. (Behavioural disturbances in breeding sows kept in boxes). Wien. Tierarzl. Mschr. 69 Jahrgang heft 6/7: 200-208.

Stolba, A., 1983. The pig park family system: housing designed according to the consistent patterns of pig behaviour and social structure. In: The Behaviour and Welfare of Farm Animals (ed. W.F. Hall). Proc. Conf. on the Human-Animal Bond. Minneapolis, Minnesota, pp. 38-65.

Stolba, A., Baker, N. and Wood-Gush, D.G.M., 1983. The characterization of stereotyped behaviour in stalled sows by information redundancy. Behaviour, 87: 157-182.

Svendsen, J. and Bengtsson, A.C., 1983. Housing of sows in gestation. Guelph Pork Symposium, Waterloo, Ontario, Canada, pp. 118-131.

Tillon, J.P. and Madec, F., 1984. Diseases affecting confined sows. Data from epidemiological observations. Ann. Rech. Vét., 15 195-199.

METABOLIC RATE OF PIGLETS BETWEEN SUCKLINGS

W. VAN DER HEL AND M.W.A. VERSTEGEN

ABSTRACT

Thirteen measurements have been made to determine the heat production of piglets (1 to 20 days of age) at various times after suckling.
Heat production was measured in an indirect calorimeter of 0.085 m^3 with
2 animals (1 to 4 days of age) or individual animals (after 4 days of
age). They were allowed to suckle about each hour. Mean metabolic rate
was 672 kJ/kg$^{0.75}$/day. It appears that at 6-14 days of age metabolic
rate was highest. There was a clear indication that at restlessness heat
production was higher (740 kJ/kg$^{0.75}$/day) compared to measurements
performed when the piglets were quiet (594 kJ/kg$^{0.75}$/day).

INTRODUCTION

Metabolic rate of young suckling piglets is not well-known. To determine maintenance requirements and efficiency of conversion of milk energy into energy deposited in the body of piglets this metabolic rate has
to be known. In the literature various investigations are reported in
which metabolic rate was estimated for sows and their piglets together
from gaseous exchange (De Lange et al., 1980). Since gaseous exchange
of sows and piglets are measured together it is necessary to partition
total exchange in the part of the piglets and of the sow. Heat production
of piglets will depend on milk intake. In addition this milk intake is very
important for piglet survival (Hacker et al., 1979). To allow such

partitioning data are needed to estimate efficiency of production of fat and protein in the body of piglets from milk consumed. Therefore 13 measurements have been made with piglets of 1 to 20 days of age at various times between sucklings.

Table 1. Metabolic rate of piglets between sucklings

Per-iod	Ani-mals	Age in days	Milk/ suckling in g	Resp. min.	Weight In g	Out g	Temp °C	RH %	H*	RQ	Re-marks**
1	2	1	8 5	70	1025 1146	1015 1135	22.6	75	616	0.83	-
2	2	1	7 10	60	1247 1760	1243 1744	23.4	80	568	0.79	0
3	2	2	-	60	1505 1165	1489 1160	23.7	73	628	0.68	-
4	2	4	19 10	60	1689 1286	1684 1282	23.5	80	587	0.72	0
5	1	6	-	65	2215	2200	23.2	70	713	0.70	-
6	1	7	20	75	2737	2726	22.7	80	769	0.76	--
7	1	13	34	56	4609	4579	24.4	79	696	0.76	-
8	1	13	36	60	4291	4210	23.5	69	717	0.84	--
9	1	14	17	60	2976	2965	23.3	78	581	0.74	0
10	1	14	-	60	4270	4250	23.7	77	705	0.78	--
11	1	14	-	40	3155	3140	23.1	80	814	0.78	--
12	1	20	-	50	4004	3990	23.2	78	699	0.82	-
13	1	20	-	105	4261	4226	24.2	78	639	0.79	0

* H in $kJ/kg^{0.75}/day$

** O=very quiet, - = quiet and -- = restless

MATERIALS AND METHODS

Large White piglets from 4 primiparous sows were ad random chosen from their litter. In the litter milk consumption was determined by weighing piglets before and after suckling according to the procedure described by Klaver et al. (1981). Animals used for measurements were between 1 and 20 days of age. When they were younger than one week two animals of one litter were selected. Otherwise one animal was used. The animals were placed into a calorimeter of 0.085 m^3 (Van der Wal et al., 1976) within 5-15 min after suckling. The measurement of CO_2 production and O_2 consumption started between 10-30 min after the animals were introduced in the chamber. Mostly the duration of measurement was around 60 min (Table 1).

RESULTS AND DISCUSSION

In Table 1 the results of measurements of metabolic rate are given. There is a large variation in metabolic rate between various measurements depending on weight, milk consumption and behaviour. If animals are restless than a much higher gaseous exchange was measured. The mean metabolic rate of all measurements was 672 $kJ/kg^{0.75}/day$. In Table 2 data are arranged by age. Metabolic rate was highest in animals of 6-14 days of age. At other ages lower values were found. As milk production and thus also milk intake was maximal at about 10 days of age (Geerse and Mesu, 1981) such a pattern can be expected. Data have also been grouped according to activity (Table 2). Animals which were quiet produced less heat than those which were active.

From these results metabolic rate between suckling can be calculated. However, there is a considerable variation which may partly be due to the fact that only one or two animals were used each time. Normally these animals are with a group together in a litter. Therefore Geerse and Mesu (1981) measured metabolic rate of a whole litter between sucklings.

One litter of 9 piglets was put in a respiration chamber directly after suckling at various ages. In the chamber no feed or water was supplied and animals remained in the chambers for 60 minutes. At an age of 6-12

Table 2. Metabolic rate of piglets $(kJ/kg^{0.75}/d)$, weight loss $(g/kg^{0.75}/d)$ and milk consumption as associated with age (days), activity and milk consumption[a].

	Period	Heat prod. $kJ/kg^{0.75}/d$	Weight loss $g/kg^{0.75}/d$	Milkconsumption $g/kg^{0.75}/d$
Age (days)				
0- 5	1,2,3,4	600	128	135/104[a1]
6- 10	5,6	741	124	147
11- 15	7,8,9,10,11	703	231/157[a2]	184
16- 20	12,13	669	132	not measured
Activity (score)				
very quiet	2,4,9,13	594	117	153/131[a1]
quiet	1,3,5,12	664	141	88
restless	6,7,8,10,11	740	229/155[a2]	185
milk consumption $(g/kg^{0.75}/d)$				
76-125	1,2	592	156	104
126-175	6,9	675	93	145
176-225	4,7,8	667/706[a1]	267/138[a2]	202

a. when two mean values are given then the first one is including an outlier.

[1] outlier period 4

[2] outlier period 8.

days they measured a mean metabolic rate of 397 $kJ/kg^{0.75}/day$ and at an age of 12-18 days 427 $kJ/kg^{0.75}/day$ was measured. It should be noted however that these piglets were extremely quiet (they went to sleep directly after being put in the chamber). Compared to the literature (Jordan, 1971; Komarek, 1972; De Goey and Ewan, 1975) metabolic rates found by Geerse and Mesu (1981) were very low. De Goey and Ewan (1975) found that metabolic rate in 20 day old piglets kept at 26°C was related to intake of metabolizable energe as:

Heat = 0.31 * ME + 365 $(kJ/kg^{0.75})$

This gave an efficiency of deposition of body energy in piglets of 69% and a maintenance requirement of 544 kJ ME per $kg^{0.75}$. Komarek (1972) found in piglets of 10 days of age a metabolic rate of 507 $kJ/kg^{0.75}$ at 28°C and 596 $kJ/kg^{0.75}$ at 30°C. Therefore it was thought that the values found by Geerse and Mesu (1981) were too low. It must be noted that suckling itself was not included in their and our measurement periods. Suckling will induce a high metabolic rate due to feed intake and due to activity at the udder of the sow. A large variation therefore can be expected in metabolic rates between piglets at various times of the day related to time of suckling.

In Table 2 metabolic rate of piglets is grouped also according to various intakes of milk. Milk intake seemed to be relatively of minor importance. This may be due to the fact that we did not record milk intake at earlier suckling(s). This would have been more appropriate since the digestion of milk intake at an earlier suckling, its absorption and use in the intermediate metabolism, will not been fully complete at the time of the next suckling and our measurement period.

The consequences of deviating results of measurements with piglets for estimation of the metabolic rate of sows is dependent on the age of the piglet. It can be calculated that in newborn piglets the correction will be small. However correction for metabolic rate of piglets of more than two weeks of age will have a considerable impact on the metabolic rate of sows. Let us consider a hypothetical sow of 160 kg with 10 piglets.

Sow 160 kg	= 45.0 $kg^{0.75}$
10 piglets 2 days old of 1,5 kg	= 13.5 $kg^{0.75}$
10 pigets 24 days old of 7 kg	= 43.0 $kg^{0.75}$

Let us assume a similar high metabolic rate of 800 kJ per $kg^{0.75}$ for sows and piglets. Two situations (A and B) will be distinguished:

A. At 2 days post partum

Sow	800 × 45.0	= 36000 kJ
Piglets	800 × 13.5	= 10800 kJ
		46800 kJ

Metabolic rate of 10 piglets is 23% of that of dam and litter together.

B. At 24 days post partum

Sow	800 × 45.0	= 36000 kJ
Piglets	800 × 43.0	= 34400 kJ
		70400 kJ

Metabolic rate of 10 piglets is 48.9% of that of dam and litter together.

If metabolic rate ascribed to piglets is under- or overestimated with 10%, then that of the sow is estimated with an inaccuracy of
- at 2 days old ± 1080 kJ = 3.0% and
- at 24 days old ± 3440 kJ = 9.6%

provided that total metabolic rate measured is the same. Therefore it depends very much on the assumed metabolic rate of piglets, which heat production of sows during lactation is found. This in turn determines the estimation of efficiency of milk production by sows. However piglets may vary in rate of activity. Therefore it is thought that our results together with those of Geerse and Mesu (1981) may give a preliminary survey of the range in which metabolic rate of piglets can be expected.

The range of metabolic rate depends on sleeping, activity, weight and milk intake of piglets. Probably the best estimate is obtained by the mean of Table 1. Moreover milk intake does have a much smaller influence on metabolic rate in piglets than in older animals. Their feeding level is much lower than that of fast growing fattening pigs or lactating sows.

REFERENCES

Geerse, C., and Mesu, J.J., 1981. De energetische efficiëntie van de melkvorming bij zeugen. MsC Thesis, Agricultural University Wageningen.

Goey, L.W. de and Ewan, R.C., 1975. Effect of level of intake and diet dilution on energy metabolism on the young pigs. J. Anim. Sci. 40: 1045-1051.

Hacker, R.R., Hazeleger, W., Poppel, F.J.J. van, Osinga, A., Verstegen, M.W.A. and Wiel, D.F.M. van der, 1979. Urinary oestrone concentration in relation to piglet viability, growth and mortality. Liv. Prod. Sci. 66: 313-318.

Jordan, J.W., 1971. Investigations into energy metabolism of bacon pigs and piglets. Agric. Progress 46: 9-25.

Klaver, J., Kempen, G.J.M. van, Lange, P.G.B. de, Verstegen, M.W.A. and Boer, H., 1981. Milk composition and daily yield of different milk components as affected by sow condition and lactation/feeding regime. J. Anim. Sci. 52: 1091-1097.

Komarek, J., 1972. Der Einfluss der Hungern und des reitlich geregelden Futterangebots auf die Entwicklung von ersten bis vier-zehnten Lebenstag. Zeitschrift Tierphysiol., Tierern. und Futtermittelkunde 29: 169-177.

Lange, P.G.B. de, Kempen, G.J.M. van, Klaver, J. and Verstegen, M.W.A., 1980. Effect of condition of sows on energy balances during 7 days before and 7 days after parturation. J. Anim. Sci. 50: 886-891.

Wal., H. van der, Verstegen, M.W.A. and Hel, W. van der, 1976. Protein and fat deposition in selected lines of mice in relation to feed intake. EAAP publ. no. 19, Proc. En. Metab. Vichy: 125-128.

INFLUENCE OF SOME ENVIRONMENTAL, ANIMAL AND FEEDING FAC-
TORS ON ENERGY METABOLISM IN GROWING PIGS

M.W.A. VERSTEGEN, A.M. HENKEN AND W. VAN DER HEL

ABSTRACT

Housing conditions and the thermal environment influence productivity of food animals by altering their heat production and thus exchange with their environment. Also feed intake and its use for maintenance and production can be changed. Housing itself may alter maintenance require-ments by influencing activity and effective environmental temperature. Many experiments have been conducted to measure environmental/nutri-tional energetics of swine. At thermoneutral conditions, heat production depends mostly on feed intake and metabolic body size. Also physical activity is important in this respect. Within the thermoneutral zone environmental temperature does not affect heat production very much. Thermoneutral heat production must be known in order to calculate or determine critical temperatures at various housing conditions. Especially systems without bedding may alter the thermal demand of the environ-ment upon the animal. Moreover housing systems may influence the level and/or pattern of activity and the related heat production. Stereotypies may have a clear effect on energy requirement for maintenance. Thus, rate of body weight gain will be affected if housing influences activity. These effects are similar to changes in feed allowance.

INTRODUCTION

Animals produce heat as a consequence of processes related to maintenance and production.

Optimum production can only be achieved if maintenance requirement is not increased by adverse environmental conditions, such as non-optimal housing and climatic conditions. The ways in which environmental factors increase metabolic rate are manifold (Blaxter, 1977; ARC, 1981). Metabolic rate is also dependent on animal factors, like bodyweight and degree of adaptation (Verhagen, 1987). All these factors affect mainly the maintenance part of metabolism.

Heat is also produced in processes related to converting feed above maintenance into production. The amount of energy which can be deposited from feed above maintenance is related to body weight and the quantity and quality of the feed provided. The feeding value describes the amount of energy which can be deposited from a feed or feed components. Feeding systems are based on the feeding value. Feed quality can be described as the capacity to sustain maintenance and to promote production (products, labor, offspring). The requirement of an animal for metabolizable energy is traditionally calculated by expressing energy gain and/or heat loss as a function of body weight and intake of metabolizable energy.

Farm animals are mostly fed for production, therefore the heat increment below maintenance will not be dealt with here. We will discuss effects of environmental, animal and some feeding factors. These factors affect energy metabolism and thus requirements for production and maintenance.

ENVIRONMENTAL FACTORS

Environmental factors determine highly the energy requirement of an animal. Among one of the first factors to be mentioned is the way in which the animal is housed. Housing conditions are often seen to affect feed conversion ratio. These conditions will also affect the variation in activity pattern of animals. Therefore it can be expected that this may lead to variation in the amount of energy retained. The magnitude of these effects is important since it determines the energetic efficiency of the production system. It may be expected that variation in physical movements will be an important factor in determining efficiency in gain.

However any interaction between feeding level and maintenance will limit the possibilities of predicting the energy gain at farming conditions. Therefore comparisons at similar feeding level may be more predictive. Another way is the determination of the heat increment at conditions which are thought equivalent to the situations they are applied for. Also if animals of different classes are to be compared, efficiency and maintenance can not be looked at independently.

Hörnicke (1971) showed that pigs considerably increase their heat production when standing. Thus, effects of the environment on activities like standing or walking will effect the heat increment and production efficiency.

Deposition of protein and fat and the heat associated with deposition may not be independent of these factors. Preferably effects on heat increments of housing and feeding should be measured at similar feeding level. Evaluation of practical conditions with regard to the energy required for activity may be an important aspect of improving predictibility of the animal's performance from a ration.

Housing factors are often associated with climatic environment. The surroundings of an animal determine the heat exchange between its body and the environment. The thermal factors together will determine the effective temperature, i.e. the temperature as encountered by the animal.

Effective temperature has been defined as compared to ESET, Equivalent standardized Environmental Temperature (Mount 1979). This is the standardized condition in which

- air and mean radiant temperature are equal to one another
- condition of free convection without forced draught of air (more or less still air)
- floor is insulated. This is important for animals which are sensitive to floor insulation.

Mount (1979) has extensively discussed the effects of various environments on the effective temperature. Especially for cold conditions these have been assessed. These effects will be described:

- the radiant temperature outdoors may decrease the loss of heat due to increased incoming short wave radiation from the sunshine or may increase heat loss due to a decrease in receiving long wave radiation at night (Blaxter, 1977). Changes in temperature indoors of surrounding walls or reflection of the walls may have similar effects on heat loss (Mount, 1964; Holmes and McLean, 1977). Radiant temperature can change heat requirements considerably as experiments of Holmes and McLean (1977) showed;

- the temperature of feed and water may also have an effect on the effective temperature, especially when large quantities of cold liquid are ingested at low air temperatures (Holmes, 1971);
- the temperature and kind of bedding may affect the effective temperature in pigs (Stephens and Start, 1970; Verstegen and Van der Hel, 1974). Conductive heat loss of a pig to the floor or other contact surfaces is determined by the nature and temperature of the contact surface. The conductive heat loss, which is generally small, may be considerable if piglets are housed on a floor with a high thermal conductivity such as concrete. Mount (1967) calculated that at 20°C the heat loss to a wooden floor was similar to that to a concrete floor at 30°C. Introduction of wood shavings, straw or woodwool on concrete floors was equivalent to a rise of 9, 15 and 19°C in ambient temperature. The benefit of keeping piglets on floors with high insulative values is increased at low environmental temperatures. Moving piglets onto straw at 10°C was similar to raising the ambient temperature with 8°C on concrete. Verstegen and Van der Hel (1974) noticed similarly that in groups of 9 pigs of about 40 kg, housed on straw, asphalt and concrete the lower border of the thermoneutral zone (critical temperature) was at 11-13, 14-15 and 19-20°C respectively.

An animal may also respond by changing its posture and thus orientation influencing in this way the area of contact. In that connection Spillman and Hinkle (1971) found that a change in air temperature more than a change in temperature of the floor itself may effect heat loss to the floor. They noticed that at an air temperature of about 33°C and a floor temperature of 24°C heat loss to the floor was about 12.7 kJ/m^2 per hour. At lower air temperatures the heat loss to the floor was reduced. Restrepo et al. (1977) on the other hand noticed that the percentage of conductive heat loss from the total heat loss increased from 34 to 48% as the floor temperature decreased from 35 to 15°C irrespective of ambient temperature. Changes in flooring can be translated into changes of effective temperature as it affects heat loss;

Table 1. Change in standardized effective critical temperature ESET in pigs at various housing, management and climatic conditions

Condition	Specifica-tion	Weight (kg)	Change in lower critical temperature (°C)	References
group vs individual		20	2 to 5	Close and Mount (1978)
		>100	2 to 6	Geuyen et al. (1984)
concrete vs straw	at 10°C	piglet	+8	Stephens and Start (1970)
concrete vs straw	at 30°C	"	+2	"
straw	group of 9	35	-4	Verstegen and v.d. Hel (1974)
concrete slats	"	35	+5	"
asphalt vs straw	"	35	+2	"
wet surface	group of 9	35	+5 to +10	Mount (1967)
Draught				
draught	insulation	-	+6	"
draught	uninsulated (winter)	-	+8	"
no draught	uninsulated (winter)	-	+2	"
no draught	uninsulated with straw	-	-4	"
draught	group, insulated floor			
	day 2	25	+6	Verhagen (1987)
	day 12	"	+2.4	Verhagen (1987)
Radiant Temperature				
+1°C	individual	piglet	-1	Mount (1964)
reflective wall	group	11	-2	Holmes and McLean (1977)

- also rainfall (Alexander, 1974) or wetting of the skin otherwise (Ingram, 1965) will affect the effective temperature;
- group size may also affect the critical temperature in pigs (Mount, 1960). Grouping in pigs may reduce lower critical temperature and

also extra heat required in the cold. Close and Mount (1978) calculated that the extra heat in the cold is reduced by 7% per increase in
group size from 1 to 9. Group sizes change social environment as a
review of Buchenhauer and Henricksen (1975) showed that 8-12 pigs
was the optimum number for minimal food conversion ratio. It can be
expected that a maximum energy gain may occur at these group sizes.
In Table 1 a survey has been given of the critical temperature at different housing conditions.

From the literature it is not clear whether and to what extent thermal
housing conditions which alter lower critical temperatures also alter extra
heat production per °C below thermoneutrality. From our own data
(Verstegen and Van der Hel, 1974) we could not clearly show an increased extra thermal heat production per °C below thermoneutrality on
a floor with little insulation. This means that apart from data with young
piglets (Mount 1967) such interactions have not yet been quantified
(Figure 1).

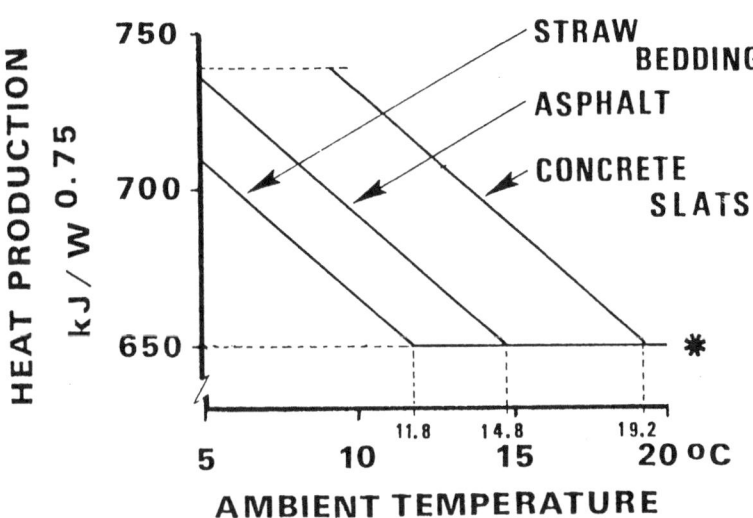

Figure 1. Relation between heat production and ambient temperature in
groups of pigs (40 kg body weight) housed on different types of floor.
* Indicates heat production at thermoneutrality and a feeding level of
1150 kJ ME/kg 0.75.

With regard to thermoneutral conditions it has been found that within the zone of thermoneutrality housing conditions can alter heat production by means of an effect on maintenance requirement. Data of such experiments are very rare. In a study with sows Cronin et al. (1986) found that sows with a high frequency of stereotypic (high) behaviour had a much higher heat production compared to sows with a low frequency (low) of steriotypic behaviour. High sows produced proportionally 0.36 more heat than low sows during the 12 hours light period. In this period 40% of heat production was associated with activity in high sows and 20% in low sows. On a daily basis 61.8 $kJ/kg^{0.75}$ of heat was associated with stereotypies in high sows and 40.6 $kJ/kg^{0.75}$ in low sows (Van Tartwijk et al., 1987). In Table 2 effects of tethering of sows on heat production are given. These data show that housing conditions resulting in an increased metabolic rate may have a similar effect on animals as decreased feed intake.

Table 2. Energy metabolism of tethered sows which had a high (H) or a low (L) activity (Cronin et al., 1986).

	ME intake $kJ/kg^{0.75}$	Heat production $kJ/kg^{0.75}$
Sows with high activity (H)		
period 1	606	572
period 2	612	590
Sows with low activity (L)		
period 1	605	484
period 2	576	472

ANIMAL FACTORS

Also factors related to the animal itself are important, i.e. age, sex, fatness, breed, being ruminant or non-ruminant, etc.

A factor which is often confounded with energy metabolism is leannes or obesitas of the animal (Johnson and Crownover, 1976; Pullar and Webster, 1974). Young animals normally contain more lean tissue than mature ones. Thorbek (1975) showed that metabolism of young animals can not be compared with that of more mature animals due to a difference in maintenance requirement. Whether this difference is due to activity or body composition can not be separated at similar feed intake. A way of overcoming this problem has been found by Pullar and Webster (1974 and 1977). They estimated from measurements of energy and protein balances in congenitally obese and lean rats efficiencies of deposition of protein and fat. Heat increment for protein is higher than for fat and the ARC (1981) adopted these values from such studies with rats and with pigs. They concluded that per kJ of metabolizable energy (ME) used for deposition of energy in protein 0.46 kJ of heat are produced and 0.54 are deposited. When fat is deposited 0.26 kJ of heat is produced and 0.74 kJ of energy are retained in fat. Their way of estimating these values was a priori free from assumption about maintenance requirements or the relation between heat production and body weight. It requires however laboratory conditions to study this.

Animal factors may decrease or increase the lower critical temperature and/or change maintenance in thermoneutral conditions.

- length of fleece or hair coat may change insulation (Blaxter, 1977). This is however not important for pigs housed indoors.
- length of duration of exposure may alter the critical temperature by changes in tissue insulation, e.g. variation in fat content and backfat thickness in pigs (Ingram, 1964; Sörensen, 1962; Irving, 1964).
- animals may have critical temperatures which vary as much as 2-7°C due to variation in condition (Young, 1975; Holmes, 1971). The lower critical temperature of an animal decreases with increasing weight. This is partly due to an increased thermal insulation and partly to the changing body composition of growing animals. A very high critical temperature of around 20°C was for instance reported for sows (Holmes and Close, 1977). Sows may, depending on their productivity stage, have a relatively high protein/fat ratio in their body.
- duration of exposure to climatic conditions. Verhagen (1987) found that the thermal demand of the environment upon growing pigs after exposure to 15, 20 or 15/25°C with draught decreased considerably from 1 to 10 days after first exposure. Verhagen (1987) found that

the overall effect of draught on thermal demand was equivalent to a lowered ambient temperature of 2.8°C at 25°C and 4.2 at 15°C. If a fluctuating temperature was applied, 25°C during day time and 15°C during night time, then draught at night resulted in an increase of the critical temperature of 4.5°C.

Table 3 summarizes effects of draught on thermal demand at various conditions as found by Verhagen (1987).

Table 3. Effect of super-imposed draught at various climatic conditions (25°C, 15°C and fluctuating 25/15) on extra thermoregulatory heat production ETH (kJ/kg$^{0.75}$) and on lower critical temperature Lct[1] (Verhagen, 1987).

Tempera-ture °C	Mean		Day 2		Day 12	
	ETH kJ/kg$^{0.75}$	Lct °C	ETH kJ/kg$^{0.75}$	Lct °C	ETH kJ/kg$^{0.75}$	Lct °C
25	35	3.2	48	4.3	-12.1	-1.1
15	66	6.0	96	8.7	45	4.1
25/15	59	5.4	61	5.5	45	4.1

1) Assuming that ETH per °C coldness is 11 kJ/kg$^{0.75}$

FEEDING

The feeding level and the quality of the feed provided is important with respect to the amount of heat produced. Each additional amount of metabolizable energy intake is assumed to result in an additional amount of heat being produced.

The increase in heat production at increasing feed intakes from ARC (1981) can be estimated as given in Table 4. It can be derived from literature (Holmes and Close, 1977) that growing animals will have an energy gain which is maximally about 50% of total metabolizable energy ingested.

If energy gain from a certain amount of feed can be predicted also heat production can be predicted. If energy gain is measured at various

Table 4. Heat increment of feed (kJ/kJ ME) and maintenance requirement ME_m in $kJ/kg^{0.75}$ in pigs from ARC (1981) and Curtis (1983).

References	ME_m	Increment
ARC (1981)	420 - 584	0.33 - 0.43
Curtis (1983)	362 - 500	0.25 - 0.35

levels of feeding the increase in energy gain is thought to be associated with the increased feed intake. It is assumed that the use of feed for maintenance is not affected. It is also assumed that activity is not affected. Heat increment is then only a function of the ration concerned. It is questionable, however, whether these prerequisites are always met. In addition heat increment as a function of intake comparable to energy gain has to be measured over periods of at least one day.

Heat production is increased considerably during eating. Charlet-Lery (1975) measured that heat production at eating is 40-80% above basal metabolism in pigs. The heat increment of feeding can be thought to be made up of energy produced as associated with eating, ingestion, digestion, adsorption, movement of digesta through the digestive tract, formation and excretion of urea or uric acid, resynthesis of complex compounds from simpler, as absorbed, compounds, etc. They all require the expenditure of energy. Curtis (1983) summarizes part of the heat increament of feeding. We will not include heat of fermentation as it is of minor importance in pigs, but in rumunants it can be as high as 5-6% of total heat production.

Cost of eating depends on the type of animal (Curtis, 1983):
- Grazing costs about 2.1 kJ per kg body weight per hour;
- Rumination costs about 1.05 kJ per kg body weight per hour;
- Sheep spend 25.2 kJ for eating 1 kg dry matter as pelleted dry grass. Eating chopped dry grass cost 6 times as much;
- Chewing in sheep was reported to cost 0.1 kJ per minute regardless of diet;
- A 2 kg hen will spent about 0.25 kJ per minute in eating.

Heat production
(Watt.kg$^{-0.75}$)

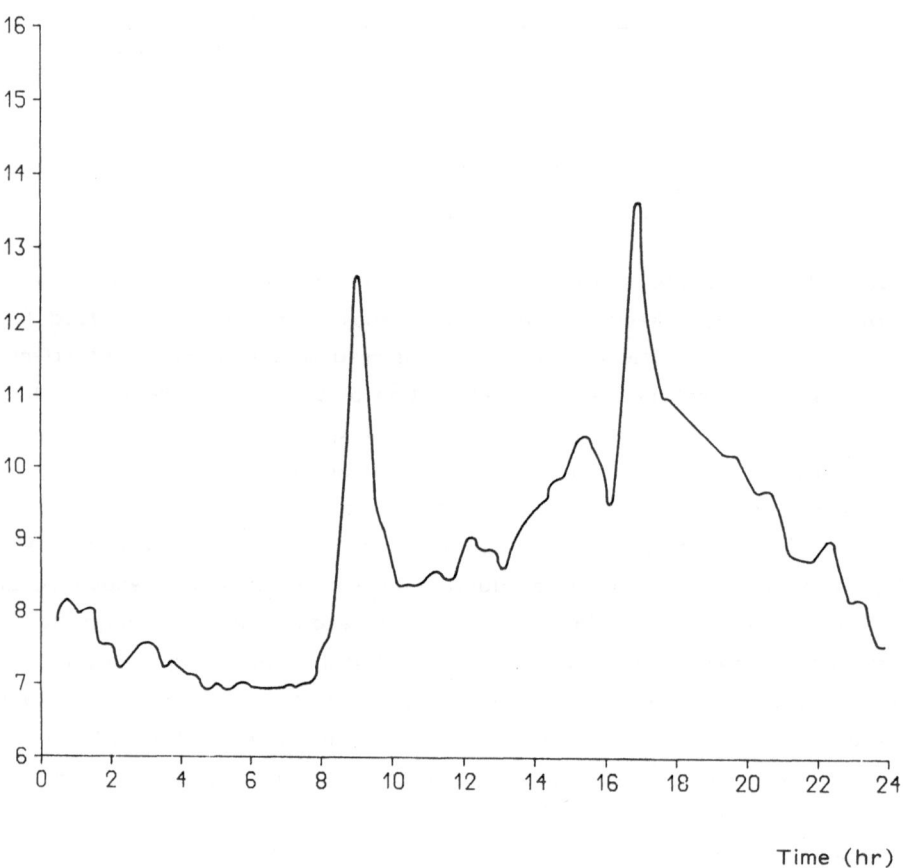

Time (hr)

Figure 2. Heat production at thermoneutrality during 24 hours (feeding at 8 am and 4 pm; lighting from 7 am to 7 pm).

The cost of eating for a non-ruminant animal can be considered constant per unit of feed intake of crum, pellets or wet mash. The effect of eating itself on the heat increment is thus only very small. Activity associated with eating is at least partly responsible for the increase as activity is associated with heat production. Hörnicke (1971) estimated the cost of standing in pigs. He found that the energy costs at standing are increased considerably above those at lying. In an experiment with animals housed in a group inside a calorimeter we noticed that the heat

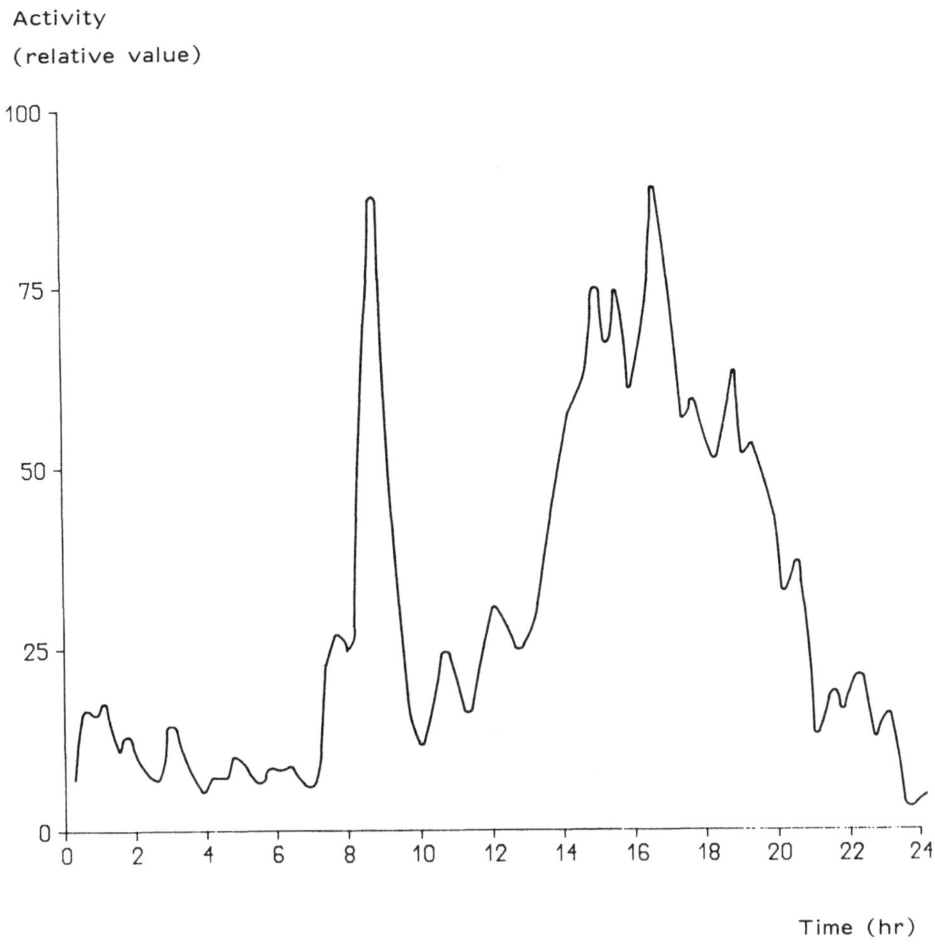

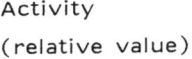

Activity
(relative value)

Figure 3. Activity pattern (relative values) during 24 hours of the same pigs and the same conditions used in Figure 2.

production level at feeding time was similar when feed intake was 74% of ad libitum or 85% of ad libitum (Verstegen et al., 1982). Figure 2+3 suggest that activity at feeding time is the main reason for the heat increment. Charlet-Lery (1976) reported a similar increase in heat production at onset of eating or at onset of drinking. Heat production in pigs after a meal was increased above maintenance for 23 hours afterwards. Charlet-Lery (1975) reported that during 7 hours after a meal 12% of ME intake was produced as extra heat irrespective of the diet. Therefore activity studies will be of increasing importance to assess correctly energy needs of farm animals. Wenk and Van Es (1976) calculated heat production associated with acticity of chickens at two feeding levels: restricted and ad libitum. They found that restriction of feed resulted in an important increase in heat production due to increased activity. They found that (all data expressed in kJ per $kg^{0.75}$) without correction of energy balance data for activity the following line could be derived:

retained energy = 0.667 ME - 337 → ME_m = 505 $kJ/kg^{0.75}$ (1)

If they subtracted energy for activity from ME data then the relation derived was:

retained energy = 0.61 ME - 185. → ME_m = 319 kJ/kg^{075} (2)

The extra heat associated with activity amounted to 15-20% of ME intake in ad libitum and to 20-25% of ME intake in restricted fed animals. In young pigs kept in groups Halter et al. (1980) observed also an increased heat production associated with increased activity. In experiments with pigs housed in a group but fed individually we found that heat associated with activity was less than that in the younger pigs of Halter et al. (1980), about 15% of ME intake. The total heat production measured at feeding time is 40-80% higher than that of basal metabolism. The effect lasts up to 23 h (Charlet-Lery, 1975 and 1976). Results of Verstegen et al. (1982) showed that heat production was associated with feed intake as:

H = 0.26 * ME + 388.8 → ME_m = 525 $kJ/kg^{0.75}$

when corrected for activity the relation was:

H = 0.30 * ME + 285.4 → ME_m = 408 $kJ/kg^{0.75}$

This means that heat increment with feeding is altered by activity related heat production. Data on actual activity related heat productive are given in Table 5.

Thus it can be estimated that normal activity will require about 15 to 20% of maintenance. However at non-optimal feeding and housing conditions this can be increased considerably.

Table 5.   Effect of feeding level (% of ad lib intake) on activity related heat production ($kJ/kg^{0.75}$) in various weight ranges (Verstegen et al., 1982)

Feeding level	Weight range (kg)			
	25-45	45-65	65-85	85-125
80%	-	71(10)*	61(8)	65(9)
70%	66(10	85(13)	93(13)	75(12)

* between brackets activity related heat production as percentage of total heat production

REFERENCES

ARC, 1981.   The nutrient requirements of pigs. Agricultural Research Council Commonwealth Agricultural Bureaux England: 341 pp.

Alexander, G., 1974.   Heat loss from sheep. In: J.L. Monteith and L.E. Mount (Editors), Heat loss from Animals and Man. Butterworths, London: 173-202.

Blaxter, K.L., 1977.   Environmental factors and their influence on the nutrition of farm livestock. In W. Haresign, H. Swan and D. Lewis (Editors), Nutrition and the climatic environment. Butterworths, London: 1-10.

Buchenauer, D. and J. Henricksen, 1975.   EAAP Meeting Warsaw.

Charlet-Lery, G., 1975.   Les dépenses énergétiques prandiales et post prandiales chez le porc en croissance. Thèse AO 11 620, Université de Paris VI.

Charlet-Lery, G., 1976.   Influence of protein feeding pattern on heat production in the growing pig. In: M. Vermorel (Editor), Energy Metabolism of Farm Animals, EAAP Publ. no. 19: 109-112.

Close, W.H. and L.E. Mount, 1978.   The effects of plane of nutrition and environmental temperature on the energy metabolism of the growing pig. 1. Heat loss and critical temperature. Br. J. Nutr. 40: 413-421.

Cronin, G.M., Tartwijk, J.M.F.M. van, Hel, W. van der and Verstegen, M.W.A., 1986. The influence of degree of adaptation to tether-housing by sows in relation to behaviour and energy metabolism. Anim. Prod. 42: 257-268.

Curtis, S.E., 1983. Environmental management in animal agriculture IOWA University Press: 409 pp.

Geuyen, T.P.A., Verhagen, J.M.F. and Verstegen, M.W.A., 1984. Effect of housing and temperature on metabolic rate of pregnant sows. Anim. Prod. 38: 477-485.

Halter, H.M., Wenk, C. and Schürch, A., 1980. Effect of feeding level and feed composition on energy utilisation. Physical activity and growth performance of piglets. In: Hughes, E.H. (Editor). Studies in Agricultural and Food Science. Butterworths, London: 195-198.

Holmes, C.W., 1971. Growth and backfat depth of pigs kept at high ambient temperature. Anim. Prod. 15: 521-530.

Holmes, C.W. and Close, W.H., 1977. The influence of climatic variables on energy metabolism and associated aspects of productivity in the pig. In: W. Haresign, H. Swan and D. Lewis (Editors), Nutrition and the climatic environment. Butterworths, London: 51-74.

Holmes, C.W. and McLean, N.R., 1977. The heat production of groups of young pigs exposed to reflective or non-reflective surfaces on walls and ceiling. Trans. ASAE 20: 527-528.

Hörnicke, H., 1971. Circadian rhythm and the energy cost of standing in growing pigs. 5th Symposium on energy metabolism. (Editors Schürch and Wenk). Vitznau: 165-168.

Ingram, D.L., 1964. The effect of environmental temperature on heat loss and thermal insulation in the young pig. Res. Vet. Sci. 5: 357-364.

Ingram, D.L., 1965. Evaporative cooling in the pig. Nature London 207: 415-416.

Irving, L., 1964. Terrestral animals in cold: birds and mammal. In D.B. Hill (Editor), Adaptation to the environment. American Physiological Society: 361-378.

Johnson, D.E. and Crownover, J.C., 1976. Maintenance energy requirements of lean vs obese growing chicks at equal age and body energy. In: M. Vermorel (Editor), Energy Metabolism of Farm Animals, EAAP Publ. no. 19: 121-124.

Mount, L.E., 1960. The influence of huddling and body size on the metabolic rate of the young pig. J. Agric. Sci. Camb. 55: 101-105

Mount, L.E., 1964. Radiant and convective heat loss from the new-born pig. J. Physiol. 173: 96-113.

Mount, L.E., 1979. Adaptation to the thermal environment. Man and his productive animals. Edward Arnold, London: 333 pp.

Mount, L.E., 1967. The heat loss from newborn pigs to the floor. Res. Vet. Sci. 8: 175-186.

Pullar, J.D. and Webster, A.J.F., 1974. Heat loss and energy retention during growth in congenitally obese and lean rats. Br. J. Nutr. 31: 377-392.

Pullar, J.D. and Webster, A.J.F., 1977. The energy cost of fat and protein deposition in the rat. Br. J. Nutr. 37: 355-363.

Restrepo, G., Shanklin, M.D. and Hahn, L., 1977. Heat dissipation from growing pigs as a function of floor and ambient temperatures. Trans. ASAE 20: 145-147.

Sörensen, P.H., 1962. Influence of climatic environment in pig performance. In: J.I. Morgan and D. Lewis (Editors), Nutrition of pigs and poultry. Butterworths, London, pp. 88-103.

Spilman, C.K. and Hinkle, C.N., 1971. Conduction heat transfer from swine to controlled temperature floors. Trans. ASAE 14: 301-303.

Stephens, D.B. and Start, I.B., 1970. The effects of ambient temperature, nature and temperature of floor and radiant heat on the metabolic rate of the newborn pig. Int. J. Biometeorol. 14: 275-281.

Tartwijk, J.M.F.M., Cronin, G.M., Verstegen, M.W.A. and Hel, W. van der, 1987. Enkele aspecten van de energiestofwisseling van aangebonden zeugen in relatie tot aanpassing. Landbk. Tijdschr. 40: 63-71.

Thorbek, G., 1975. Studies on energy metabolism of growing pigs. No. 434 Beretning fra stat Husdyr brugs forsog Kobenhavn.

Verhagen, J.M.F., 1987. Acclimation of growing pigs to climatic environment. PhD Thesis Agric. Univ. Wageningen: 128 pp.

Verstegen, M.W.A. and Hel, W. van der, 1974. Effects of temperature and type of floor on metabolic rate and effective critical temperature in growing pigs. Anim. Prod. 18: 1-11.

Verstegen, M.W.A., Hel, W. van der, Brandsma, H.A. and Kanis, E., 1982. Heat production in groups of growing pigs as affected by weight and feeding level. In: A. Ekern and F. Sundstøf (Editors), Proc. 9th Symposium on Energy Metabolism of Farm Animals, EAAP Publ. no. 29:218-221.

Wenk, C. and Es, A.J.H. van, 1976. Energy metabolism of growing chickens as related to their physical activity. In: M. Vermorel (editor), Proc. 7th Symposium on Energy Metabolism of Farm Animals, EAAP Publ. no. 19: 189-192.

Young, B.A., 1975. Some physiological costs of cold climates. Univ. Mo. Agric. Exp. Stn. Special Rep. 175.

THE EFFECTS OF HOUSING CONDITIONS ON ENERGY UTILIZATION OF POULTRY

E.H. KETELAARS

ABSTRACT

In two balance respiration experiments with medium weight laying hens a cage system was compared to a floor system, at 21°C and 65-70 percent humidity (RH). Feed conversion and body weight gain tended to be better in the cage system. Energy utilization was more efficient in cages, presumably due to a (calculated) lower maintenance energy requirement in this system of 6,7%.

Introduction of a high ambient temperature (to 32°C) revealed a different response to the heat treatment in these systems. The hens in cages showed a much more pronounced reduction in production and body weight than hens in the floor system. However, there could have been a relation with the higher body weights of the caged hens at the start of the experiment.

In another experiment a comparison was made between ventilated and non-ventilated cages, with White Leghorn hens, under three different climate regimes, i.e. at 21°C and 65 percent relative humidity, 28°C at 60 percent relative humidity, and a temperature fluctuating from 35°C in the light and 28°C in the dark period. It could be concluded that cage ventilation under normal climatic conditions might lead to a lower effective temperature, and that only at very high ambient temperatures an alleviation of heat stress may be observed.

Finally it was concluded from an experiment with broilers that intermittent lighting compared to continuous lighting contributes to a higher efficiency of energy utilization through a less total activity of the birds.

The general conclusion is that housing and climatic conditions affect performance of fowl by influencing intake and utilization of dietary energy.

INTRODUCTION

The energy metabolism of the bird is often described as the balance of intake of metabolizable energy (ME) on one side and energy output in the form of egg energy (EE), energy retention (ER) and heat production (H) on the other side, so:

$$ME = EE + ER + H$$

All components of this equation, but in particular energy intake (ME) and heat production (H), are influenced by ambient factors through their impact on energy requirements for maintenance. Energy metabolism is therefore also affected by individual characteristics of the bird itself such as body weight, feather cover and activity. Furthermore there is a strong interaction between the environment and bird characteristics. That is the reason why housing systems can play such an important role in the energy utilization of the fowl.

CAGES VS FLOOR SYSTEM

Hens in cages have a better feed efficiency than hens in floor systems. From field observations we can conclude that this efficiency is apparently due to the higher ambient temperature in battery houses in relation with the higher total heat production of the flock in this high density housing system. However there might be also an effect of the system itself as one might suppose that activity in cages is lower and therefore the energy requirement for maintenance may be less.

Order to investigate the impact of the housing system two experiments were carried out in two respiration chambers (Ketelaars et al., 1985a). In each experiment 192 medium large laying hens were used. In one chamber 96 hens were housed in 24 flat deck cages and in the other one 96 hens in 2 wire floor pens. In both chambers the ambient temperature

Table 1. Egg production, feed conversion, growth and mortality in two experiments with laying hens in two housing systems: cage and wire floor, under equal climatic conditions.

| | Experiment 1 | | Experiment 2 | |
	cages	floor	cages	floor
Experimental period (age in weeks)	19-27	19-27	19-42	19-42
Eggs produced per hen day (%)	54.2	55.0	80.8 91.7[1]*	80.5 90.9[1]*
Eggs collected per hen day (%)	51.0*	46.7*	79.5 90.5[1]*	78.0 88.6[1]*
Average egg weight (g)	53.3	51.9	60.5	59.2
Feed consumption per hen per day (g)	119.9	118.9	128.7	127.0
Feed conversion (kg/kg eggs produced) (kg/kg eggs collected)	3.80 4.04	3.82 4.51	2.61 2.65	2.64 2.72
Growth (g)	562	490	825	676
Mortality (%)	5.2	2.1	6.3	1.0
Egg characteristics at 26 weeks of age:				
Egg shell (%)	9.26	9.19	-	-
Egg albumen (%)	67.07	66.52	-	-
Egg yolk (%)	23.56*	24.26*	-	-
Cracked eggs (%)	1.74	2.13	-	-

* $p < 0.05$

[1] from 29 to 42 weeks of age

was maintained at 20°C (± 3°C) with a relative humidity of 65-70 percent. In addition to performance records such as egg production, egg
quality, feed and water consumption, body weights and mortality, also
energy balances were determined including respiratory gaseous exchange. From these respiratory data heat production can be calculated
(Verstegen et al., 1987).

As far as production efficiency is concerned it can be concluded that
feed conversion in the cages tended to be slightly more efficient than in
the floor system, mainly as a result of a larger egg weight, because egg
production and feed consumption did not differ, whereas body weight of
the hens in cages was higher (Table 1). A number of eggs got lost because of damaging or egg eating, particularly in the floor system. On
basis of eggs actually collected egg production was 1.4 to 3 percent
larger in the cages. Egg characteristics were not significantly different,
except egg yolk percentage (p < 0.05).

These egg production data are in agreement with field observations
and other comparative investigations, but the difference in feed conversion is in many observations larger. Most probably this is due to the

Table 2. ME intake, heat production (H), egg energy (EE) and energy
retention (ER) in kJ per $kg^{0.75}$ per day.

	Experiment 1		Experiment 2	
	cages	floor	cages	floor
Experimental period				
(age in weeks)	19-27	19-27	29-42	29-42
ME intake	757	741	876*	900*
Heat production	566	572	615***	650***
Egg energy	98	94	265	264
Energy retention	93	75	-3	-13

*   p < 0.05
*** p < 0.001

fact that in temperate climates in winter the cage system is causing a higher ambient temperature through the high bird density of the system. In these experiments however the ambient temperature was maintained at the same level in both systems. The non-significant difference in feed conversion, particularly on the basis of produced eggs, suggests that the effect on the efficiency of feed utilization in cages is not as large as often is assumed in practice, although the energy metabolism remains more efficient if we take into consideration the higher body weights of the birds in cages.

The energy balance data indicate a relatively higher efficiency of the cage system. Per kg of metabolic weight ($kg^{0.75}$) the intake of ME was significantly lower in the cage system in experiment 2, together with a lower heat production (Table 2), whereas egg energy output was the same in both systems with a tendency to a somewhat higher energy retention component in the cage system.

The higher energy efficiency on metabolic weight basis can be illustrated by the metabolizable energy required for maintenance in both systems. Assuming that the efficiency of energy conversion into egg and growth energy will be the same in both systems, namely at a level of 75 percent, maintenance energy requirement (ME_m) can be derived from the equation:

$$ME_m = ME \text{ intake} - EE/0.75 - ER/0.75$$

In Experiment 1 the thus calculated ME_m was 502 kJ ME per $kg^{0.75}$ per day for the caged birds and 516 kJ for the birds on the wire floor, and in Experiment 2 these estimates were 527 and 565 kJ per $kg^{0.75}$ per day respectively. This difference of 6.7% is probably due to a difference in heat production as a consequence of a difference in activity of the hens in these systems.

HIGH AMBIENT TEMPERATURES IN TWO HOUSING SYSTEMS

High ambient temperatures can have a detrimental effect on egg production, especially when temperature rises above 30°C (Mowbray and Sykes, 1971). According to Scheele and Musharaf (1978) and Sykes (1979) this effect is associated with a reduced energy retention. As

Table 3.   Performance of medium large laying hens with rapidly increasing and decreasing ambient temperatures in two housing systems.

Age (days)	Temp. (°C)	Eggs/hen-day		Egg Weight (g)		Body Weight (g)		Egg shell perc. *	
		cages	floor	cages	floor	cages	floor	cages	floor
302	20	0.91	0.83	64.8	62.9	2484	2269	8.9	8.4
305	23	0.89	0.87	64.8	62.7	-	-	8.5	9.0
308	26	0.88	0.86	64.8	63.1	-	-	8.4	8.9
311	29	0.83	0.87	63.8	62.6	2481	2307	8.0	8.5
314	32	0.72	0.82	63.1	62.5	-	-	7.6	8.2
317	32	0.71	0.80	61.0	61.4	2372	2276	7.3	8.1
320	29	0.68	0.83	60.5	62.7	-	-	7.8	8.4
323	26	0.68	0.81	61.1	62.5	2319	2277	8.4	8.5
326	23	0.68	0.83	62.3	6.27	-	-	8.8	9.0
329	20	0.72	0.78	62.2	63.0	2387	2322	9.1	8.5

*   measured 2 days after temperature change

pointed out there may be an interacting effect on the energy metabolism of temperature level and housing system. In order to test this possibility the birds of the before mentioned second experiment were, at the end of this experiment, at an age of 42 weeks exposed to rapid increasing ambient temperatures (every 3 days 3°C) from 20°C to 32°C in 12 days whereafter the temperature was lowered to again 20°C, also in 12 days. At the same time the relative humidity was lowered from 70 percent to 50 percent at 32°C (Ketelaars et al., 1985b). Again performance and energy metabolism were measured.

At the start of the experiment the body weight and egg weight of the hens in cages were higher than in the floor system, whereas production was about the same. With increasing temperature egg production, egg weight and body weight of the caged hens dropped significantly (Table 3), together with a sharp decrease in feed consumption (Table 4). The

Table 4. Production efficiency of medium large laying hens with rapidly increasing and decreasing ambient temperatures in two housing systems.

Age (days)	Temp. (°C)	Feed cons. per hen per day (g)		Feed conversion	
		cages	floor	cages	floor
302	20	130	120	2.25	2.29
305	23	129	130	2.25	2.38
308	26	124	126	2.19	2.31
311	29	111	121	2.10	2.22
314	32	89	109	1.98	2.16
317	32	76	104	1.75	2.11
320	29	84	113	2.06	2.16
323	26	99	121	2.41	2.38
326	23	113	122	2.65	2.36
329	20	118	127	2.65	2.56

effect however was much more pronounced in the cage system. There was also a remarkable difference in response in egg shell percentage (Table 3).

These observed differences in response to high temperatures in rela-tion to the applied housing systems were also observable in the energy metabolism (Table 5). Intake of ME decreased in both systems. In the cage system it decreased by 133 kJ when the temperature rose from 26-29 to 32°C, whereas heat production fell only by 62 kJ. In the floor system ME intake decreased by only 53 kJ and heat production by 31 kJ. The reduction in egg energy output and energy retention therefore was the greatest in the cage system, indicating a more severe heat stress to the hens at 32°C than in the floor system.

However it needs to be pointed out that the cage hens were heavier at the start and therefore probably more susceptible to heat stress. Therefore we can only conclude that under the conditions of the present

Table 5. Energy balance data (in $kJ \cdot kg^{-0.75} \cdot day^{-1}$) as affected by changes in ambient temperature in laying hens in two housing systems.

Age (days)	Temp. (°C)	ME intake		Heat production		Egg energy		Energy retention	
		cage	floor	cage	floor	cage	floor	cage	floor
302	20	816	859	590	619	252	254	- 26	- 14
311	26-29	631	770	517	577	213	248	- 99	- 55
317	32	498	717	455	546	186	234	-143	- 63
326	26-23	722	844	535	590	197	227	- 10	27

experiment the effect of a short period of heat stress appeared to be more pronounced in the cage system. This could be due to fewer possibilities for heat dissipation in cages, but there could have been also a relation with the higher body weights of the cage hens at the start of the experiment.

EFFECTS OF CAGE VENTILATION

In order to facilitate the transport of poultry manure surpluses the manure is pre-dried in the poultry house by blowing air through tubes, in between or along battery cages, on the droppings underneath these cages. Field observations gave evidence to the existence of air movements in such a system near the hens in the cages which may affect the microclimate around the birds.

In each of two climate respiration chambers three cage sections, each consisting of two rows of four cages, containing five birds, were placed, so that in each chamber 120 birds were kept. In one of the chambers a metal air flow tube was installed at the inner side of the cages. At one end of the tube a small ventilator was mounted to blow air through the tube with a maximum of 0,5 m³ air per bird per hour. The air escaped through holes in the bottom of the tube. The dried droppings under-

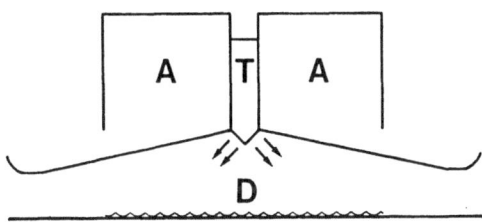

Figure 1. Cages (A), tube (T), and metal tray for droppings (D) in the experiment for cage ventilation.

neath the cages were collected in metal trays (Figure 1.)

Three experiments were carried out. Experiment 1 lasted four weeks, preceeded by one week of acclimation. During this experiment the ambient temperature in both chambers was maintained at 21°C, with a relative humidity (RH) of 65 percent.

Experiment 2 consisted of one week of acclimation and five experimental weeks, during which a temperature level of 29°C was maintained with a RH of approximately 60 percent in both chambers.

Experiment 3 consisted of one week of acclimation and two experimental weeks, during which the ambient temperature in both chambers fluctuated from 28°C in the dark period of the day (9 hours) to 35°C in the light period (15 hours).

In all three experiments White Leghorn laying hens were used. In Experiment 1 the birds were reared on the floor, housed in laying cages at 18 weeks of age, and transferred to the respiration chambers at 26 weeks of age. In Experiment 2 the birds were reared in cages, transferred to laying cages at 18 weeks of age, and placed in cages in the respiration chambers at an age of 25 weeks. The same birds were used for Experiment 3. So, at the start of the experiments the birds were respectively 27, 26 and 36 weeks old.

Performance data were collected and balance measurements carried out in the usual way. To estimate the energy requirement for maintenance (ME_m) the egg energy (EE) and energy retention for weight gain (ER) were subtracted from the ME intake. It was assumed that conversion of ME into egg energy takes place with an efficiency of 60 percent (De

Grootte, 1974). It is further assumed that conversion of ME available for weight gain has an efficiency of 80 percent (Hoffman and Schiemann, 1973).

Under the climatic conditions of Experiment 1 (21°C and 65 percent RH) there was no difference in weight gain, production and feed conversion, but the air flow caused a highly significant difference in percentage dry matter of the droppings (Table 6).

Table 6. Effect of cage ventilation on the performance of WL laying hens.

	Experiment 1		Experiment 2		Experiment 3	
	vent.	non-vent.	vent.	non-vent.	vent.	non-vent.
Growth (g per hen per day)	1.3	1.4	1.5*	1.0*	-2.2	1.7
Egg production (%)	93.1	94.1	91.3	92.9	86.5	85.7
Egg weight (g)	58.2	58.4	55.0	54.9	52.9*	51.1
Egg mass (g per hen per day)	54.2	54.9	50.2	50.9	45.8	43.8
Feed consumption (g per hen per day)	117.4	117.6	102.6	101.0	82.6***	76.6***
Feed conversion (g/g)	2.17	2.14	2.05**	1.98**	1.81	1.75
Dry matter of droppings (%)	29.0***	22.1***	39.9***	19.7***	41.6**	23.7**

* $p < 0.05$
** $p < 0.01$
*** $p < 0.001$

In Experiment 2 (28°C and 60% RH) the ventilated group showed a better growth, even under these conditions there was no difference in performance and feed consumption. Feed conversion however was significantly worse in the group with cage ventilation. Again there was a clear increase of dry matter percentage of the manure, rising to almost 40 percent under the ventialted cages.

In Experiment 3 the increased temperature during daytime caused a decrease in live weight, production and feed intake in both groups, but as a result of the relatively strong decrease in feed intake, feed conversion was more efficient than in Experiment 2.

There was no difference in growth and egg production between the two groups. Egg weight however was clearly less affected by the high temperature in the ventilated group in Experiment 3 (Table 6). Therefore daily egg mass was higher, but not significantly ($p = 0.16$). As a result of the significantly better maintained level of feed intake, feed conversion tended to be worse in the ventilated group in Experiment 3. Dry matter contents of the droppings under the ventilated cages were higher again.

The ME intake per kg metabolic weight decreased, as expected, with the applied increasing ambient temperatures in the experiments, and so did heat production, egg energy and energy retention (Table 7). However egg energy deposition was decreasing at a slower degree than ME intake, so that the conversion of energy into eggs (EE/ME) is tending to grow with higher temperature levels.

From Table 7 it appears that there was no clear effect of the applied ventilation in Experiment 1. In Experiment 2, with the higher ambient temperature, cage ventilation caused a slightly lower ME intake. In Experiment 3, with a still higher temperature, ME intake was significantly higher in the ventilated cages. At the same time egg energy deposition tended to be higher in this group ($p = 0.08$). In none of the experiments however a significant difference was found in the efficiency of conversion of ME into EE (Table 7).

Heat production can be divided into heat production due to activity and heat produced during resting (activity free heat production). It appeared that heat production caused by activity was significantly higher in the ventilated cages, at every investigated temperature level, but in particular in Experiment 1 (Table 7). This suggests that hens in venti-

Table 7. Effects of cage ventilation on the energy metabolism of WL laying hens.

	Experiment 1		Experiment 2		Experiment 3	
	vent.	non-vent.	vent.	non-vent.	vent.	non-vent.
ME intake ($kJ/kg^{0.75}$)	1029	1033	903*	919*	739*	712*
Heat production ($kJ/kg^{0.75}$)	683	684	596	600	535	531
Egg energy (kJ)	429	436	405	413	359	336
Energy retention (kJ)	46	43	16	23	- 91	-100
EE/ME (%)	44	45	49	50	59	57
Heat production by activity (kJ)	136***	112***	93*	85*	90*	69*
Energy for maintenance (kJ)	466	464	396	394	371	378

* $p < 0.05$
*** $p < 0.001$

lated cages are more active. The total heat production, however, did not differ between the experimental and the control groups.

There seemed to be no difference in calculated energy requirements for maintenance (ME_m) between control and treatment groups (Table 7), except a tendency to a lower value for ME_m requirement in ventilated cages in Experiment 3 ($p = 0.11$).

The results indicate that under normal field conditions (Experiment 1) ventilation of cages does not significantly affect growth and production of light laying hens. Yet from an energy point of view it seems that cage ventilation may increase activity and therefore increase heat production.

At higher temperature levels (Experiment 2) hens in ventilated cages had a slightly higher weight gain, compared to not ventilated cages. There was no difference in egg production and egg weight (Table 6). Their feed intake tended to be higher (p = 0.08), whereas their daily egg mass tended to be lower (p = 0.13). This is probably related to their lower ME intake (Table 7). The latter effect however is difficult to explain. At still higher temperature levels (Experiment 3) the first positive effect of cage ventilation was observed, in that egg weight was less negatively affected as in the control group. A tendency to a better egg mass however could not compensate for the higher feed intake (Table 6). The higher ME intake per kg metabolic weight with about the same heat production was leading to a strong tendency (p = 0.08) to higher egg energy deposition in the ventilated group.

From the above mentioned data it can be concluded that cage ventilation at very high temperature levels may contribute to a better energy efficiency in the long run. However, this is apparently not caused by a lower energy requirement for maintenance, because there is a strong indication to a higher activity of the ventilated birds. There is therefore a strong indication that in ventilated cages under normal temperature conditions the effective temperature is decreased, and that only at very high ambient temperatures an alleviation of heat stress may be observed.

EFFECT OF INTERMITTENT LIGHTING ON PERFORMANCE AND ENERGY METABOLISM OF BROILERS

Lighting regimes may also interfere with the energy metabolism of poultry through their impact on activity. In windowless broiler houses intermittent lighting (IL) is practiced, i.e. alternating short periods of light and dark periods, e.g. 1 hour light and 3 hours of darkness (IL:3D), instead of continuous lighting (23L: 1D). To test the possibility of a greater efficiency of IL an experiment was designed to investigate the effect of IL on performance and energy metabolism of broilers (Ketelaars et al., 1987).

For this experiment 960 1-week old broiler chickens were allotted to cages in two respiration chambers, each containing 24 cages in a 3-tier battery system. In one respiration chamber the light regimen remained at 23L:1D (CL treatment) and in the other chamber the regimen was 1L:3D (IL treatment). Climatic conditions and feeding were the same as under field conditions and equal for both chambers. Performance data were registered and also energy data according to the usual procedures (Ketelaars et al., 1987).

Maintenance energy requirement (ME_m) was computed by subtracting ME for growth from the total ME intake. It was assumed that ME for growth is converted into ER with an efficiency of 65 percent. Thus ME_m = ME - ER/0.65.

Table 8. Effect of an intermittent lighting regimen (IL: 1 hr light, 3 hrs darkness) compared to a continuous lighting regimen (CL: 23 hrs light, 1 hr darkness) on performance and energy metabolism of broilers.

		IL	CL
Growth (g/bird/day)		45.8	43.7
Feed consumption (g/bird/day)		75.3	73.7
Feed conversion		1.79	1.93
ME intake	$(kJ/kg^{0.75}/day)$	1395	1433
Heat production	$(kJ/kg^{0.75}/day)$	786	817
Energy retention	$(kJ/kg^{0.75}/day)$	619	615
Energy for maintenance	$(kJ/kg^{0.75}/day)$	458	486

Growth per bird per day was higher in the IL group compared to the CL group (Table 8). This was also found by Van Voorst (1982) together with a lower incidence of leg abnormalities. Total activity in the IL birds was less, but peaks of activity were higher.

Feed conversion in the IL group was better than in the CL group. The balance data show that ME intake and heat production per kg metabolic weight of the IL group were lower, whereas energy retention was equal to that of the CL group. Maintenance energy requirement was lower in the IL group which may be explained by less total activity in this group.

The higher efficiency of energy utilization of IL birds compared to CL birds can be attributed to a lower maintenance energy requirement.

CONCLUSION

From the above described effects of a difference in housing systems (cage vs. floor), the different response at high temperatures in different housing systems, the effect of cage ventilation, and the effect of intermittent lighting on the performance of broilers, we can conclude that housing and climate conditions affect performance of fowl by influencing intake and utilization of dietary energy.

REFERENCES

Groote, G. de, 1974. Utilization of metabolisable energy. In: Energy requirements of Poultry, 1974: 113-134.
Hoffmann, L. and Schiemann, R., 1973. Die Verwertung der Futterenergie durch die legende Henne. Archiv für Tierernährung 23: 105-132.
Ketelaars, E.H., Arets, W.L.M., Hel, W. van der, Wilbrink A.J. and Verstegen, M.W.A., 1985a. Effect of housing systems on the energy balance of laying hens. Netherlands Journal of Agricultural Science 33: 35-43.
Ketelaars, E.H., Brandsma, H., Hel, W. van der, Linden, J.M. van de, Verstegen, M.W.A. and Wilbrink, A.J., 1985b. Effect of high ambient room temperatures on metabolic rate and performance of laying hens in two housing systems. Netherlands Journal of Agricultural

Science 33: 235-240.

Ketelaars, E.H., Verbrugge, M., Hel, W. van der, Linden, J.M. van de and Verstegen, M.W.A., 1986. Effect of intermittent lighting on performance and energy metabolism of broilers. Poultry Science 65: 2208-2213.

Mowbray, R.M. and Sykes, A.H., 1971. Egg production in warm environmental temperatures. British Poultry Science 12: 25-29.

Scheele, C.W. and Musharaf, N.A., 1978. Balance experiments with laying hens at high environmental temperature. Reports 188-78, 189-78. Poultry Research Institute "Spelderholt", Beekbergen, The Netherlands.

Sykes, A.H., 1979. Environmental temperature and energy balance in the laying hen. In: K.N. Boorman and B.M. Freeman (eds.), Food intake regulation in poultry: 207-229. British Poultry Science Ltd. Edinburgh.

Verstegen et al., 1987. The Wageningen Respiration Unit for Animal Production Research: A Description of the Equipment and its possibilities. In: M.W.A. Verstegen and A.M. Henken (eds.), Energy metabolism of farm animals with special reference to effects of housing, stress and disease. Martinus Nijhoff Publications, Dordrecht.

Voorst, A. van, 1982. Draaipoten, groei, voederefficiëntie en vetgehalte van slachtkuikens bij verschillende lichtbehandelingen. IPS-onderzoekverslag no. 95. Instituut voor Pluimveeonderzoek, "Het Spelderholt", Beekbergen, The Netherlands.

CHAPTER III. CLIMATIC CONDITIONS AND ENERGY METABOLISM

SURFACE TEMPERATURES AS PARAMETERS

R. GEERS, W. VAN DER HEL AND V. GOEDSEELS

ABSTRACT

With respect to quantification of the energy balance of a mammal, determination of activity related heat production is important. The change of posture and the movement of the body of the mammal is related to physical and behavioural thermoregulation. The surface temperature plays a part in the physical and behavioural thermoregulation. The study of the surface temperature of a mammal can be helpful to understand the relationship between climatic conditions and energy metabolism.

In order to study surface temperatures in relation to climatic conditions, the accuracy of the measurements is very important. Especially point to point variations, differences according to direction and stimula created by the measuring instruments themselves have to be taken into account.

INTRODUCTION

The surface temperature is an important parameter with respect to the heat exchange of homeotherm animals.

The first part of this paper will deal with the surface temperature in relation to physical and behavioural thermoregulation. Special attention is paid to the physiological mechanisms dealing with the surface temperature.

In order to evaluate the role of the surface temperature, it is very important to measure the temperature very accurately. In the last part therefore some methods will be described to do this.

THE SURFACE TEMPERATURE OF HOMEOTHERM ANIMALS IN RELATION TO PHYSICAL AND BEHAVIOURAL THERMOREGULATION

The surface temperature

The temperature field at the surface of a tissue is determined by heat conduction and convection, metabolic heat generation, thermal energy transferred to this tissue from an external source or the surrounding tissue, and by the tissue geometry.

The thermal conduction is characterized by a thermal conductivity at steady state and by a thermal diffusity in transient states. The thermal convection is characterized by the topology of the vascular bed and the blood flow rate, which is subject to thermoregulation (Jain, 1985).

Obviously, the surface temperature of an animal depends on the climatic conditions of its environment.

A drop in surface temperature of growing pigs (± 25 kg) of about 2°C was observed after the air temperature had been lowered from 25 to 15°C. The resulting surface temperature was about 28°C (Geers et al., 1987). A temperature difference of 2°C between measurements under hair and over hair is realistic (Kelly et al., 1954). Then, the measurements of Geers et al. (1987) correspond to the air temperature related skin temperature measurements of Ingram (1964). He measured in pigs of a comparable body weight a skin temperature of 30°C at an air temperature of 15°C.

With respect to draught (1 m.s.$^{-1}$; 10 °C) a drop of about 1,5°C was stabilized after 4 minutes exposure of the surface of growing pigs (Table 1; Geers et al., 1987). Ingram and Mount (1965) found a drop of 4°C when the skin was exposed to 1 m.s.$^{-1}$ at 25°C during 15 minutes. Since the equilibrium temperature of the surface is determined by the conduction of heat through the coat (Cena, 1974), a decrease of heat loss as a result of vasoconstriction can be the result as well as the origin of the initial lower surface temperature.

Interactions with respect to feed intake have also to be taken into account. A drop of skin temperatures in relation to the time after the

Table 1. The mean surface temperature (°C) per skin region as a function of the night periods with draught sequences every 8 min during 4 min (1 m·s^{-1}; 10°C).

night-period	neck	chest	abdomen
21.00-23.00	30.1-28.2	30.6-28.8	31.1-29.4
01.00-03.00	29.0-28.1	30.0-28.8	30.5-28.9
05.00-07.00	28.1-27.2	29.1-27.7	29.4-28.0

last meal was shown for men by Dauncey et al. (1983).

Acclimation as measured by skin temperature was also observed, because at 8 weeks of age skin temperature of legs and ears was higher when pigs were kept at 10°C versus 35°C from birth on when measured at 10, 20 and 27°C (Macari et al., 1983).

Taking into account the complex of factors and processes which determine the surface temperature of animals, the question remains whether or not the observed circadian rhythm is a pure endogenic process (Geers et al., 1987).

Thermal comfort

The thermal comfort of homeotherm animals is situated in fairly narrow intervals of skin temperature and sweat secretion rate. Comfort is usually "sensed" by a subject at skin temperatures between 32 and 35°C (Fanger, 1972).

The conscious sensation of thermal comfort is one feature which distinguishes behavioural thermoregulation from physiological thermoregulation. But thermal comfort may ultimately be given a physiological interpretation based on the accommodation of a wider range of sensory inputs than the definition based on a single hypothalamic thermoreceptor-thermostat (Eberhart, 1985).

Physical thermoregulation and skin temperature

Important for thermoregulation is the quantitative analysis of a local heat balance, of which skin or surface temperature is a parameter. Because of the number of influences which can act synaptically upon any sensor-to-effector pathway, the response to a stimulus can depend upon much more than on the stimulus itself only. In the case of thermoregulation, the response to a thermal stimulus can vary according to what other central nervous activities are exerting influences upon the thermosensor to thermoregulatory pathways at that time. Hence, as in many biological systems, no clear distinction between the principal and the modulating components may be seen. The relative influences may change, due to differing environmental and physiological conditions (Bligh, 1973).

Within thermoneutrality internal heat transport is still influenced by ambient temperature (vasomotor control of blood heat convection; alterations in insulation capacity, sub- and supracutaneous; posture). Careful calorimetric studies of the dynamics of heat loss and heat production by the body under thermoneutral conditions indicate that the daily fluctuation in deep body temperature is primarily due to changes in thermal conductance rather than to change in heat production. It is generally accepted that the control of peripheral blood flow is the principal thermoregulatory effector process in a thermal neutral environment, and this is controlled by local skin temperature change and core temperature differences as well (Bligh, 1973; Heller and Glotzbach, 1985).

However, a simple averaging over places of skin temperature without consideration of local conductance variations leads to major inconsistencies. Even if one knew the distribution of tissue conductivity, one would still not know how to weigh the regional influence of skin temperature on the central nervous system. That is, thermoreceptor density will also vary regionally (Rowell and Wyss, 1985). Indeed, studies showing sensitivity only to heat or only to cold in discrete areas of skin, indicate that heat and cold might be sensed by different neural structures in the skin, which are purely thermosensitive (Bligh, 1973).

Hence, attention should be shifted to specific skin region temperature measurements (Geers et al., 1987). Especially the estimation of local exchange and partitioning of energy at the surface is necessary to understand the behavioural thermoregulation of homeotherms in terms of heat exchange (Cena, 1974).

Behavioural thermoregulation and skin temperatures

The initial responses are partly behavioural, from a simple reflex (e.g. change of posture) to a complex sequence of muscle activities to construct a shelter. But it is still unknown whether behavioural and autonomic thermoregulatory processes are wholly integrated or separated (Bligh, 1973).

Rats began to press a lever to obtain radiant heat when the subcutaneous temperature dropped with 8°C. The rate at which the rats worked for reductions in hypothalamic temperatures was increased above neutrality (Baldwin, 1974).

But the interactions between animals kept individually or in group, as well as acclimation mechanisms have to be taken into account. Baldwin (1974) found that growing pigs preferred to huddle during the night, instead of operating a switch to produce radiant heat. During the night piglets also preferred a lying area with low air velocity and tended to lay proportionally more side by side (Geers et al., 1986). When applying enhanced air velocities growing pigs, using operant supplemental heating, preferred air temperatures which were about 5°C lower during night time (Verstegen et al., 1986). At 8 weeks of age, the shivering response to a change of skin temperature occurred sooner in cold-reared pigs (Heath and Ingram, 1983).

The change of posture introduces a change of curvature of the surface, a change of convective heat loss and a redistribution of skin surface temperatures. But interactions with the time of the day, the part of the body and even species differences, were observed. Hence, the same mean skin temperature can be accompanied by different amounts of overall heat loss, irrespective of the possible role of evaporation (Aschoff, et al., 1974).

Future studies

Although progress has been made defining the channels of heat loss in the homeothermic animal, and in understanding some of the control mechanisms, there are still considerable gaps in our knowledge: the factors which elicit autonomic as opposed to behavioural control of heat loss, or of the ways in which these alternatives may change over a 24-hour period. We are even more ignorant of the mechanism by which the animal adapts to long term changes in climatic conditions (Ingram, 1974).

The study of local heat balances by means of measuring specific skin temperatures can be very helpful within this respect.

METHODS FOR MEASURING THE SURFACE TEMPERATURE

Problems

When measuring thermal properties of tissues the following problems have to be taken into account: individual variability; point to point variations; differences according to direction; stimuli created by the measuring instruments themselves (Chato, 1985).

It is not easy to construct a stable, accurate and convenient instrument that measures only temperature. Probe thermometers sense their own temperature, which is determined by the exchange with the medium to be measured. The temperature sensed is an average over a finite volume, that depends on the presence of gradients, the thermal properties of the medium and those of the probe. Heat is exchanged by conduction, convection and radiation, each of which must be considered especially when measuring surface temperatures. Because of the small temperature range within biological systems, an improved accuracy is obtained when measuring temperature differences (Cetas, 1985).

Thermocouples

Precautions have to be given to eliminate conduction errors along the lead wires and to minimize the thermal disturbances created by the wire itself (Chato, 1985).

An ordinary thermocouple junction just touching the skin will produce considerable errors ($\pm$ 25%), unless the air temperature is very close to that of the skin, the wire is led along an isotherm away from the junction in good contact with the skin, the wire is as thin as possible without causing mechanical problems. The wires themselves can distort the temperature pattern of the tissue, because of their high thermal conductivity. The distortion will increase with higher temperature gradients (Chato, 1985).

The good contact with the skin can cause an enhanced resistance to heat flow and to evaporation. The warming of the surface can introduce vasodilatation and the pressure vasoconstriction (Chato, 1985).

Hence, non-invasive techniques, such as infrared (IR)-thermography, which measures the whole temperature distribution of the surface seem more appropriate.

Infrared thermography

The IR-technique depends on the classic principle that all bodies emit radiation according to the law of Stefan-Boltzmann. The physical nature of human skin is such that the reflection of radiation from the skin is less than 1%. Thus, the energy sensed by thermography is essentially that emitted by the skin surface. All non-contact heat measurement devices use the radiation of an object to measure its temperature. This energy is emitted in the form of electromagnetic waves which travel with the velocity of the light. If a body is allowed to come to equilibrium with its surroundings, the emission and absorption will become equal and the body will be neither hot or cold.

The infrared spectrum band is part of the electromagnetic spectrum and is broken into four pieces: the near, middle, far and extreme far infrared. The middle and far infrared are the domain of remote temperature measuring instruments (8 to 14 micrometer).

Many of the objects that we would like to observe over normal temperature ranges are virtually (± 90%) blackbodies, or objects whose emissivity is one, reflectivity zero and transmittance zero. All of the energy surrounding is emitted or absorbed. Using an objects' radiation to measure its temperature is more difficult than attaching a thermometer or thermocouple (e.g. change of geometry, temperature related emissivity). So how does one use these instruments to obtain a practical, accurate temperature reading, which still has a relative value: take temperature readings from things that are dull in the visible spectrum; focus on the object; move around to eliminate reflections; the higher the temperature above ambient, the better the reading; watch out for spectral absorption and emission (Wolfe and Zissis, 1978).

The application of thermography allows the mapping of skin temperature patterns and the interpretation of that information to assess physiological status. The temperature levels may be displayed by various shades of grey or by colours representing discrete temperature levels.

The advantages of the technique are numerous. First, the surface temperature distribution is seen almost instantaneously and in detail.

Second, it allows the measurement of the radiative temperature of objects which have such a small heat capacity that measurements with solid probes give false readings (e.g. animal coats). Third, measurement of the temperatures of inaccessible subjects is often possible, Fourth, surface temperatures can be determined in situations where the proximity of an observer would disturb the object of measurement.

The main disadvantages are in the low precision of point measurements and relatively poor accuracy in determining absolute temperatures. Also the careful control of environmental conditions and the instrument calibration that are required (Cena, 1974; Love, 1985).

CONCLUSIONS

A lot of effort has still to be given to study the fundamental mechanisms of heat exchange between the animal and its environment. Especially the value of heat balances of specific parts of body tissues and the dynamic aspect of the heat transfer seem very important in order to clarify the relationship between physical and behavioural thermoregulation.

New techniques with respect to data acquisition and mathematical evaluation of the results (finite element analysis, system analysis) as well, may be useful in this respect.

REFERENCES

Aschoff, J., Biebach, H., Heise, A., Schmidt, T., 1974. Day-night variation in heat balance. In: J.L. Monteith and L.C. Mount (Editors), Heat loss from animals and man, Butterworths, London: pp. 147-172.

Baldwin, B.A., 1974. Behavioural thermoregulation. In: J.L. Monteith and L.C. Mount (Editors), Heat loss from animals and man, Butterworths, London: pp. 97-118.

Bligh, J., 1973. Temperature regulation in mammals and other vertebrates, North-Holland, Amsterdam: 300 pp.

Cena, K., 1974. Radiative heat loss from animals and man. In:

J.L. Monteith and L.C. Mount (editors), Heat loss from animals and man, Butterworths, London: pp. 33-58.

Cetas, T.C., 1985. Analysis and application of thermography in medical diagnosis. In: A. Shitzer and R.C. Eberhart (Editors), Heat Transfer in Medicine and Biology, Plenum Press, New York: pp. 373-392.

Chato, J.C., 1985. Measurement of thermal properties of biological materials. In: A. Shitzer and R.C. Eberhart (Editors), Heat Transfer in Medicine and Biology, Plenum Press: pp. 167-192.

Dauncey, M.J., Haseler, C., Page Thomas, D.P., Parr, G., 1983. Influence of a meal on skin temperatures estimated from quantitative IR-thermography. Experientia 39: 860-862.

Eberhart, R.C., 1985. Thermal models of single organs. In: A. Shitzer and R.C. Eberhart (Editors), Heat Transfer in Medicine and Biology, Plenum Press, New York:pp. 261-324.

Fanger, P.O., 1972. Thermal comfort. McGraw-Hill, New York: 244 pp.

Geers, R., Goedseels, V., Parduyns, G., Vercruysse, G., 1986. The group postural behaviour of growing pigs in relation to air velocity, air and floor temperature. Appl. Anim. Beh. Sci. (in press).

Geers, R., Van der Hel, W., Verhagen, J., Verstegen, M., Goedseels, V., Brandsma H., Henken, A., Schöller J., Berckmans, D., 1987. Surface temperatures of growing pigs in relation to the duration of acclimation to air temperature or draught. J. Thermal Biol. (in press).

Heath, M. and Ingram, D.L., 1983. Thermoregulatory heat production in cold-reared and warm-reared pigs. Am. J. Physiol. 244: R273-278.

Heller, H.C. and Glotzbach, S.F., 1985. Thermoregulation and sleep. In: A. Shitzer and R.C. Eberhart (Editors), Heat Transfer in Medicine and Biology, Plenum Press, New York: pp. 107-136.

Ingram, D.L., 1964. The effect of environmental temperature on body temperature, respiratory frequency and pulse rate in the young pig. Res. Vet. Sci. 5: 348-356.

Ingram, D.L., 1974. Heat loss and its .control in pigs. In: J.L. Monteith and L.C. Mount (Editors), Heat loss from animals and man, Butterworths, London: pp. 233-254.

Ingram, D.L. and Mount, L.C., 1965. The metabolic rates of young pigs living at high ambient temperatures. Res. vet. Sci. 6: 300-306.

Jain, R.K., 1985. Analysis of heat transfer and temperature distribu-

114

tions in tissues during local and whole-body hyperthermia. In: A. Shitzer and R.C. Eberhart (Editors), Heat Transfer in Medicine and Biology, Plenum Press, New York: pp. 3-54.

Kelley, C.F., Bond, T.E., Heitman, H., 1954. The role of thermal radiation in animal ecology. Ecology 35: 562-569.

Love, T.J., 1985. Analysis and application of thermography in medical diagnosis. In: A. Shitzer and R.C. Eberhart (Editors), Heat Transfer in Medicine and Biology, Plenum Press, New York: pp. 333-352.

Macari, M., Ingram, D.L., Dauncey, M.J., 1983. Influence of thermal and nutritional acclimatization on body temperatures and metabolic rate. Comp. Biochem. Physiol. 74a: 549-553.

Rowell, L.B. and Wyss, C.R., 1985. Temperature regulation in exercising and heat-stressed man. In: A. Shitzer and R.C. Eberhart (Editors), Heat Transfer in Medicine and Biology, Plenum Press, New York: pp. 53-78.

Verstegen, M., Siegerink, A., Van der Hel., W., Geers, R., Brandsma, C., 1986. Operant supplemental heating in groups of growing pigs in relation to air velocity. Journal of Thermal Biology: in press.

Wolfe, W.L. and Zissis, G.J., 1978. The Infrared Handbook, Office of Naval Research, Department of the Navy, Washington DC.: 1500 pp.

THE INFLUENCE OF CLIMATIC ENVIRONMENT ON SOWS

B. KEMP AND M.W.A. VERSTEGEN

ABSTRACT

The influence of environmental temperature on metabolism of sows is discussed. First of all principles like heat production, insulation, heat loss, regulation of heat loss and thermoneutrality are discussed. The following items are discussed for sows:
- the level of thermoneutral heat production
- the lower and upper critical temperature
- the extra thermoregulatory heat production at low ambient temperature
- the consequences of not feeding sows according to their increased metabolic rate below the thermoneutral zone.

A sow should be fed at least 420 kJ ME/kg$^{0.75}$ for maintenance. At ambient temperatures below 21°C she should be fed an extra 17 kJ/°C/day/kg$^{0.75}$ to account for the extra thermoregulatory demand.

If a sow is kept below the zone of thermoneutrality or not fed to compensate her increased metabolic rate below thermoneutrality there is a rise of piglets born with a reduced birth weight. Moreover, the sow will need to use her body stores. An increased maintenance requirement and a lower digestibility and metabolizability of nutrients and energy contribute to this.

INTRODUCTION

The most important effect of climatic factors on pigs is on the exchange of heat between the pig and its environment. This exchange is
highly influenced by ambient temperature. This is one of the most important climatic factors. Other climatic factors which might influence the
heat exchange are housing condition, wind and radiation. The influences
of wind, radiation, relative humidity and insulation of the housing on
sows is hardly investigated. As far as these factors are investigated and
reported untill 1976 these have been discussed sufficiently by Holmes
and Close (1977). New data on influences of ambient temperature on sows
are summarized in this paper. A sow is a homeothermic animal which
means that she tries to maintain a constant body temperature. A sow will
be successful in maintaining such a constant body temperature if she is
able to balance heat production with heat loss.

HEAT PRODUCTION

A sow produces heat because of several processes associated with
maintenance and synthesis of body tissue. Heat is produced with synthesis because of the inefficiency of the processes involved. The rate of
synthesis is influenced by food intake. Heat production associated with
this metabolic process is therefore dependent on food intake (Table 1).

Table 1. Heat production (MJ per $kg^{0.75}$ per day) at thermoneutral
conditions of sows fed different amounts of metabolizable energy (ME).

| References | ME intake ($kJ/kg^{0.75}$/day) | | |
	420	840	1260
Holmes and MacLean (1974)	0.393	0.531	0.699
Verstegen et al. (1971)			
- 60 days pregnant	0.456	0.594	0.732
- 112 days pregnant	0.510	0.653	0.787

A sow will produce extra heat in a cold environment to keep her body temperature constant. This extra heat is called extra thermoregulatory heat production (ETH). This heat production is quite distinct from other types of heat production since it is produced especially to meet the environmental demand for heat. The non ETH heat production is produced as an inevitable byproduct of the metabolic activities within the body.

HEAT LOSS

The amount of heat loss from an animal is dependent on two factors:
1. the insulation of the animal, and
2. the possibility to regulate heat loss.

Insulation

Blaxter (1977) distinguished two types of insulation to restrict the heat flow to the environment: the tissue insulation (also called internal insulation) and the external insulation. The internal insulation is the resistance against heat loss from the internal body to the skin surface and can be described in the following way:

$$\text{Internal insulation} = I_i = \frac{T_r - T_s}{H}$$

I_i = Internal insulation ($°C/kJ/m^2/day$)

T_r = rectal temperature ($°C$)

T_s = skin temperature ($°C$)

H = heat loss ($kJ/m^2/day$)

The external insulation is the insulation which includes that of the coat and that of the boundary layer in the coat. Pigs have only a limited amount of hair. Moreover, the insulative value of the hair coat can not be altered very much (Hovell et al., 1977). The external insulation can be described as:

$$\text{External insulation} = I_e = \frac{T_s - T_a}{H - E}$$

I_e = External insulation (°C/kJ/m²/day)
T_s = skin temperature (°C)
T_a = environmental temperature (°C)
H = heat loss (kJ/m²/day)
E = heat loss by evaporation (kJ/m²/day)

The tissue insulation of an adult pig was estimated at 1.67°C per MJ heat loss per m² per day by Irving et al. (1956). The external insulation of a sow was estimated to be 1.63°C per MJ heat loss per m² per day which is relatively low compared to other animals because of the poor coat (Blaxter, 1977). The tissue insulation and the external insulation will change with time when pigs are kept for long periods in cold environments. Animals will probably get a larger coat. When fed enough, also a larger fat layer develops compared to their lean tissue (Verstegen et al., 1985).

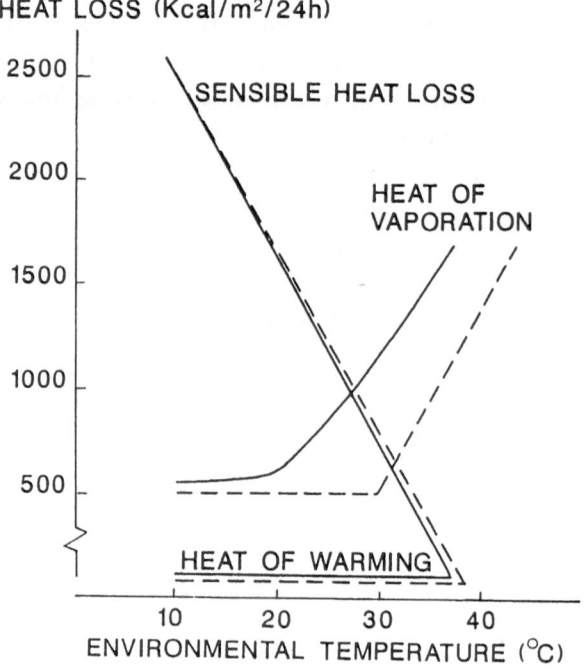

Figure 1. The partitioning of heat loss of sows in sensible and eveporative heat loss at various ambient temperatures at a very low (---) or a high (——) feeding level.

Regulation of heat loss

The heat of animals is exchanged through different channels. These channels can be classified into two main types:

- non-evaporative, or sensible, heat transfer by convection, conduction and radiation,
- evaporative heat transfer by sweating and panting.

The partitioning of heat losses over the non-evaporative and evaporative heat transfer depends on temperature as shown in Figure 1.

There is a linear relationship between heat loss via sensible channels and environmental temperature. If the environmental temperature increases, the sensible heat loss reduces. At low temperatures heat of evaporation is minimal and constant. At high ambient temperatures especially above the zone of thermal comfort (Mount, 1974) an animal will need to evaporate (to pant or to sweat), which rises heat loss by evaporation.

Because of the increase of evaporative heat loss above a given temperature and the constant decrease in sensible heat loss, there will be a temperature range in which total heat loss is constant (Figure 2).

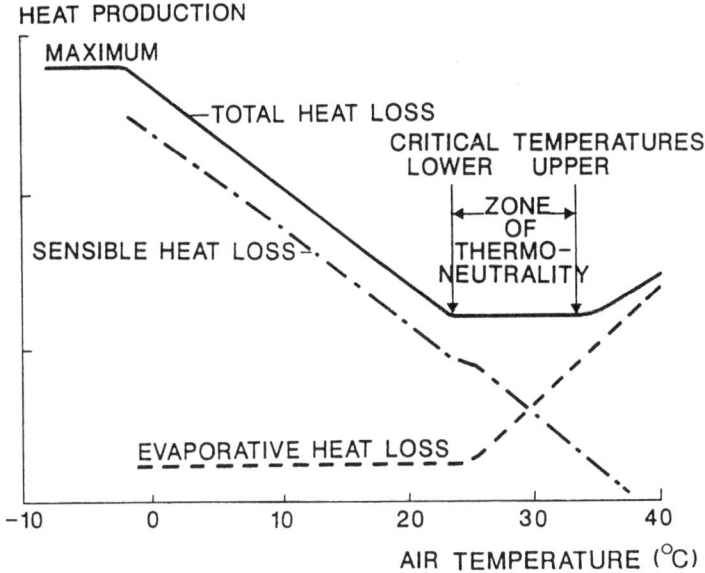

Figure 2. A diagrammatic representation of the relation between air temperature and the components of heat loss in a pig of 60 kg live weight with a ME intake of 2 × maintenance (Holmes and Close, 1977).

When body temperature remains constant heat loss is equal to heat production. The temperature range of constant heat production is called the thermoneutral zone. The lower boundary of this zone is called the lower critical temperature (Lct) or the temperature of maximum insulation. The upper boundary is called the upper critical temperature (Uct) or the temperature of minimum insulation. Below the Lct an animal will have to produce extra heat to keep the body temperature at a constant level. This extra heat is called extra thermoregulatory heat production. Above the Uct an animal must get rid of heat actively (panting), which in itself will cause heat production to rise. Critical temperatures and extra thermoregulatory heat production are influenced by many factors such as feeding level, degree of adaptation of the animal and type of housing.

In this chapter the following will be discussed for sows:
- heat production in the zone of thermoneutrality,
- the upper and lower critical temperatures,
- the extra thermoregulatory heat production at lower ambient temperatures,
- the effect of low environmental temperatures on sow productivity.

HEAT PRODUCTION IN THE ZONE OF THERMONEUTRALITY

The values of heat production in the zone of thermoneutrality as far as published in the literature are given in Table 2.

Heat production of sows fed at maintenance and kept at thermoneutrality is thus about 420 kJ/kg$^{0.75}$/day (Holmes and Close, 1977). It also seems likely that metabolic rate increases with progress in pregnancy (Verhagen et al., 1986).

Thermoneutral heat production of lactating sows fed above maintenance is about twice as high compared to non-lactating or pregnant sows because of the higher metabolic rate associated with milk synthesis (Verstegen et al., 1985). Lactating sows are mostly fed 2 to 2.3 x maintenance.

Table 2. Thermoneutral maintenance heat production in sows.

Stage of pregnancy (days)	Temperature (°C)	Maintenance heat production $(kJ/kg^{0.75}/day)$	References
90	18	418	Verstegen et al. (1971)
n.p.*	23	385	Holmes and MacLean (1974)
n.p.	18	435	Holmes and MacLean (1974)
n.p.	20	476	Hovell et al. (1977)
72	20	459	Geuyen et al. (1984)
72	20	431	Geuyen et al. (1984)
46	21	419	Verhagen et al. (1986)
83	21	448	Verhagen et al. (1986)

*   non-pregnant

UPPER AND LOWER CRITICAL TEMPERATURES

Upper critical temperature

The upper critical temperature or minimum level of insulation for sows has not been experimentally researched. Holmes and Close (1977) calculated the Uct, proposing a maximum heat loss of 1000 $kJ.m^{-2}.°C^{-1}$. The maximum heat loss from the animal can be calculated by multiplying the minimum level of insulation with the surface area of a pig as calculated by the formula of Brody (1945): surface area of a pig (m^2) = 0.097 * (live weight)$^{0.633}$. The Uct can be calculated as:

$$Uct = rectal\ temperature - \frac{thermoneutral\ heat\ production}{minimum\ level\ of\ insulation}$$

Some Uct values for sows weighing 140 kg are given in Table 3.

Table 3. Upper border of thermoneutrality (Uct) in °C (M = maintenance = 420 kJ ME per kg $M^{0.75}$, live weight of the sows was 140 kg).

Stage of pregnancy (days)	Feeding level*			References
	M	2M	3M	
0	32	29	27	Holmes and Close (1977)
112	30	27	25	Holmes and Close (1977)

* times maintenance

The upper boundary of thermoneutrality seems to be lowered by about 2°C in highly pregnant sows compared to non-pregnant sows. High feeding levels will lower the upper critical temperature due to extra heat production. Pregnant and non-lactating sows are kept on feeding levels lower than two times maintenance. In moderate climatic conditions (Western Europe) the upper boundary of thermoneutrality for pregnant sows is not reached very often. However, lactating sows should receive at least 2.5 to 3.5 times maintenance to account for the need for their milk production. At high temperatures sows will depress feed intake in order to prevent reaching the upper limit of thermoneutrality. This is only a limited possibility since metabolic rate arrives from synthesis of milk components. However, milk from body stores will be accompanied by less heat compared to milk synthesis from feed (Verstegen et al., 1985). Lactating sows in a warm environment will loose therefore more body stores compared to sows housed in the cold (Van der Klis and De Bie, 1987).

Lower critical temperature

Holmes and Close (1977) calculated that sows fed at about maintenance will have a lower critical temperature of about 20°C. More investigations have been done on calculating to Lct In Table 4 data on Lct are summarized.

Most authors estimate the Lct for individually housed sows at a feeding level of 1.0 to 1.35 times maintenance at about 18 to 21°C.

Table 4. The lower border of thermoneutrality (Lct) in °C (M = maintenance = 420 kJ ME/kg$^{0.75}$).

Weight (kg)	Stage of pregnancy (day)	Housing system	Feeding* level (×M)	Lct	References
140	0	indiv.	1.0	23	Holmes and Close (1977)
	60	indiv.	1.0	21	Holmes and Close (1977)
	112	indiv.	1.0	19	Holmes and Close (1977)
164 - 192	46 - 97	group	1.2	14	Geuyen et al. (1984)
168 - 186	46 - 97	indiv.	1.2	20	Geuyen et al. (1984)
140	0	indiv.	1.0	18.2	Hovell et al. (1977)
162 - 185	43 - 79	indiv.	1.35	18	Verhagen et al. (1986)
162 - 185	44 - 80	indiv.	1.1	20	Verhagen et al. (1986)

* times maintenance

Holmes and Close (1977) found that Lct was decreased by about 4°C for
sows in late pregnancy compared to sows at an early stage of pregnancy. Geuyen et al. (1984) found that the type of housing (in groups
or individually) influenced the Lct considerably. Sows housed in groups
showed a 5-6 K lower Lct compared to individually housed sows. This
was explained by huddling of sows housed in groups. Verhagen et al.
(1986) showed that even small changes in feeding level could influence
the Lct significantly. Increasing feed intake from 1.1 to 1.35 times maintenance lowered the Lct by 2 K. Holmes and Close (1977) found similar
effects.

EXTRA THERMOREGULATORY HEAT PRODUCTION

For an assessment of feed requirements at low temperatures the extra
heat in the cold above the thermoneutral level need to be known. The

extra thermoregulatory heat (ETH) has been calculated in various studies (Table 5).

Table 5. Extra thermoregulatory heat (ETH: $kJ/kg^{0.75}/day$) per °C below Lct (M = maintenance requirement at thermoneutrality = 420 $kJ/kg^{0.75}/day$).

Weight (kg)	Condition	Housing Level	Feeding* (xM)	ETH kJ/°C/ d/kg$^{0.75}$	References
140	fat	indiv.	1.0	10.0	Holmes and Close (1977)
140	thin	indiv.	1.0	17.4	Holmes and Close (1977)
180	normal	indiv.	1.2	12.7	Geuyen et al. (1984)
180	normal	group	1.2	6.1	Geuyen et al. (1984)
180	normal	indiv.	1.35	13.4	Verhagen et al. (1986)
180	normal	indiv.	1.1	17.6	Verhagen et al. (1986)
140	normal	indiv.	1.0	17.4	Hovell et al. (1977)

* times maintenance

It can be calculated that the increase per K coldness is about 4% when expressed as part of thermoneutral maintenance heat production. Therefore, if the temperature is 5 K below thermoneutrality at least 15 or 20% more feed is required for maintenance. Holmes and Close (1977) estimated that fat sows have a 57% lower ETH production below thermo-neutrality compared to thin sows. This might be explained by the higher rate of insulation in fat sows. Geuyen et al. (1984) found a 52 percent lower ETH production in sows kept in group housing compared to individual housing. Also feeding level seems to effect the Lct (Verhagen et al., 1986).

From data on ETH and heat production of lactating sows the critical temperature can be derived. We may further assume that lactating sows have a high ETH (about 17.4 $kJ/°C/day/kg^{0.75}$) due to poor insulation of the udder. We may assume heat production to be about 200 $kJ/kg^{0.75}/$

125

day above the maintenance level of pregnant sows. The lower border of thermoneutrality of lactating sows would be about 22-200/17.6 = 10°C as compared to 22°C for a thin non-lactating sow at pregnancy.

THE EFFECT OF LOW ENVIRONMENTAL TEMPERATURES ON SOW PRO-
DUCTIVITY

Rate of gain and piglet production

In Figure 3 the mean rate of live weight gain is given for 10 individually housed pregnant sows on two feeding levels kept at various temperatures (Verhagen et al., 1986).

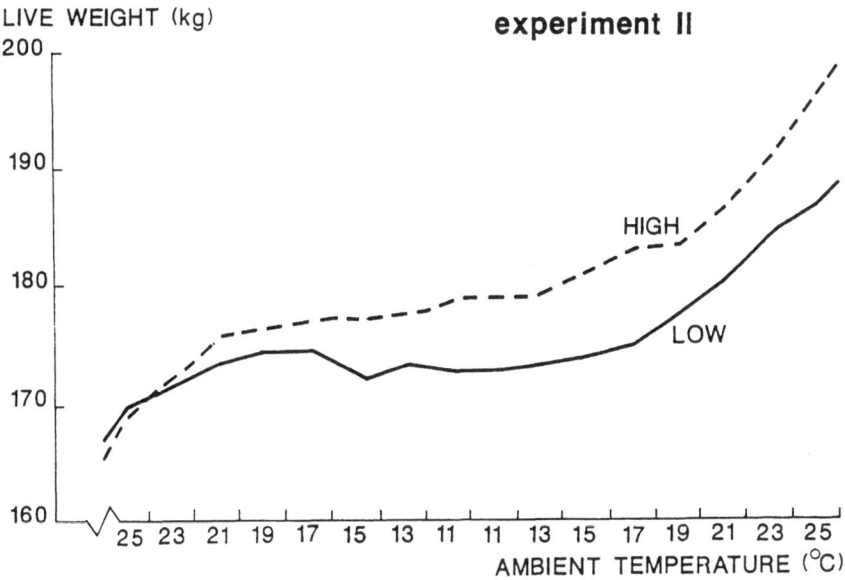

Figure 3. Development of individually housed pregnant sows in the cold. (Exp. II of Verhagen et al., 1986).

Low ambient temperatures depress the rate of gain significantly. On average, rate of gain at temperatures above 20°C was 765 grams per day and below 20°C it was 241 grams per day.

Kemp et al. (1987) found that sows kept for a part of their parity in the cold at a feeding level of 1.1 times maintenance produced piglets

with a 18% reduced birth weight compared to sows on a 1.35 times maintenance feeding level. Noblet et al. (1985) found similar results in sows
kept at lower feeding levels compared to higher feeding levels. However
there is no evidence that the number of piglets can be reduced by low
temperatures or low feeding levels (Vanschoubroek and Van Spaendonck,
1973).

Digestibility and metabolizability

Hovell et al. (1977) found a significant decrease in digestibility of
dry matter, energy and protein in sows kept at 5°C compared to sows at
20°C. Dry matter, energy and protein digestibility were decreased with
5, 4 and 3 percent (absolutely), respectively.

Kemp et al. (1987) found a reduced metabolizability of GE-intake (77%
at 20°C compared to about 74% at 12-14°C). These data suggest that
thermal effects on digestibility and metabolizability are partly responsible
for less efficient feed conversion at lower ambient temperatures.

Energy balance, protein and fat retention

In Table 6 energy balances and protein and fat retentions in sows as
found in various investigations at different ambient temperatures are
given. In each study one temperature is assumed to be thermoneutral
(20, 21 or 24°C), the other is obviously below thermoneutrality. Data on
lactating sows are not yet available.

Sows at feeding levels of 1.1 to 1.35 times maintenance kept at thermoneutral conditions will maintain a positive energy balance. At low temperatures, e.g. at 5 to 12°C, energy balances became negative in all
cases except for the sows (64 days pregnant) kept at 12°C at a 1.35
times maintenance level. Sows at a later stage of pregnancy (85 days) at
the same conditions had a negative energy balance. The fat retention,
the amount of fat deposited in the sow, followed a similar pattern as the
energy balance. Hovell et al. (1977) showed that sows at 20°C accumulated 111 g body fat per day while sows kept at 5°C lost about 151 g fat
per day at similar intake. Kemp et al. (1987) found that fat gain at a
low feeding level was depressed by about 154 g per animal per day at a
temperature of 12°C compared to 21-24°C.

From data of Kemp et al. (1987) we calculated that protein gain was
depressed by about 5.1 g per °C below thermoneutrality in Exp. I and

6.9 g per °C in Exp. II. Similar data were obtained by Hovell et al. (1977).

Table 6. Energy balance, protein and fat retention at different temperatures.

Temperature (°C)	Condition	Stage of pregnancy (days)	Feeding level* (×M)	Energy balance (kJ/kg$^{0.75}$/ day)	Protein retention (g/an/d)	Fat retention (g/an/d)	References**
20	thin	0	1.3	136	− 3	109	1
5	thin	0	1.3	− 232	− 106	− 106	1
20	normal	0	1.3	116	34	113	1
5	normal	0	1.3	− 180	− 24	− 196	1
21	normal	63	1.1	65	79	28	2
12	normal	63	1.1	− 92	46	− 135	2
21	normal	64	1.35	150	102	118	2
12	normal	64	1.35	50	57	25	2
24	normal	70	1.1	49	103	− 2	3
12	normal	70	1.1	− 160	28	− 208	3
24	normal	85	1.35	108	123	61	3
12	normal	85	1.35	− 68	58	− 120	3

* times maintenance (M = 420 kJ/kg$^{0.75}$/day)

** 1 Hovell et al., 1977

2 Kemp et al., 1987 (exp. I)

3 Kemp et al., 1987 (exp. II)

They used non-pregnant sows. Their animals lost protein at low temperatures (in addition to loss of fat). There is some discrepancy between the data of Kemp et al. (1987) and those of Hovell et al. (1977). The pregnant sows of Kemp et al. (1987) still deposited protein even at low feeding levels while fat loss was considerably. The sows of Hovell et al. (1977), however, lost protein in some cases where fat retention remained positive.

Protein retention in pregnant sows seems to vary less with temperature than fat. Protein is accumulated in reproductive tissue (uterus
tissue, mammary tissue and piglets). Moustgaard (1962) calculated the
deposition of protein and energy in intra-uterine tissue (placenta, fluids
plus piglets) of pregnant sows as a function of stage of pregnancy. We
assume that daily deposition of protein in placenta, fluids and piglets is
not influenced by level of feeding or by low ambient temperature. With
the formula of Moustgaard (1962) it was calculated which part of protein
and energy retained in the sows is deposited in intra-uterine in relation
to the pregnancy stages of the animals of Kemp et al. (1987). The remaining protein gain is then deposited the in maternal body. In Table 7
the results of this partitioning of protein is given for the data of Kemp
et al. (1987).

Table 7. Calculated protein and fat deposition in maternal and intra-
uterine tissue

Ex-peri-ment	Feeding* level (×M)	Mean temper-ature (°C)	Stage of preg-nancy (days)	Protein deposition		Fat deposition	
				intra-uterine** (g/an/d)	maternal (g/an/d)	intra-uterine** (g/an/d)	maternal (g/an/d)
I	1.35	17.1	44- 87	17	65	3	74
	1.1	17.1	43- 86	16	46	3	-36
II	1.35	18.0	56-109	34	66	4	-12
	1.1	18.0	51-107	32	35	5	-87

*   times maintenance (M = 420 $kJ/kg^{0.75}/day$)
** placenta, fluids and piglets

The protein free energy fraction is assumed to be deposited as fat.
In the body of piglets there is also some glycogen storage, but this
fraction is neglected. Thus, about 20 to 48% of the total protein retention is deposited intra-uterine. Intra-uterine energy retention is mostly

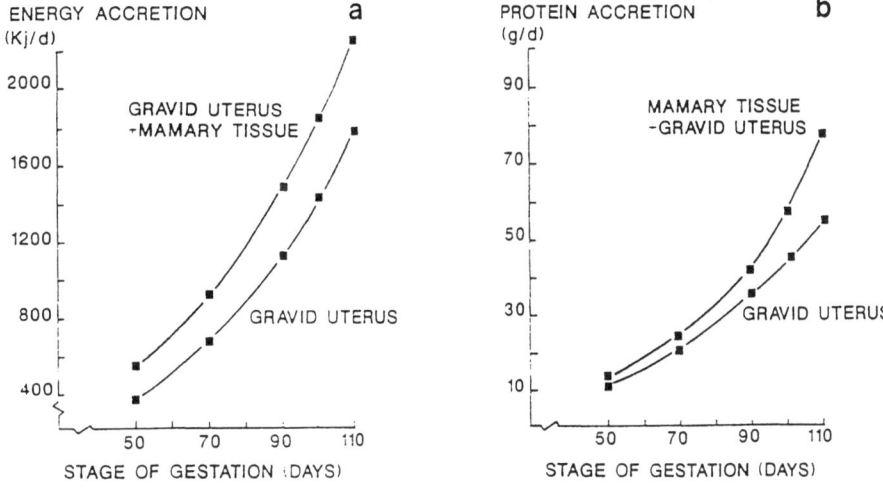

Figure 4. Predicted rates of energy and protein accretion in the reproductive tissues of pregnant gilts according to Noblet et al. (1985).

determined by protein retention.

The protein and energy deposited in the maternal tissue will be present partly in muscle tissue, etc., and partly in reproductive tissue (udder). Noblet et al. (1985) dissected udders from pregnant gilts in different stages of pregnancy. After analysis they calculated the rate of energy and protein deposition in mammary tissue. In Figure 4 the predicted rates of energy (kJ/d) and protein (g/d) accretion in the reproductive tissues (uterus tissue, placenta, fluids, piglets and mammary tissue) of pregnant gilts according to Noblet et al. (1985) are given. ·Values have been calculated for a litter size of twelve piglets and at a ME intake of 20 or 30 MJ/day for the sow. The protein deposition in the mammary tissue was estimated at 1 g/day at day 50 of pregnancy and 21 g/day at day 110 of pregnancy. The energy deposition in the mammary tissue is 0.192 MJ/day at day 50 of pregnancy and 0.488 MJ/day at day 110 of pregnancy. The deposition of energy and protein in uterus tissue is not calculated here, but from these data it is clear that pregnant sows do not deposite very much less protein at low temperature compared to non-pregnant sows. There is no doubt that this "extra" protein is deposited mainly in reproductive tissue. It is obvious from these results that sows which are able to accumulate both protein and fat at thermoneutral conditions may loose large amounts of body fat, even at the cost of fat stores, when they are kept below thermoneutral conditions. Increased maintenance requirements and a lower digestibility and metabolizability of nutrients and energy contribute to this.

CONCLUSIONS

Maintenance requirements of a sow at thermoneutral conditions is about 420 kJ ME per kg metabolic weight. When using this estimate, the zone of thermoneutrality for pregnant and non-lactating sows is bordered by a lower critical temperature of about 18 to 20°C and an upper critical temperature of about 30 to 32°C at normal feeding levels of 1.1 to 1.3 times maintenance. Extra thermal heat requirements will be about 4% of the thermoneutral heat production per K coldness. For lactating sows the upper limit of thermoneutrality will be about 25°C. The lower border of thermoneutrality for lactating sows is not known, but can be estimated at about 10°C or lower.

Critical temperatures, thermoneutral heat production and ETH depend heavily on factors as food intake, housing conditions and animal fatness. From the point of view of energy saving housing in groups is much more economical than individual housing. Lower critical temperatures are clearly increased, when sows are fed below maintenance.

If a pregnant sow is not sufficiently nourished at lower ambient temperatures, this can not only influence her productivity (risk of lower birth weight of piglets and lower gain) but it can also seriously exhaust the body reserves. It is clear that sows normally store body reserves in the period of pregnancy to have sufficient ability to use part of it during the subsequent lactation period. In practice sows are housed individually and fed on low feeding levels because of assumed economic benefits. If we desire that a sow should store sufficient energy and protein in her body reserves in addition to reproductive gain, we must provide good environmental conditions.

REFERENCES

Blaxter, K.L., 1977. Environmental factors and their influence on the nutrition of farm livestock. In: W. Haresign, H. Swan and D. Lewis (editors). Nutrition and the climatic environment of pigs. Butterworths, London: 1-16.

Bond, T.E., Kelly, C.F. and Heitman, H., 1952. Heat and moisture loss from swine. Agric. Engin. 33: 148-152.

Brody, S., 1945. Bio-energetics and growth. Reinhold, New York.

Geuyen, T.P.A., Verhagen, J.M.F. and Verstegen, M.W.A., 1984. Effect of housing and temperature on metabolic rate of pregnant sows. Anim. Prod. 38: 477-485.

Holmes, C.W. and Close, W.H., 1977. The influence of climatic variables on energy metabolism and associated aspects of productivity in the pigs. In: W. Haresign, H. Swan and D. Lewis (editors). Nutrition and the climatic environment of pigs. Butterworths, London: 51-73.

Holmes, C.W. and MacLean, N.R., 1974. The effect of low ambient temperatures on the energy metabolism of sows. Anim. Prod. 19: 1-12.

Hovell, F.D. DeB, Gordon, J.G. and MacPherson, R.M., 1977. Thin sows. 2. Observations on the energy and nitrogen exchanges of thin and normal sows in environmental temperatures of 20 and 5°C. J. Agric. Sci., Camb. 89: 523-533.

Irving, L., Peylon, L.J. and Manson, M., 1956. Metabolism and insulation of swine as bare-skinned mammals. J. appl. Physiol. 9: 421-426.

Kemp, B., Verstegen, M.W.A., Verhagen, J.M.F. and Hel, W. van der, 1987. The effect of environmental temperature and feeding level on energy and protein retention of individual housed pregnant sows. Anim. Prod. 44: 275-283.

Klis, J.D. van der and Bie, G. de, 1987. De aanzet en mobilisatie van lichaamsweefsel van zeugen gedurende het laatste deel van de dracht en de daarop volgende lactatie. MSc Thesis, Agricultural University Wageningen, The Netherlands: 94 pp.

Mount, L.E., 1974. The concept of thermal neutrality. In: Heat loss from animals and man. Eds. Monteith and Mount, London, Butterworths: 425-439.

Moustgaard, J., 1962. Foetal nutrition in the pig. In: Nutrition of pigs and poultry (Eds. Morgan and Lewis), Butterworths, London: 189-206.

Noblet, J., Close, .W.H., Heavens, R.P. and Brown, P., 1985. Studies on the energy metabolism of the pregnant sow. 1. Uterus and mammary tissue development. Brit. J. Nutr. 53: 251-265.

Vanschoubroek, F. and Spaendonck, R. van, 1973. Faktorieller Aufbau des Energiebedrafs tragender Zuchtsauen. Z. Tierphysiol., Tierernähr. u. Futtermittelkde 3: 1-21.

Verhagen, J.M.F., Verstegen, M.W.A., Geuyen, T.P.A. and Kemp, B., 1986. Effect of environmental temperature and feeding level on heat production and lower critical temperature of pregnant sows. Z. Tierphysiol., Tierernähr. u. Futtermittelkde 55: 246-256.

Verstegen, M.W.A., Es, A.J.H. van and Nijkamp, H.J., 1971. Some aspects of energy metabolism of the sow during pregnancy. Anim. Prod. 13: 677-683.

Verstegen, M.W.A., Mesu, J.J., Kempen, G.J.M. van and Geerse, C., 1985. Energy balances of lactating sows in relation to feeding level and stage of lactation. J. Anim. Sci. 60-3: 731-740.

THERMAL REQUIREMENTS OF GROWING PIGS FROM BIRTH TO SLAUGHTER

M.W.A. VERSTEGEN, A.M. HENKEN, W. VAN DER HEL AND
H.A. BRANDSMA

ABSTRACT

The relations between climatic environment and energy metabolism of growing pigs have been studied intensively during the last two decades. From a number of studies thermal requirements have been derived. These requirements have been quantified in terms of lower critical temperature, extra thermal heat requirement and depression in rate of gain.

Critical temperatures of young piglets are reduced from about 31-34°C at birth to 20°C at 20 kg. Weaning requires special attention for thermal requirement. During the fattening period the lower critical temperature is diminished further from about 20°C to about 10-12°C depending on weight, feed intake and housing (groups vs individual). Lower critical temperature depends clearly on feeding level, since high feed intake reduces this critical temperature. Extra heat requirement can be translated into extra feed requirement or depression in rate of gain.

Studies of variation in thermal requirement within days have revealed that thermal requirement is not constant within a day. It appears that animals require higher temperatures in the evening compared to day time and night.

INTRODUCTION

Animals produce heat as a result of processes related to maintenance and production. Efficient production is only possible if heat production from maintenance and production processes is minimal and not affected by housing conditions and climatic environment. Heat production can then be thought to be related to level of feeding and the ration provided. In its simplest way a linear increase of heat increment with increase in feeding level can be assumed. For evaluation of feeding values of rations or feed requirements of animals for practical conditions this assumption is normally made. Efficiencies of conversion of metabolizable energy into energy gain are determined and then applied to practice. Environmental factors are important because they determine whether extra heat is to be produced for homeothermia. Intereaction of feeding level and thermal environment is important with respect to deviations in spending feed energy for heat production and for energy gain. Lower critical temperatures depend on feeding level, body weight and body composition. Some environmental factors can also be corrected towards equivalent standard conditions (Mount, 1975). Some practical conditions on the farm may be simulated in the laboratory. It is therefore important to describe climatic conditions on a farm and to assess zones of thermoneutrality for conditions and animals which are present on farms.

THERMONEUTRALITY

The principles of thermoneutrality have been extensively described (Mount, 1974; Curtis, 1982). Within the zone of thermoneutrality heat production is not affected by climatic conditions. Thus heat production occurs primarily at a rate which depends on level of feeding and on live weight of the animal. Below the thermoneutral zone (in the cold) the animal may be found to increase its heat production in order to maintain homeothermia. This increase is called extra thermoregulatory heat production (ETH). The lower border of the thermoneutral zone is called the lower critical temperature (LCT) and is defined as:
- that temperature below which heat production has to increase in order to maintain homeothermia.
- that temperature below which the insulation of the body (fat, skin and hair) is maximal (and about constant).
- that temperature at which normal (thermoneutral) heat production is

just the same as heat required to maintain homeothermia.

If animals are housed below thermoneutrality heat production is primarily determined by heat loss. Many experiments have shown that both environmental temperature and plane of nutrition influence, through their effect on heat production, the extent to which protein in the diet and the metabolizable energy is converted within the body into protein and fat gain. Thus there are two ways of controlling energy gain and heat loss of animals:

- manipulating the feed intake,
- manipulating the thermal environment.

Table 1.   Calculated lower critical temperatures (°C) for young piglets at three feeding levels.

Live weight	Housed	Feeding level		
		M*	2M*	3M*
2 kg	individually	31	29	29
	group	27	25	25
5 kg	individually	29	23	22

*   M = maintenance, i.e. 0.4-0.5 MJ ME per $kg^{0.75}$ (2M and 3M represent two and three times the maintenance requirements)

Within the zone of thermoneutrality heat production is minimal and thus energy retention can be maximal at a given feed intake. In this zone estimates of heat increment in dependency of feeding level can be made. However if housing conditions or feeding level are such that animals are more active on one level than on another heat production will be no longer minimal. Apparently some results given in Table 1 may have been influenced by such an effect. Therefore it may be advantageous to compare energy retained at various diets at similar feeding level. For assessing thermal requirements this is beneficial because of interaction between lower critical temperature and feeding level. Numerous experiments (see Holmes and Close, 1977) have shown that at a low level of feeding the risk of being below thermoneutrality is much higher than at a higher feeding level. Theoretically the relation between feeding level, ambient temperature and heat production can be visualized as follows.

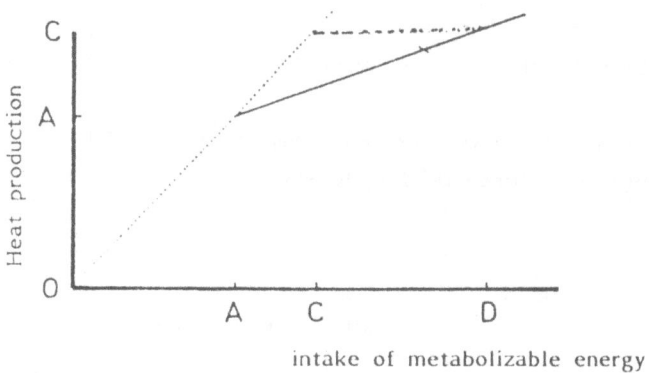

Figure 1. Heat production as affected by feed intake (metabolizable energy). The dotted line indicates the level of heat production at a specific temperature below thermoneutrality (c kJ heat).

At thermoneutrality heat production is related linearly to intake of metabolizable energy (ME). This is shown in Figure 1. In Figure 1 the straight line represents the increase in heat production with increase of metabolizable energy (ME)-intake (equation 1). From A kJ heat at A kJ ME-intake (maintenance) there is an increase to C kJ heat at D kJ of ME-intake. The NRC (1981) proposed the following equation for relating heat production (H) to ME-intake (both in kJ/kg$^{0.75}$).

$$H = 270 + 0.32 * ME \tag{1}$$

When using also the relation ME = EB + H the other component of use of metabolizable energy, i.e. energy balance (EB), can be computed:

$$EB = -270 + 0.68 * ME \tag{2}$$

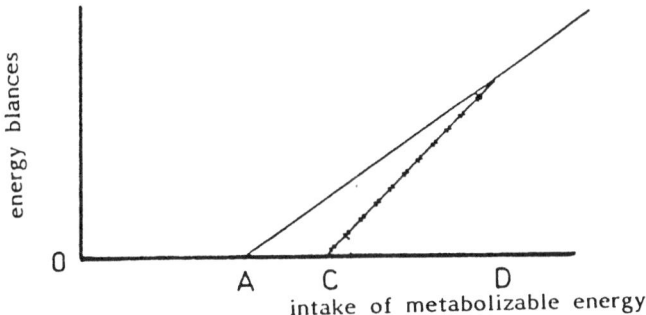

Figure 2. Effect of ME-intake on the energy balance at thermoneutrality (straight line). The dotted line indicates the increase in energy balance with an increase in ME when the thermal demand is c kJ heat.

This is depicted in Figure 2. The straight line from A represents the relation given in equation (2).

When temperatures are below thermoneutrality the temperature determines heat production. This means that production can be similar at two different feeding levels. This is shown in Figure 1 with the dotted horizontal line. The consequences for the energy balance are given in Figure 2. It means that in formula 1 0.32 will change to zero while in formula 2 0.68, as partial efficiency of conversion of ME above maintenance into energy gain, will change towards unity (=1). There is sufficient evidence from the literature that the partial efficiency as derived in formula 2 does not become exactly unity below thermoneutrality (Close, 1980). However nearly all data do show that the partial efficiency is higher in the cold compared to thermoneutral conditions.

THERMAL REQUIREMENTS FROM BIRTH TO 20 KG

There are numerous studies on climatic requirements of newborn piglets (see Mount, 1979 and Curtis, 1982). Most studies agree on thermal requirements of 30-35°C for piglets of a few days old. New-born piglets may even have a lower critical temperature of about 34°C. The tempera-

138

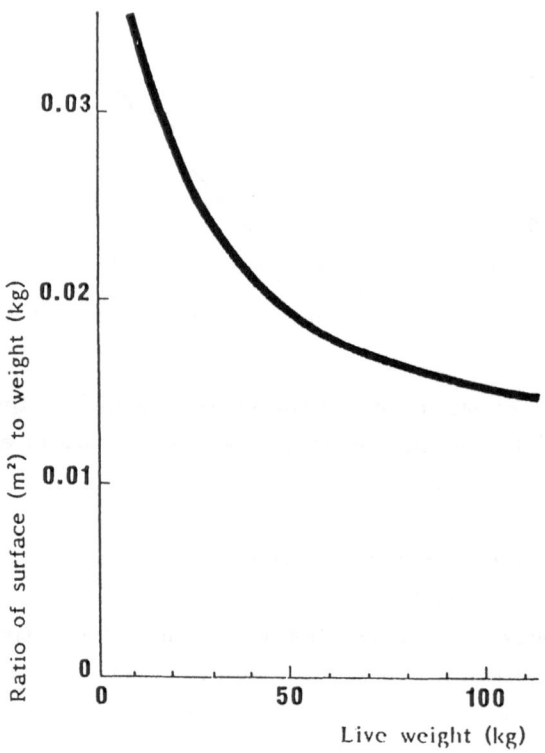

Figure 3. Ratio of surface area to weight as a function of weight (derived from Curtis, 1982).

ture requirement, when expressed as lower critical temperature, falls when animals grow older. Young pigs have a much larger surface to weight area than older pigs (Figure 3). Therefore homeothermia is also more difficult to maintain for young pigs if heat production does not take place at an equivalent higher rate. In addition a baby pig has much less body fat and accordingly a much higher lower critical temperature.

The lower critical temperature of young pigs of 2 kg fed at maintenance (0.4-0.5 MJ ME per $kg^{0.75}$) is about 31-33°C (Holmes and Close, 1977). When more feed is consumed the critical temperature is lowered. In Table 1 the lower critical temperatures are given at three levels of ME-intake. Below this lower critical temperature pigs of 2 kg produce about 47 kJ/kg$^{0.75}$ per °C too cold. This extra heat per °C coldness (ETH) is related to the overall conduction of the animals. Young pigs have a low insulative value, which increases when they grow older (Mount, 1979). The ETH will therefore be greater in older pigs.

As animals grow older their lower critical temperature will diminish. However, the lower critical temperature rises during a few days after weaning as a result of the withholding of milk. Le Dividich (personal communication) found that the fat content of piglets diminishes sharply after weaning. As a result the thermal requirement (lower critical temperature) is suddenly increased from about 22°C at weaning to 27°C a few days after weaning. This principle has been given in Figure 4.

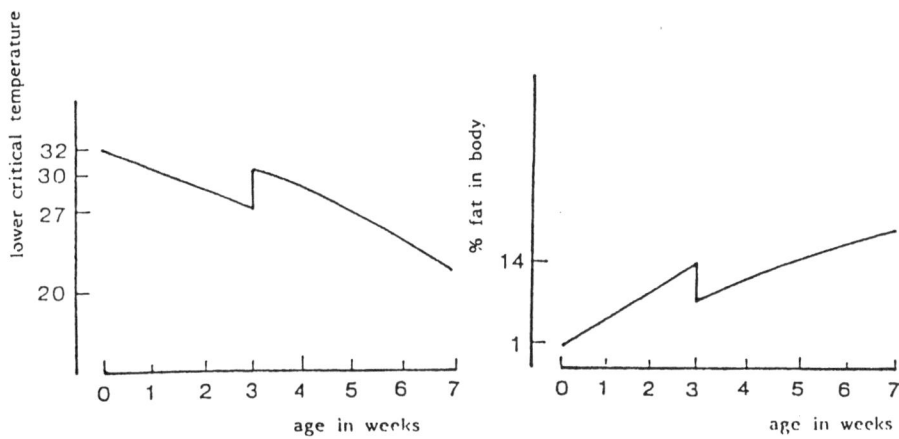

Figure 4.   Effect of age and weaning on the lower critical temperature of piglets.

THERMAL REQUIREMENTS FROM 20 KG TO SLAUGHTER WEIGHT

In areas with a moderate climate cold conditions are generally important for a greater part of the year than hot conditions. In the laboratory thermal requirements are determined mostly at standardized climatic conditions which do not always reflect the conditions at farms.

It is obvious that the energy requirements of animals at cold condi- tions (below thermoneutrality) will be increased due to ETH. The magni- tude of this ETH depends on the insulation of the animal's surface (Blaxter, 1977; Mount, 1979) and on the temperature difference between the animal and its environment. The extra heat produced is a measure for the extra amount of metabolizable energy required by the animal for every degree that the temperature is lower than the lower border of thermoneutrality. By supplying that amount of extra feed the animal can maintain its energy balance. The extra feed requirement can thus be only computed if data are available on the lower critical temperature and on the extra thermal heat production.

The lower critical temperature can be determined in various ways:
- by measurements of heat production at various climatic conditions,
- by measurements of the insulation of the animal.

With both methods some assumptions must be made, e.g. that at low tem- perature heat loss due to evaporation is minimal and independent of tem- perature or that the surface area and the insulation of the animal's sur- face does not change below thermoneutrality. The assumptions are not fully justified, but are necessary (Mount, 1979). The lower critical tem- perature will depend on several factors:
- level of feeding,
- other climatic factors than temperature.

Factors like air movement (Bond et al., 1965; Mount and Ingram, 1965; Verstegen and Van der Hel, 1976; Close et al., 1981) may change heat loss by forced convection.

In order to make critical temperatures comparable Mount (1975) pro- posed to use the effective temperature or the Equivalent Standardized Environmental Temperature (ESET) for pigs. This is the air temperature with a radiant temperature equal to the air temperature, with a RH of 50% and with air flowing at such a low level that no forced convection occurs. Moreover, the animals have to be kept on a dry, well insulated, floor. Any combination of climatic factors like air temperature, radiant temperature, air velocity and kind of bedding can be expressed in terms of effective temperature or ESET.

At standardized conditions there are a lot of data on critical temper- atures in growing fattening pigs. These have been reviewed often in recent years (Holmes and Close, 1977; ARC, 1981; NRC, 1981; Close, 1982).

Table 2. Lower critical temperatures (°C) at ESET conditions in individually and group housed pigs during fattening in relation to body weight (kg) and feeding level (Holmes and Close, 1977).

Kind of animal	Housed	Live weight (kg)	Feeding level		
			M*	2M*	3M*
growing pig	individually	20	26	21	17
	group	20	24	19	15
finishing pig	individually	60	24	20	16
	group	60	23	18	13
finishing pig	individually	100	23	19	14
	group	100	22	17	12

* M = maintenance, i.e. 0.42 MJ ME per kg$^{0.75}$ (2M and 3M represent two and three times the maintenance requirements)

In Table 2 some data on critical temperatures of single and group housed pigs are given. In general the lower critical temperature falls by about 0.8 to 1°C per 10 kg weight increase. All these data have been derived from studies measuring heat production. Various studies have been made to check whether or not the depression in rate of gain agrees with the values derived from measurement of ETH. Measurements of extra thermal heat production are to be preferred since they can be done within a short time. Rate of gain can only be measured over a longer period due to the relatively large contribution of weighing errors in case weight at the beginning and at the end are not very much different. Weight gain depression can also be calculated from measurement of ETH and energy balance (EB). Holmes and CLose (1977) calculated weight gain depressions assuming:
a. depression in rate of gain only at the cost of fat, or
b. composition of the depression in rate of gain similar to the normal

composition of gain.

In Table 3 the theoretical reductions are given. Verstegen et al. (1978) concluded on basis of results reported in the literature up to 1977 that the mean depression was about 14 g per °C below thermoneutrality.

Table 3. Depression in rate of gain (g/°C coldness) from energy balance studies with groups of pigs (Holmes and Close, 1977).
(a) = 1°C below thermoneutrality (only fat deposition affected),
(b) = 1°C below thermoneutrality (all tissues equally affected by coldness).

| | Live weight (kg) | | |
	20	60	100
Depression (a)	4	8	11
Depression (b)	28	16	18

Table 4. Depression in rate of gain and change in feed conversion at similar feed intake (93 $g/W^{0.75}$/day) per °C below LCT (Verstegen et al., 1979).

| | Body mass range (kg) | | | |
	20 - 60		60 - 100	
Depression in gain	9.9		22.4	
Change in feed/gain	0.14		0.13	
Coldness (°C below LCT)	2°C	6°C	2°C	6°C
Lean to fat ratio	1.99	2.04	1.57	1.57
Backfatthickness (mm)	9.3	9.3	16.2	17.1

In Table 4 results of studies designed to compare growth rate and slaughter quality at various degrees below thermoneutrality are reported (Verstegen et al., 1979). The data in Tables 3 and 4 show that depressions in rate of gain per °C coldness are high in animals of high liveweight compared to animals of low weight. As an average about 16-18 g per °C coldness between 20 and 100 kg may be adopted as depression in rate of gain assuming that the animals are not exposed to extreme coldness or exposed to coldness during a very long time. The data show that the depression in rate of gain is somewhat less than as calculated with the assumption that the composition is unaffected (see Table 2). The data therefore suggest that fat gain is somewhat more affected than protein gain. This is supported by the lean to fat ratio's as given in Table 4.

Table 5. Extra heat (ETH) produced per °C coldness ($kJ.°C^{-1}.d^{-1}$) and meal equivalents ($g.d^{-1}.°C^{-1}$) to compensate for this ETH, assuming that 1 g of concentrates contains 12-13 kJ of ME

Housing	Weight (kg)	Meal equivalent ($g \cdot °C^{-1} \cdot d^{-1}$)	References
indiv.	20	14	Holmes and Close, 1977
group	20	13	Holmes and Close, 1977
indiv.	100	36	Holmes and Close, 1977
group	100	35	Holmes and Close, 1977
group	20 - 60	27	Verstegen et al., 1982
group	60 - 100	38	Verstegen et al., 1982
indiv.	20 - 100	35	LeDividich et al., 1985

From the data on the lower critical temperature and ETH it can be calculated how much extra feed is required per °C coldness to maintain the same rate of gain (Blaxter, 1977). In Table 5 some estimates are given as derived from various sources. It can be derived that during the fattening period about 30-40 g of extra feed are required per °C coldness.

Other climatic factors like air velocity and radiant temperature are also important. They may affect the ESET of the environment and consequently also the extra heat required per °C below thermoneutrality. The air velocity acts mainly on the small air boundary layer within the hair coat of the animal. When this insulation layer has been broken the lower critical temperature of the animal housed individually may rise with as much as 10°C (Close et al., 1981). Various studies have been made to evaluate the rise in heat production due to air velocity, especially with very young pigs. Mount and Ingram (1965) estimated the rise in heat loss due to air velocity to be equivalent to the square root of the air speed. However a linear relationship between increase in heat loss and increase in air velocity has also been found (Bruce and Clark, 1979). Huddling behaviour of pigs makes them less sensitive to air movement (Mount, 1960 and 1966; Boon, 1981). Verstegen and Van der Hel (1976) found a rise in the lower critical temperature of only 1.4°C when air velocity was 0.45 m/s instead of 0.15 m/s. This means that the effect of air velocity on heat loss and lower critical temperature is much less pronounced in group housed animals than in individually housed animals. In addition Siegerink (1986) found that group housed animals of about 15-20 kg choose a somewhat higher temperature at a higher air velocity (0.08, 0.25 and 0.4 m/s were compared). At 0.4 m/s the pen temperature chosen by means of operant supplemental heating was 1.2°C higher than that at 0.25 m/s and 3.5°C higher than that at 0.08 m/s. Between 0.08 and 0.25 m/s the largest effect was found. Data of Mount (1979) suggested a much larger effect of air velocity on the lower critical temperature. If we assume that differences in preferred temperatures are a measure for differences in lower critical temperatures then Siegerink's (1986) data suggest similar changes in LCT as data of Verstegen and Van der Hel (1976).

VARIATION IN CLIMATIC REQUIREMENT DURING THE DAY

In recent years sufficient evidence has been collected to suggest that thermal requirements are not constant during the day. The thermal preference of pigs was studied by the operant method (Baldwin and Ingram, 1967; Ingram, 1975; Baldwin, 1979). Individually housed animals regulated a fan in relation to air temperature and air velocity (Baldwin and Ingram, 1967). At a low ambient temperature a lower air speed was chosen by the animals and vice versa. Interactions of preferred thermal environment with the level of feed intake have been observed by Baldwin and Ingram (1968). Preferred conditions were dependent on time of day (Baldwin and Ingram, 1967; Heath and Ingram, 1983). Indeed, since the activity and the heat production are higher during day-time (Van der Hel et al., 1984) the operant activity is also higher (Ingram et al., 1975; Van der Hel et al., 1986). In data of Siegerink (1986) also a large

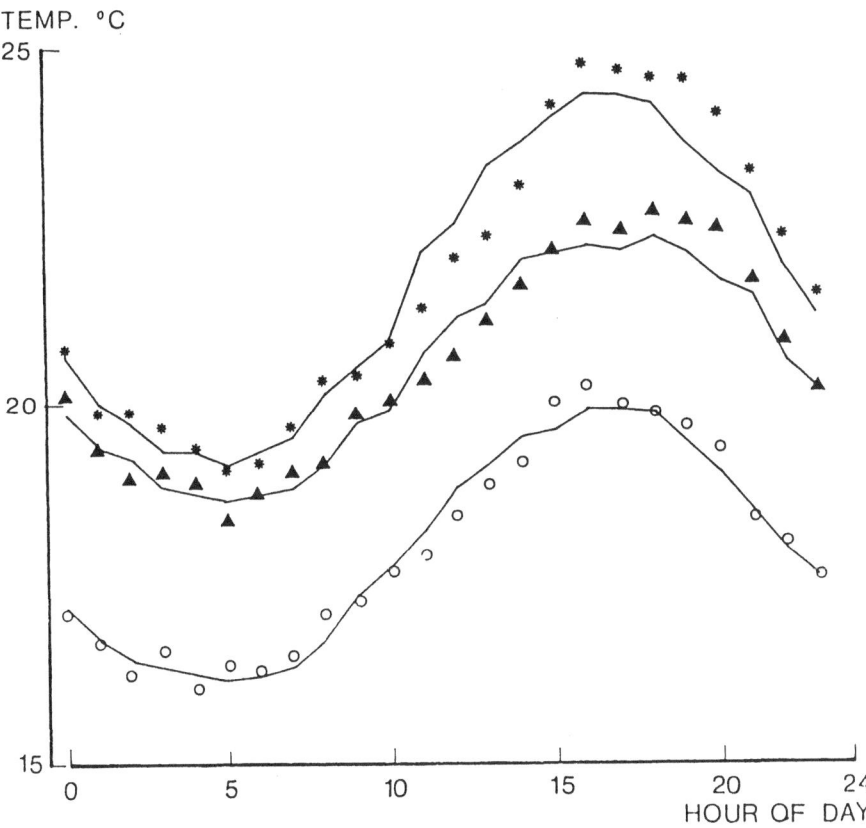

Figure 5. Effect of air velocity on temperatures chosen by pigs with operant supplemental heating (0 = 0.08 m/s; ▲ = 0.25 m/s; * = 0.4 m/s)

variation in temperature preference within a day was found. In Figure 5 this has been depicted for three air velocities.

Thermal demand may change therefore with time of day in relation to activity and behaviour of animals. Balsbaugh and Curtis (1979) and Curtis and Morris (1982) reported that pigs preferred a higher temperature during daytime than at night. Pigs have an innate diurnal rhythm in activity with a small peak in the morning and a great peak in the evening. In these periods pigs became more active at lower temperatures (Van der Hel et al., 1986), this being associated with their greater heat production (heat loss). Van der Hel et al. (1984) calculated the thermal demand from the data of heat production at various times of the day. It was calculated that differences in LCT within a day may be about 2.5 to 3°C. The highest LCT was found at night time. The thermal demand on one part of the day however was not independent from another part since a change in activity pattern over the day at various ambient temperatures was found.

REFERENCES

A.R.C., 1981. The nutrient requirement of pigs. Agricultural Research Council. Commonwealth Agricultural Bureaux, England: 1-30.

Baldwin, B.A., 1979. Operant studies on the behaviour of pigs and sheep in relation to the physical environment. J. Anim. Sci. 49: 1125-1134.

Baldwin, B.A. and Ingram, D.L., 1967. Behavioural thermoregulation in pigs. Physiol. Behav. 2: 15-21.

Baldwin, B.A. and Ingram, D.L., 1968. The effect of food intake and acclimatization to temperature on behavioural thermoregulation in pigs and mice. Physiol. Behav. 3: 395-400.

Balsbaugh, R.K. and Curtis, S.E., 1979. Operant supplemental heating by pigs. Abstract no. 91, 71st Annual Meeting, University of Arizona, Tucson. American Society of Animal Science.

Blaxter, K.L., 1977. Environmental factors and their influence on the nutrition of farm livestock. In: W. Haresign, H. Swan and D. Lewis (Editors). Nutrition and the climatic environment. Butterworths, London: 1-16.

Bond, T.E., Heitman, H. and Kelly, C.F., 1965. Effects of increased air velocities on heat and moisture loss and growth of swine. Trans. Am. Soc. Agric. Eng. 8: 167-172.

Boon, C.R., 1981. The effect of departures from lower critical temperature on group postural behaviour of pigs. Anim. Prod. 33: 71-79.

Bruce, J.M. and Clark, J.A., 1979. Models of heat production and critical temperature for growing pigs. Anim. Prod. 28: 353-369.

Close, W.H., Heavens, R.P. and Brown, D., 1981. The effects of ambient temperature and air movement on heat loss from the pig. Anim. Prod. 32: 75-84.

Close, W.H., 1980. The significance of the environment for energy utilisation in the pig. Proc. Nut. Soc. 39: 169-175.

Close, W.H., 1982. The climatic requirement of the pig. In: J.A. Clark (editor). Environmental aspects of housing for animal production. Butterworths, London: 149-166.

Curtis, S.E., 1982. Environmental management in animal agriculture. Iowa University: 409 pp.

Curtis, S.E. and Morris, G.L., 1982. Operant supplemental heat in swine nurseries. In Livest. Environm. Vol. II: 295-297. ASAS, St. Joseph, Michigan.

Heath, M. and Ingram, D.L., 1983. Thermoregulatory heat production in cold-reared and warm-reared pigs. Behav. Neural. Biol. 31: 273-278.

Hel, W. van der, Verstegen, M.W.A., Baltussen, W. and Brandsma, H., 1984. The effect of ambient temperature on diurnal rhythm in heat production and activity in pigs kept in groups. Int. J. Biometeor. 28: 303-315.

Hel, W. van der, Duyghuisen, R. and Verstegen, M.W.A., 1986. The effect of ambient temperature and activity on the daily variation in heat production of pigs kept in groups. Neth. J. Agric. Sci. 34: 164-173.

Holmes, C.W. and Close, W.H., 1977. The influence of climatic variables on energy metabolism and associated aspects of animal productivity in the pig. In: W. Haresign, H. Swan and D. Lewis (Editors). Nutrition and the climatic environment. Butterworths, London: 51-74.

Ingram, D.L., 1975. The effect of operant thermoregulatory behaviour in pigs as determined from the rate of oxygen consumption. Pflügers Arch. 353: 139-149.

Ingram, D.L., Walters, D.E. and Legge, K.F., 1975. Variations in behavioural thermoregulation in the young pig over 24 hour periods. Physiol. Behav. 14: 689-695.

LeDividich, J., Desmoulin, B. and Dourmad, J.Y., 1985. Influence de la temperature ambient sur les performances du porc en croissance, function en relation avec le niveau alimentaire. Journées Rech. Porcine en France 17: 275-282.

Mount, L.E., 1960. The influence of huddling and body size on the metabolic rate of the young pig. J. Agr. Sci. 55: 101-105.

Mount, L.E., 1966. The effect of wind-speed on heat production in the newborn pig. Quart.J. Exp. Physiol. 81: 18-26.

Mount, L.E., 1974. The concept of thermal neutrality. In: J.L. Monteith and L.E. Mount (Editors). Heat loss from animals and man. Butterworths, London: 425-439.

Mount, L.E., 1975. The assessment of thermal environment in relation to pig production. Livest. Prod. Sci. 2: 381-392.

Mount, L.E., 1979. Adaptation to the thermal environment, man and his productive animals. Arnolds, London: 333 pp.

Mount, L.E. and Ingram, D.L., 1965. The effect of ambient temperature and air movement on localized sensible heat-loss from the pig. Res. Vet. Sci. 6: 84-91.

NRC, 1981. The effect of environment on nutrient requirements of domestic animals. National Academy of Sciences, National Research Council, Washington D.C.: 152 pp.

Siegerink, A., 1986. Effect van luchtsnelheid op temperatuurskeuze van varkens via operant conditioning. Internal report. Department Animal Husbandry, Wageningen: 55 pp.

Verstegen, M.W.A. and Hel, W. van der, 1976. Energy balances in groups of pigs in relation to air velocity and ambient temperature. In Energy Metabolism of Farm Animals (edited by Vermorel, M.) EAAP publ. no. 19: 347-350.

Verstegen, M.W.A., Brascamp, E.W. and Hel, W. van der, 1978. Growing and fattening of pigs in relation to temperature of housing and feeding level. Can. J. Anim. Sci. 58: 1-13.

Verstegen, M.W.A., Mateman, G., Brandsma, H.A. and Haartsen, P.I., 1979. Rate of gain and carcass quality in fattening pigs at low ambient temperature. Livest. Prod. Sci. 6: 51-60.

Verstegen, M.W.A., Brandsma, H.A. and Mateman, G., 1982. Feed requirement of growing pigs at low environmental temperatures. J. Anim. Sci. 55: 88-94.

A FORMULA TO DESCRIBE THE RELATION BETWEEN HEAT PRODUC-
TION AT THERMONEUTRAL AS WELL AS BELOW THERMONEUTRAL
TEMPERATURES SIMULTANEOUSLY

G.F.V. VAN DER PEET, M.W.A. VERSTEGEN AND W.J. KOOPS

ABSTRACT

The relation between heat production and ambient temperature is mostly described by two separate linear models; one for data obtained in the
zone of thermoneutrality and one for data of conditions below the thermoneutral zone. In this study a continuous model is developed to describe the relation in the thermoneutral zone and below the critical temperature (Tcr) simultaneously. Data on energy balances as reported by
Close (1970) and Verstegen (1971) were used to test the formula. These
data were obtained in experiments on 54 animals at various temperatures
and housed in groups of 3 to 5. In the formula heat producton is described as a function of intake of metabolizable energy (ME), metabolic
weight (WP) and ambient temperature (T):

$$H = (1-a) * ME + a * k_m * W^p + \ln (1+e^{(-c*(Tcr-T)*W^p)})$$

Results show that the residual error with this model is lower as compared to the error with linear functions used so far.

INTRODUCTION

Part of the metabolizable energy (ME) consumed by animals is used
for maintenance (ME_m), the remainder being used for growth (ME_p). The
maintenance part of ME is dissipated as heat (H). Heat production from

energy for growth is dependent on the efficiency of the conversion of ME_p into energy gain. The relation between heat production (or heat loss) and productivity (energy gain: EB) can be summarized as: H = ME - EB (all in $MJ \cdot day^{-1}$).

In research of Verstegen (1971), Verstegen et al. (1973) and Holmes and Close (1977) the heat production has been described at two different temperature ranges. These temperature ranges are separated by a so called lower critical temperature (T_{cr}). The lower critical temperature is defined as that temperature limit below which metabolic rate must rise if body temperature is to be maintained (Mount, 1974). This means that heat production below T_{cr} is mostly determined by weight of the animal and temperature of the environment. Verstegen (1971) assumed a linear relationship between the heat production and the number of degrees cen- tigrade below the critical temperature. However according to Holmes and Close (1977) there is experimental evidence which suggests that the rate of increase of heat production at temperatures below T_{cr} increases pro- gressively at lower temperatures. Above T_{cr} in the thermoneutral zone (TNZ) the heat production has a more or less constant level primarily determined by feeding level and live weight of the animal and is nearly independent of ambient temperature.

Up to now the general method has been to combine these two linear models mentioned into one figure resulting in two intersecting straight lines. The point of intersection is called the critical temperature.

The aim of this study was to develop a continuous model which de- scribes the heat production in the zone of thermoneutrality and below the critical temperature at the same time.

The second aim was to develop the formula in such a way that an as- sessment of the critical temperature can be made from weight, feeding level and temperature. The formula and the assessment are checked on heat production data of growing pigs from 20-100 kg from two series of experiments reported in the literature.

DEFINING THE MODEL

Above Tcr

The thermoneutral heat production at maintenance can be described by the following well-known formula:

$$H_m = k_m * W^p \tag{1}$$

in which k_m is a constant in MJ per unit of metabolic weight. It describes the maintenance part of the heat production per W^p. W^p is the mean metabolic weight per day and H_m is the total maintenance requirement in MJ per day. In this formula p is the power to which weight (W) is raised to express metabolic weight (Kleiber, 1965). The maintenance part of the heat production (H_m) is considered constant and independent of feeding level.

The thermoneutral heat production of growing animals is strongly determined by feeding level:

$$ME = H + EB, \text{ and}$$

$$EB = a * ME - a * k_m * W^p \qquad \text{(Holmes and Close, 1977)}$$

this results in:

$$H_{TNZ} = ME - a * ME + a * k_m * W^p$$

$$= (1-a) * ME + a * k_m * W^p \tag{2}$$

where H_{TNZ} is the heat production in the zone of thermoneutrality. The coefficient a is the efficiency (dimensionless) of the utilization of metabolizable energy above maintenance for energy deposition.

Below Tcr

The heat production in the thermoneutral zone is considered independent of temperature, however heat production is considered to become increasingly dependent on ambient temperature below T_{cr} (Close, et al., 1973; Verstegen et al., 1973). Assuming a linear relationship heat production (H) in this temperature traject can be described as:

$$H = H_{TNZ} - (c * (T_{cr} - T) * W^p) \tag{3}$$

where H_{TNZ} is the heat production in the thermoneutral zone, c is the regression coefficient describing the heat increment per °C below T_{cr} (in $J \cdot kg^{-p} \cdot {}^\circ C^{-1}$) and T is the air temperature (Verstegen, 1971; Van der Hel, et al., 1984).

<u>Combining above and below Tcr</u>

To describe the heat production in a continuous model from below the critical temperature into and including the zone of thermoneutrality, the equation has to meet the following requirements:

1. a constant heat production in the TNZ according to equation 2;
2. below T_{cr} an increasing additional heat production per degree with decreasing temperature;
3. result in an estimation of the critical temperature.

Above the critical temperature the part of equation (3) ($c*(T_{cr}-T)* w^p$) has to become zero. This can be fulfilled by the following mathematical equation:

$$y = \ln (1+e^{(-c*(T_{cr}-T)*w^p)})$$ (4)

For values of $T > T_{cr}$, $1+e^{(-c*(T_{cr}-T)*w^p)} \cong 1$ (c is negative)

so $\ln (1+e^{(-c*(T_{cr}-T)*w^p)}) \cong \ln (1)$

and therefore $\ln (1+e^{(-c*(T_{cr}-T)*w^p)}) \cong 0$

As $T < T_{cr}$, $e^{(-c*(T_{cr}-T)*w^p)} \gg 1$

so $\ln (1+e^{(-c*(T_{cr}-T)*w^p)}) \cong \ln(e^{(-c*(T_{cr}-T)*w^p)})$

and therefore $y \cong (-c*(T_{cr}-T)*w^p)$

To meet the requirements mentioned above, the mathematical transformation (4) is implanted into equation (3) resulting in

$$H = H_{TNZ} + \ln (1+e^{(-c*(T_{cr}-T)*w^p)})$$ (5)

Within the thermoneutral zone the second part of this equation is negligible (at a temperature $T > T_{cr} + 0.5°C$, $\ln (1+e^{-c*(T_{cr}-T)*Wp})$ is smaller than 7 J).

Substituting (2) in (5) results in the final formula

$$H = (1-a)* ME + a * k_m * w^p + \ln (1+e^{-c*(T_{cr}-T)*w^p})$$ (6)

Data used to test the formula

The equation (6) was tested on data originating from a study of Verstegen (1971). He used two respiration chambers to measure the heat production in growing pigs. Fifty-four pigs were fattened from 20 to 100 kg and were housed in groups of 4-5 animals. The pigs used were castrated male animals of the Dutch Landrace (NL). All pigs used were fed restrictedly in relation to their metabolic weight (about 93 g per $kg^{0.75}$ per day per animal). The temperature trials were carried out as follows: the animals were placed in a chamber at a weight of about 20 kg maintaining an ambient temperature of 22-24°C. After a short period of adaptation to the chamber, the temperature was decreased by 2-3°C every 2-3 days. The measurements in the chamber were restarted a few hours after each temperature change. When a certain minimum temperature was reached (about 8°C), the temperature change was reversed until a certain maximum temperature was reached etc. At each temperature heatproduction was determined from measurements of gaseous exchange of CO_2 and O_2 over 48 hours. For a more detailed description of the experiments see Verstegen (1971).

Secondly the data of experiments reported by Close (1970) were also analyzed with equation 6. He used pigs of the breed Large White. The most important difference between these trials and those of Verstegen (1971) is that all groups of pigs were kept at constant temperature and various feeding levels over longer periods of time. Rates of heat loss from 54 pigs divided in 12 groups were measured in a direct heat sink calorimeter over periods of 4 weeks. The calorimeter temperatures were kept constant at 7, 12, 20 and 30°C with 3 groups of animals per temperature. The levels of feeding were 34, 39, 45 and 52 g per kg body weight. For a detailed description of these experiments see also Close et al. (1971).

PARAMETERIZATION

The data on weight, ambient temperature, amount of ME fed to the animals and the heat production or heat loss were incorporated into a non-linear regression procedure to estimate the parameters a, k_m, p, c and Tcr in model (6). The sum of the squared differences between heat production estimated and heatproduction actually measured provides a measure for the error in the parameters estimated. Using the routine BMD computer program (Dixon, 1973), the parameter estimates were subsequently adjusted by iteration to minimize the residual sum of squared deviations. Parameterization was done for both sets of data (Verstegen, 1971; Close, 1970) separately.

RESULTS AND DISCUSSION

The data of Verstegen (1971) and Close (1970) were analyzed. In Table 1 the structure of both materials is presented by the minimum, the maximum and the mean value with standard deviation. The parameters obtained from the data are:

Table 1. The minimum, maximum and mean value with S.D. of the variables: metabolizable energy intake (ME) in MJ per day per animal, heat production (H) in MJ per day per animal, weight (W) in kg and temperature (T) in °C. 1 = material of Verstegen (1971); 2 = material of Close (1970).

		minimum	maximum	mean	S.D.
ME (MJ/day)	1	8.70	30.31	18.10	6.40
	2	11.70	27.42	18.03	3.69
H (MJ/day)	1	5.03	21.73	11.56	3.99
	2	7.08	13.87	10.25	1.31
W (kg/animal)	1	15.80	96.60	45.40	21.20
	2	21.00	47.00	32.30	6.00
T (°C)	1	4.20	24.00	15.10	4.10
	2	7.00	30.00	17.30	8.70

a   = partial efficiency for conversion of metabolizable energy above
       maintenance into energy gain;

k_m = maintenance requirement in kJ ME per kg metabolic weight;

p   = power to express weight as metabolic weight;

c   = regression coefficient describing the extra thermoregulatory heat
       (per °C below Tcr);

T_{cr} = lower border of thermoneutrality, i.e. the critical temperature.

   The results of regressing heat production on various independent
variables according to equation 6 are given in Table 2. It appears that
in the material of Verstegen (1971) the estimated value for a and p agree
very well with the value of these parameters generally accepted or found
in other studies (Kleiber, 1965; Blaxter, 1972). The estimated values for
the other parameters are more trial and animal dependent. However they

Table 2.   Parameter estimates with S.D. and the measure of goodness of
fit (R^2) (formula 6). 1 = material of Verstegen (1971); 2 = material of
Close (1970).

partial efficiency		maintenance				extra thermoregulatory heat production		critical temperature		
a		k_m (kJ)		p		c(kJ/ °C/w^P)		Tcr (°C)		
mean	S.D.	mean	S.D.	mean	S.D.	mean	S.D.	mean	S.D.	R^2
1 0.70	0.04	640	64	0.69	0.02	-10.84	3.36	11.0	1.2	0.98
2 0.78	0.02	1091	202	0.56	0.06	39.22	6.95	12.0	0.7	0.75

are within the ranges expected (Close and Verstegen, 1981). The estimated values of the parameters obtained from the material of Close (1970)
are to be interpreted on its own. In the latter material the animals were
fattened at constant temperature and various feeding levels. The results
were therefore obtained with animals which were adapted to low temperatures for a longer period. This may influence the values of the para-

meters estimated. It may also make the estimates of some parameters (e.g. T_{cr} and c) more dependent on the ME intake. At the lowest ambient temperatures only 3 instead of 5 animals were housed in one group. This may also influence the factor c (Mount, 1975). Due to a reduced effect of huddling in a smaller group an increased rate of heat production per °C below T_{cr} can be expected.

The results presented in Table 2 are shown in Figure 1 where the heat production is plotted against temperature using mean values for weight and metabolizable energy intake.

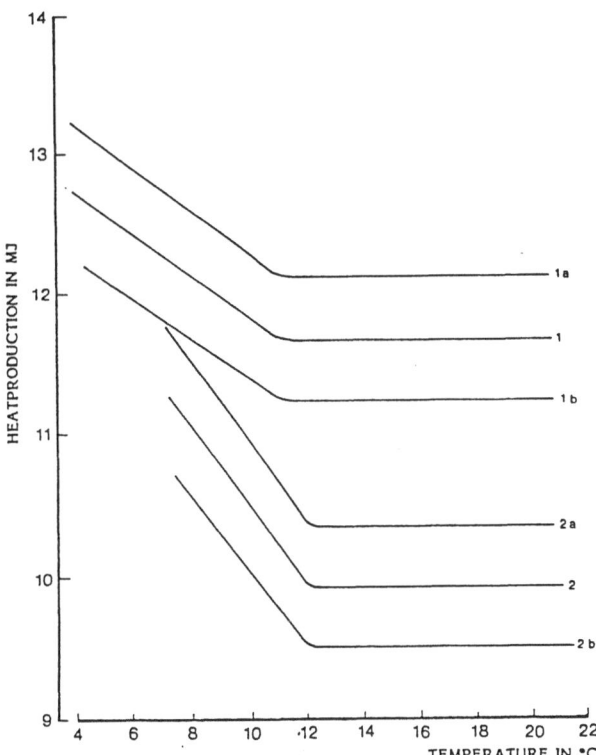

Figure 1. Heat production at various temperatures plotted according to equation 6. 1 = material of Verstegen (1971); 2 = material of Close (1970). Mean ME: 18.1 (1) and 18.0 (2) MJ; mean weight: 45.4 kg (1) and 32.3 kg (2). a,b is resp. weight + or - 2.5 kg and feeding level according to (metabolic) weight.

It is important to realize that the general direction of the curves computed from the material of Verstegen (1971) is not mutually comparible with the one of Close (1970). The performance is not the same, i.e. the weight of the pigs and the feed given to the animals differ from one another. Therefore means the general direction of the curve and the critical temperature is only applicable to the material from which it is derived.

The advantage of the method followed in this study is that all data are combined for the whole temperature range simultaneously. Analysis of data with this formula (6) gives estimates of the parameters according to all data obtained and not for a subset of data, i.e. one for above T_{cr} and one for below T_{cr}. This results in a lower residual error as compared with the traditional methods using linear functions above and below the critical temperature. The small error is demonstrated in Figure 2a and 2b where the residuals are plotted against the fitted heat production. The more accurate description of the heat production is shown in Figure 3a and 3b where the dotted line represents the estimated heat production and the solid lines give the 95% confidence interval for these estimates. In the method followed by Verstegen (1971) it was necessary to separate the data in the parts below and above the critical temperature. Therefore the regression coefficient c could only be computed with data in the part below the critical temperature. This may result in an underestimation of c in case T_{cr} is overestimated.

The advantage of the formula presented here is that all data are incorporated to estimate the relation between metabolic rate, feed intake and temperature without grouping data in an arbitrary manner as done by Verstegen (1971).

The formula may be used only under the condition that the variables weight and temperature are independent of each other.

As is shown in the derivation of equation 4 at temperatures below T_{cr} equation 6 can be written in the following form:

$$H = (1-a) * ME + a*k_m * W^p - c*(T_{cr}-T) * W^p \tag{7}$$

In this form it is easy to show that the multiplicative addition of W^p to the regression coefficient c and the critical temperature allows estimation of the two parameters in relation to metabolic weight. At increasing weight of growing animals a decrease of the critical temperature is ob-

a. Material of Verstegen.

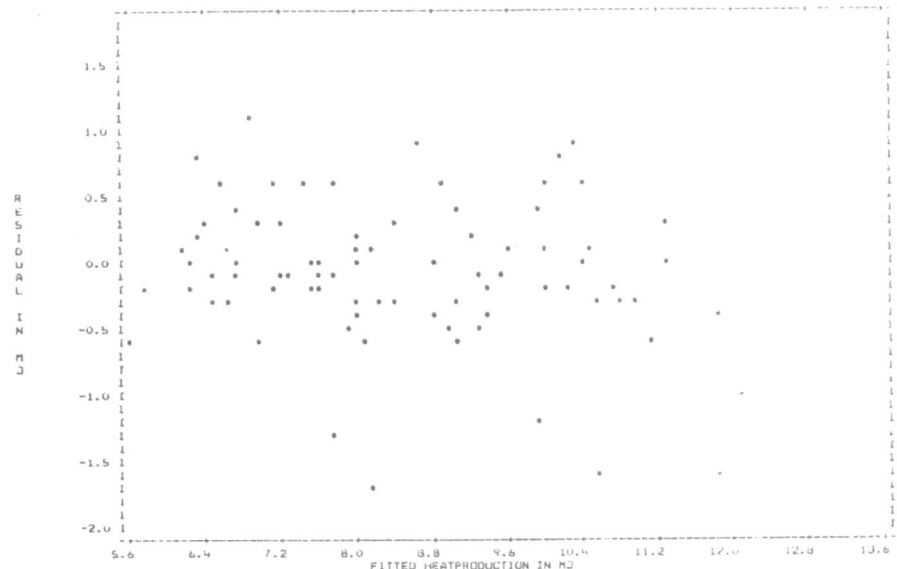

b. Material of Close.

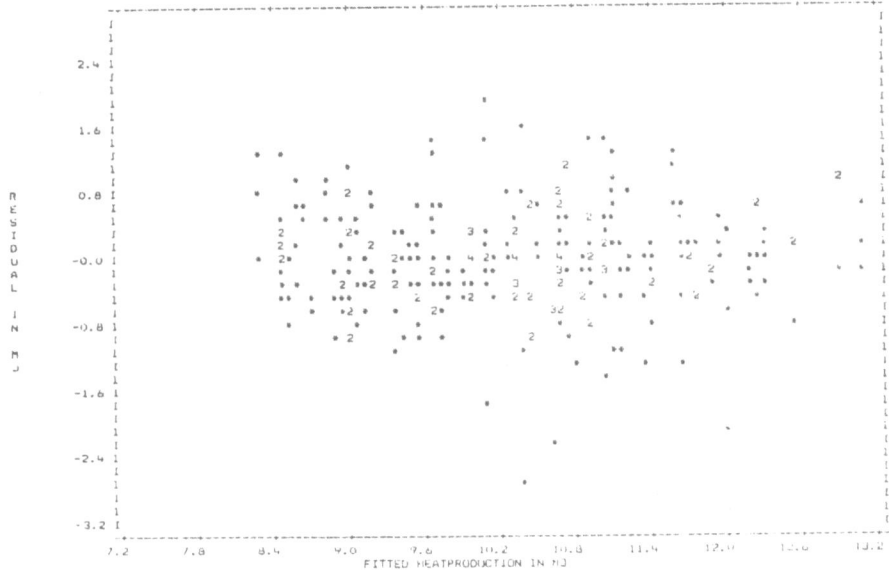

Figure 2. Residuals plotted against fitted heat production.

a. material of Verstegen.

b. material of Close.

160

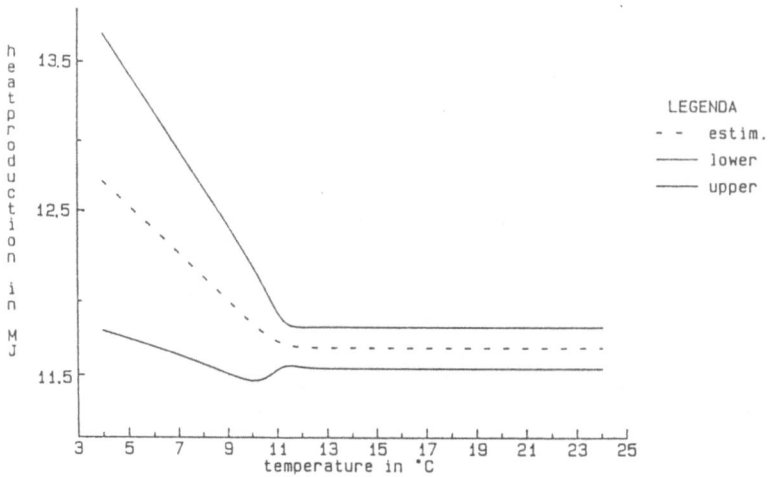

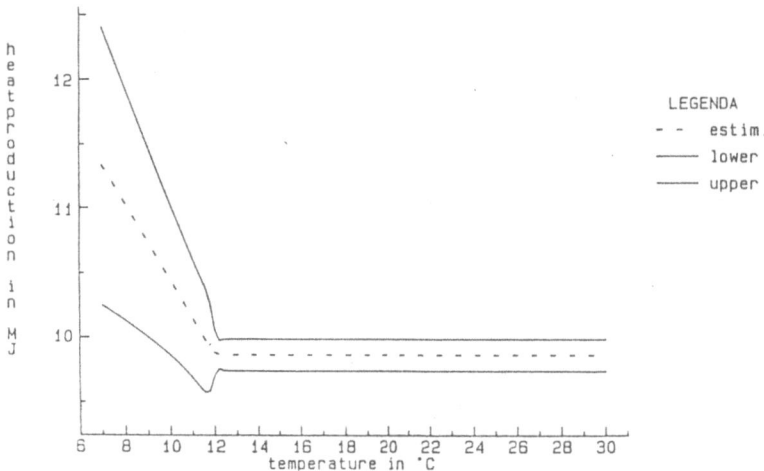

Figure 3.   Estimated heat production according to equation 6 and 95% confidence interval for the estimates.

a. material of Verstegen.

b. material of Close.

served. The accuracy of the estimation of T_{cr} is highly dependent on the value of c. The change below T_{cr} in rate of decrease in heat production with rising temperature to a constant level at thermoneutrality will become more abrupt at higher levels of c. This results in a smaller standard error of the estimated T_{cr}.

In an earlier report by Close and Verstegen (1981) it was concluded that below T_{cr} the heat increment is less dependent upon ME than within the TNZ. Especially with data obtained at varying levels of ME-intake the model can be further improved making the critical temperature a function of metabolizable energy and of metabolic weight:

$$T_{cr} = f(ME, W^p) \tag{8}$$

It may be concluded from the formula developed and tested that a continuous model can be used with more accuracy than the approaches used so far in the literature. The relation between metabolic rate and various animals' and environmental factors can be used to test such a continuous model. The dependency of metabolism on temperature has been quantified in a way which differs from that generally used for data obtained in the zone of thermoneutrality (model 2). The formula was developed and described to estimate the dependency less arbitrarily. From the two sets of data used here it can be concluded that the formula can be used to calculate these relations more accurately. Moreover the parameters of interest will be estimated with greater accuracy.

REFERENCES

Blaxter, K.L., 1972. Fasting metabolism and the energy required by animals for maintenance. In: Festskrift Knut Breirem. Mariendals Boktrykkeri, A.s. Gjøvik, Norway (Eds. Lars S., Spildo, Thor Homb, Harold Huidsten): 19-36.

Close, W.H., Mount, L.E. and Start, I.B., 1971. THe influence of environmental temperature and plane of nutrition on heat loss from groups of growing pigs. Anim. Prod. 13: 285-294.

Close, W.H., 1970. Nutrition-environmental interactions of growing pigs. Ph.D. Thesis. Queens University, Belfast, U.K.: 238 pp.

Close, W.H. and Verstegen, M.W.A., 1981. Factors influencing thermal losses in non-ruminants: a review. Livest. Prod. Sci. 8: 449-463.

Dixon, W.J. (editor), 1973. Biomedical computer programs (BMD). University of California press. Berkely, USA (3rd edition): 772 pp.

Hel, W. van der, Verstegen, M.W.A., Baltussen, W. and Brandsma, H., 1984. The effect of ambient temperature and activity on diurnal rhythm in heat production by group kept pigs. Int. J. Biometeor. 28: 303-315.

Holmes, C.W. and Close, W.H., 1977. The influence of climatic variables on energy metabolism and associated aspects of productivity in the pig. In: K.H. Menke, H.J. Lantzch and R.J. Reichl (editors), Energy Metabolism of Farm Animals, EAAP publ. no. 14: 170-180.

Kleiber, M., 1965. Metabolic body size. Third Symposium on Energy Metabolism, Troon, Scotland, EAAP publ. no. 11: 427-435.

Mount, L.E., 1974. The concept of thermoneutrality. In: J.L. Monteith and L.E. Mount (editors), Heat loss from animals and man. Butterworth, London: 425-439.

Mount, LE., 1975. The assessment of thermal environment in relation to pig production. Live Stk. Prod. Sci. 2: 381-392.

Verstegen, M.W.A., 1971. Influence of environmental temperature on energy metabolism of growing pigs housed individually and in groups. Meded. Landbouwhogeschool, Wageningen: 115 pp.

Verstegen, M.W.A., Close, W.H., Start, I.B. and Mount, L.E., 1973. The effects of environmental temperature and plane of nutrition on heat loss, energy retention and deposition of protein and fat in groups of growing pigs. Br. J. Nutr. 30: 21-35.

APPENDIX

Meaning and dimensions of symbols.

H   = heat production in MJ per day

a   = coefficient of efficiency of the utilization of metabolizable energy for energy deposition.

ME = feed intake in MJ metabolizable energy per day.

k_m = the maintenance part of the heat production in MJ per kg metabolic weight.

W   = weight in kg.

p   = power to which weight (W) is raised to express metabolic weight.

c   = regression coefficient describing the heat increment per °C below T_{cr} in MJ d^{-1} °C^{-1} . kg^p.

T_{cr} = the critical temperature in °C.

T   = the air temperature in °C.

EFFECT OF ENVIRONMENTAL TEMPERATURE AND AIR VELOCITY TWO DAYS PRESLAUGHTERING ON HEAT PRODUCTION, WEIGHT LOSS AND MEAT QUALITY IN NON-FED PIGS

E. LAMBOOY, W. VAN DER HEL, B. HULSEGGE AND H.A. BRANDSMA

ABSTRACT

The heat production of non-fed slaughterpigs was determined at environmental temperatures of 8, 16 and 24°C and 24°C with intermittent showering. Each of these four thermal conditions was tested at two air velocities (0.2 and 0.8 m/s, respectively).

Live weight loss in pigs housed at 24°C was significantly higher than that in pigs housed at 16°C. During their stay of 44 hrs in the calorimeters the pigs drank on average 4.6 ± 2.4 l/pig. while they drank only 1.2 ± 0.6 l/pig at 24°C with showering ($p \leq 0.05$).

The heat production at 16°C was significantly lower than that at 8 and 24°C (551 ± 22 vs 603 ± 35 and 584 ± 15 kJ/kg$^{0.75}$/day, respectively). Showering at 24°C decreased the activity related heat production. The heat production at an air velocity of 0.8 m/s was significantly higher at any environmental temperature than that at 0.2 m/s.

The pH of the musculus semimembranosus in carcasses of pigs housed at 24°C was significantly higher than the pH in those at 8 and 16°C. The pH of the musculus longissimus in carcasses of pigs housed at 16°C and 0.2 m/s was lowest, while the pH was nearly highest in carcasses of pigs housed at 16°C and 0.8 m/s.

From the results it was calculated that the pigs at an air velocity of 0.2 m/s lost less body fat (on average 56 g/pig less) than at 0.8 m/s. The lowest body fat loss (824 g/pig) was at 16°C and 0.2 m/s as compared with the other treatments (874 to 944 g/pig).

It may be concluded that an environmental temperature of 16°C and an air velocity of 0.2 m/s may be efficient for slaughterpigs during transport.

INTRODUCTION

Within the European Economic Community a considerable number of animals is imported, exported or exchanged between member countries. Yearly about 252 million of animals pass across European frontiers (Lockefeer, 1982). During transport the animals are exposed to many stressors such as environmental, physical (vehicle design, noise) and metabolic (deprivation of food and water) factors. Environmental factors are: temperature, humidity, air velocity, loading density and duration of transport (Hails, 1978; Connell, 1984). Moreover loading and unloading may impose stress (Augustini, 1976). Transport conditions may affect post mortem meat quality by provoking stress or fatigue of the animals (Sybesma and Van Logtestijn, 1967; Eikelenboom, 1972; Lambooy et. al, 1985).

The variation in temperatures encountered by the pigs during transport may increase up to approximately 20 degrees Celsius. This variation in temperature within the vehicle is related to variation in air temperature outside (Lambooy, 1987). Therefore ventilation rate during transport should be adapted to the inside temperature which is the resultant of heat flowing from the outside to the inside and heat produced by the animals. However, data of heat production at climatic conditions that occur during international transport of slaughterpigs are not known or estimated indirectly. It may be assumed that in most conditions during transport pigs are in their upper level for thermal tolerance. Therefore additional cooling may be beneficial. However this is not investigated in literature. Especially data at high stocking densities are needed, because of the high number of animals per unit surface of the vehicle during transport.

The aim of the present research was to determine the effect of different thermal conditions at high stocking density on heat production of non-fed slaughterpigs. Moreover effects on some meat quality parameters of slaughterpigs at these pre-slaughter conditions were determined.

166

MATERIAL AND METHODS

Animals

Slaughterpigs (females and castrated males) of about 110 kg live weight were used. The pigs were deprived of feed from 16 hours before entrance in the calorimeter (at 11.00 am) onwards. They were loaded at the farm and transported (about 10 km) to the experimental facilities, unloaded and weighed.

Facilities

Two identical climate controlled calorimeters with a volume of 80 m³ each, equipped with two pens of 9m², were available at the departments of Animal Production of the Agricultural University, Wageningen (Verstegen et al., 1987). The pigs were confined in one pen (stocking density $\approx$ 225 kg/m²) in each calorimeter.

Treatments

The pigs were housed during 44 hours at 8, 16, 24°C and 24°C with showering. Showering took place with water of 15°C every half hour during half a minute (Vajrabukka et al., 1986). In one calorimeter the air velocity was 0.2 m/s and in the other 0.8 m/s. The air velocity of 0.8 m/s was obtained by extra ventilators in the pen which blow the air of the chamber to the pigs. The experimental design is given in Table 1.

The light schedule was 12 hours light (7.00 am to 7.00 pm) and 12 hours dark. The pigs were not fed during their treatment but had free access to water at room temperature provided by a water bowl.

Parameters

Heat production was calculated from measurements of gaseous exchange of CO_2 and O_2 determined continuously during successive 3 hour periods during the day (Van der Hel et al., 1984). Activity was measured at 6-minute intervals with a burglar alarm device (Messl Spaceguard, type 15X). Activity related heat production and activity free heat production were determined according to the method described by Wenk and Van Es (1976). The measurements of gaseous exchange started immediately after unloading and the door of the calorimeter had been closed.

Table 1. Thermal conditions applied.

Experiment	Calorimeter 1 (0.2 m/s)	Calorimeter 2 (0.8 m/s)
1	8°C	8°C
2	16°C	16°C
3	24°c	24°C
4	24°C with shower	24°C with shower
5	8°C	8°C
6	16°C	16°C
7	24°C	24°C
8	24°C with shower	24°C with shower

After exposure to the treatment in the calorimeter the pigs were weighed again, loaded and transported to a slaughter-house (about 35 km). They were slaughtered within one hour after arrival.

The meat quality parameters pH, rigor (Sybesma and Van Logtestijn, 1967) and temperature of the musculus semimembranosus (SM) and musculus longissimus (LD) and the backfat thickness at the shoulder (thickest place over the shoulder), at the midback (over the last rib) and the loin (over the last lumbar vertebra) were measured 45 min post mortem.

Statistics

The data were subjected to analysis of variance (Genstat: Alvey et al., 1982). Temperature (8, 16, 24°C and 24°C with showering) and air velocity (0.2 and 0.8 m/s) were included as factors in the model. When interaction between temperature and air velocity was significant (P $\leq$ 0.10), the eight treatment combinations were compared pairwise using a Student's t-test.

The liberal procedure of comparing treatments pairwise with Student's t-test was used in preference to more conservative multiple comparison procedures, since otherwise in an experiment of this size interesting indications for future research or effects which agree with other experiments may be overlooked.

RESULTS

General

All pigs survived during all conditions; transport from the farm to the calorimeters, housing in the calorimeters and transport from the calorimeters to the slaughter-house. The water consumption was on average 4.6 ± 2.4 l/pig/44 hr based on means per trail.

Effects of environmental temperature

The mean heat production of the pigs at 8, 16 and 24°C was 603 (s.d. = 35), 551 (s.d. = 22) and 584 (s.d. = 15) $kJ/kg^{0.75}/day$, respectively. The heat production was lower ($p \leq 0.01$) at 16°C compared with 8 and 24°C (Figure 1; Table 2). The mean (± s.d.) activity related heat production was 141 (s.d. = 8), 111 (s.d. = 52) and 123 (s.d. = 12) $kJ/kg^{0.75}/day$, respectively. The activity related heat production was not affected by environmental temperature (Figure 1; Table 2).

Table 2. Heat production and activity related heat production of non-fed slaughterpigs housed at different thermal conditions.

Temperature (°C)	8		16		24		24 with showering	
Air velocity (m/s)	0.2	0.8	0.2	0.8	0.2	0.8	0.2	0.8
Heat production ($kJ/kg^{0.75}/day$)	574	633	537	566	573	594	568	595
Activity related heat production ($kJ/kg^{0.75}/day$)	141	140	98	124	122	123	99	83

After the start of measuring at the respective conditions the heat production of the pigs increased during the second and third 3 hour period

Heat production
kJ/kg$^{0.75}$ /day

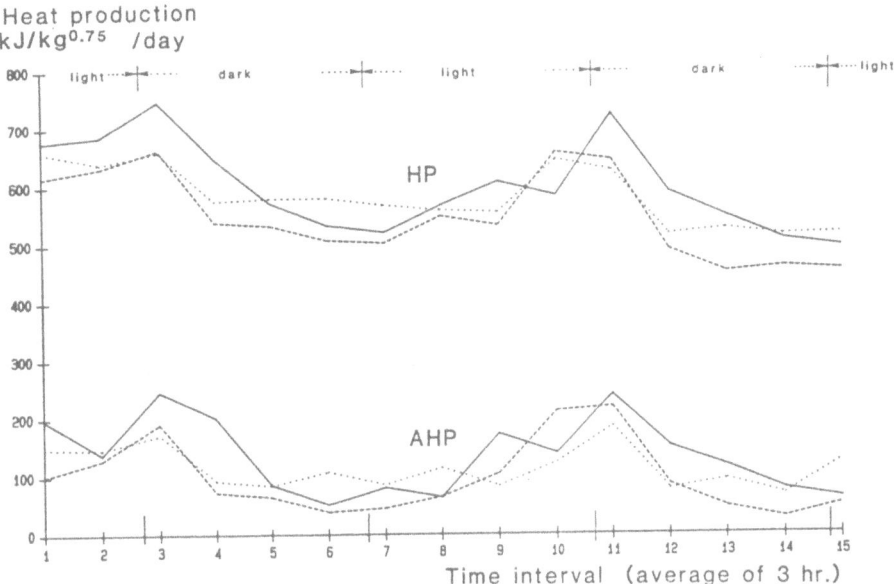

Figure 1. Heat production (HP) and activity related heat production (AHP) (kJ/kg$^{0.75}$/day) of non-fed slaughterpigs of each 3 hour periods in a calorimeter at an environmental temperature of 8°C (———) 16°C (-----) and 24°C (.....) during 44 hours.

(between 2.00 and 8.00 pm) and decreased during the following three to four 3 hour periods (dark period). During the next day the heat production increased again between 2.00 and 8.00 pm and decreased again at the dark period (Figure 1).

Live weight loss of pigs housed at 24°C was significantly (p $\leq$ 0.05) higher than that of pigs housed at 16°C. Between pigs housed at 8 and 16°C no significant difference in weight loss was observed (Table 3).

The pH of the LD in carcasses of pigs housed at 16°C was lower (p $\leq$ 0.10) than that of pigs at the other temperatures at an air velocity of 0.2 m/s (Table 3 and 4). The rigor mortis value in carcasses of pigs at 24°C was significantly ($\leq$ 0.01) higher than that in carcasses of pigs at 8, 16 and 24°C with showering. Backfat was significantly (p $\leq$ 0.05) thicker in carcasses of pigs housed at 24°C (including showering) than in carcasses of pigs houses at 8 and 16°C before slaughter.

Table 3. The mean values (± s.d.) of weight, heat production, meat quality and backfat thickness (average of shoulder, midback and loin) of non-fed slaughterpigs during different environmental temperatures ($p \leq 0.05$ with different superscript).

	Environmental temperature							
Characteristic	8°C		16°C		24°C		24°C + showering	
	mean	s.d.	mean	s.d.	mean	s.d.	mean	s.d.
Live weight:								
at start (kg)	107.2	4.1	111.1	5.9	109.5	2.9	110.4	1.8
at finish (kg)	100.7	3.7	104.6	5.3	102.5	2.5	103.4	1.8
Weight loss (%)	6.1^{ab}	0.3	5.8^a	0.5	6.4^b	0.3	6.3^b	0.5
Slaughter weight (kg)	83.4	2.8	86.6	5.1	84.8	2.7	85.4	1.4
Water use[1]	4.8^a	0.6	6.1^a	1.4	6.4^a	3.2	1.2^b	0.6
pH_{SM}	6.58^a	0.07	6.60^a	0	6.65^{ab}	0.05	6.70^b	0.07
pH_{LD}	6.61	0.08	6.59	0.11	6.67	0.04	6.67	0.06
Rigor mortis	6.8^a	1.3	6.4^a	1.2	9.4^b	0.6	6.1^a	1.5
Temperature$_{SM}$	41.8	0.6	41.1	0.3	41.4	0.2	41.6	0.5
Temperature $_{LD}$	41.2	0.9	40.3	0.1	40.9	0.5	41.0	0.7
Fat thickness (mm)	26.7^a	1.8	24.1^a	0.6	31.0^b	1.7	29.3^{bc}	3.5

[1] (l/pig/44 hr)

Table 4. The mean values of the pH of the LD at the eight thermal conditions tested ($p \leq 0.10$ with different superscript).

	Environmental temperature			
Air velocity	8°C	16°C	24°C	24°C with showering
0.2 m/s	6.66^a	6.51^b	6.66^a	6.73^c
0.8 m/s	6.57^a	6.67^c	6.68^c	6.62^b

Effects of showering at 24°C

Animals subjected to intermittent showering drank less (p < 0.05) water compared to pigs in the other treatments (Table 3). Showering pigs at 24°C had no effect on weight loss.

The activity related heat production was decreased at showering compared to not showering (Table 2). Total heat production was not different, but showed a different pattern over night (Figure 2).

The pH of the SM in carcasses of pigs housed at 24°C with showering was higher (p ≤ 0.05) than the pH of the SM in carcasses of pigs housed at the other temperature conditions. The pH of the LD in carcasses of pigs housed at 24°C with showering was higher (p ≤ 0.10) than that at other treatments at an air velocity of 0.2 m/s. However, at an air velocity of 0.8 m/s the pH of the LD at 24°C with showering was lower (p ≤ 0.10) than that of pigs at 16 and 24°C (Table 4).

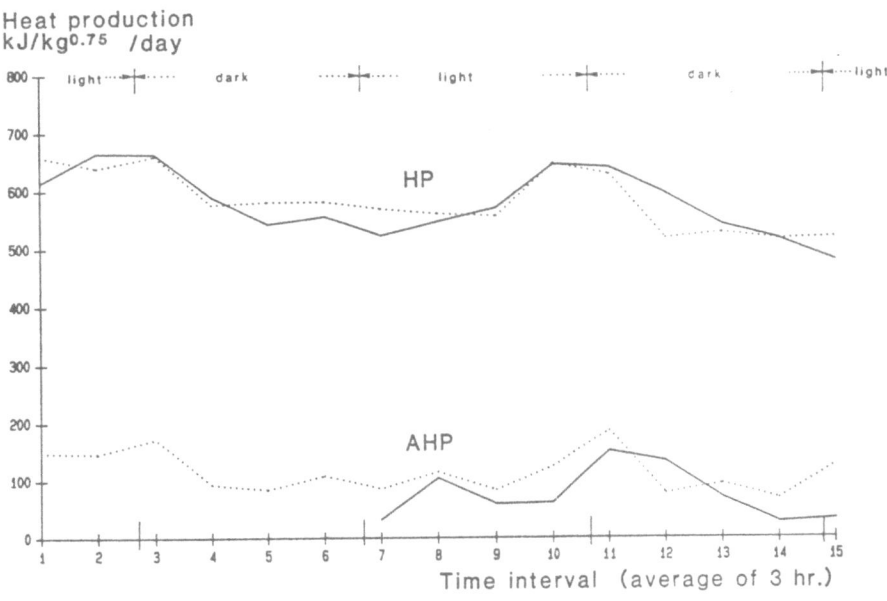

Figure 2. Heat production (HP) and activity related heat production (AHP) $(kJ/kg^{0.75}/day)$ of non-fed slaughterpigs of each 3 hour periods in a calorimeter at an environmental temperature of 24°C with (————) or without (.....) intermittent showering during 44 hours.

Effects of air velocity

The mean heat production at 0.2 and 0.8 m/s was 563 (s.d. = 19) and 597 (s.d. = 27) $kJ/kg^{0.75}/day$ while the activity related heat production was 115 (s.d. = 25) and 118 (s.d. = 41) $kJ/kg^{0.75}/day$, respectively. The heat production at 0.8 m/s was 6.1% higher (p $\leq$ 0.01) than at 0.2 m/s (Figure 3; Table 2). The activity related heat production was not affected by air velocity.

Live weight loss in pigs housed at 0.8 m/s was significantly (p $\leq$ 0.05) higher than that of pigs at 0.2 ms (Table 5).

The effect of environmental temperature and air velocity showed a significant (p $\leq$ 0.10) interaction for pH of the LD. The effect of both factors is specified in Table 4.

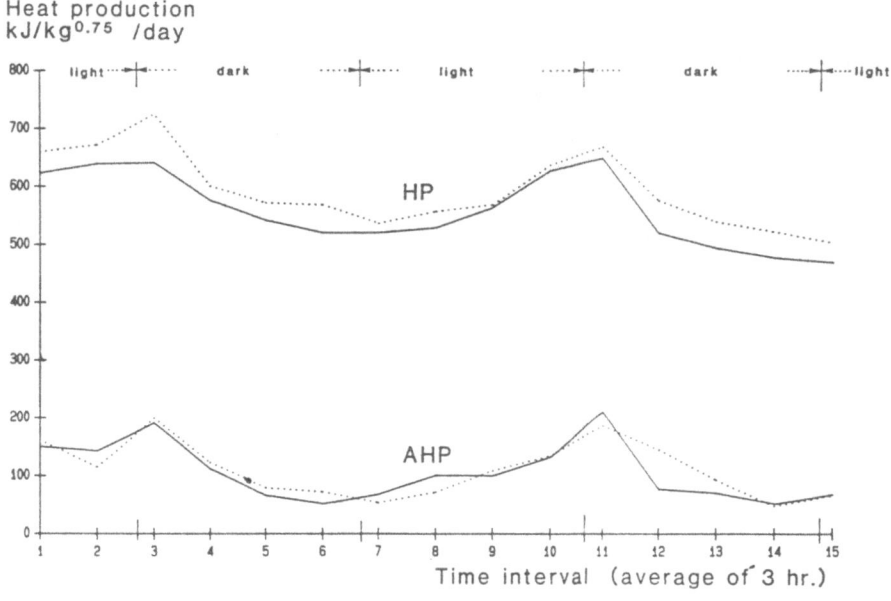

Figure 3. Heat production (HP) and activity related heat production (AHP) $(kJ/kg^{0.75}/day)$ of non-fed slaughterpigs of each 3 hour periods in a calorimeter at air velocities of 0.2 (———) and 0.8 (.....) m/s during 44 hours.

Table 5. The mean values (± s.d.) of weight, heat production, meat quality and backfat thickness of non-fed slaughterpigs housed at two air velocities (p ≤ 0.05 with different superscript).

| Characteristic | Air velocity | | | |
| | 0.2 m/s | | 0.8 m/s | |
	mean	s.d.	mean	s.d.
Live weight at start (kg)	109.1	3.2	110.0	4.6
at finish (kg)	102.6	2.9	103.0	4.2
Weight loss (%)	5.9[a]	0.4	6.4[b]	0.3
Slaughter weight (kg)	84.7	2.7	85.4	3.7
Water use (l/pig/44 hr)	5.2	2.9	4.1	2.5
Heat production (kJ/kg$^{0.75}$/day)	563	19[a]	597	27[b]
Activity related heat production (kJ/kg$^{0.75}$/day)	115	25	118	41
pH_{SM}	6.65	0.05	6.62	0.09
pH_{LD}	6.66	0.08	6.63	0.06
Rigor mortis	6.8	1.8	7.5	1.9
Temperature$_{SM}$	41.5	0.6	41.6	0.3
Temperature$_{LD}$	40.8	0.8	41.0	0.6
Fat thickness (mm)	28.6	3.1	29.0	3.6

DISCUSSION

Effects of environmental temperature

Environmental factors that may stress pigs during transport are e.g. low and high temperatures and air velocity (Augustini, 1976). The effects of these stressors are related to death, weight loss, heat production and meat quality. In The Netherlands in 1980 the mean transport death rate was 0.3%, while in the summer death rate was approximimately 0.4% (Van Logtestijn et al., 1982). During our experiments none of the pigs died. Also Markov (1981) and Lambooy (1983) observed a low death rate (approximately 0.1%) during transports over a long distance.

In our experiments the pigs lost more weight at both 24°C and 8°C environmental temperature compared to 16°C. This means that 16°C can be considered more towards the optimum than other conditions. Weight loss during transport is increased during periods of high temperatures and low relative humidity (Dantzer, 1970). During very cold weather pigs also loose more weight (Hails, 1978). Live weight loss during transport might be mainly associated with evaporation and respiratory exchange (Dantzer, 1982). Also part of loss will be associated with excreta voided by the animals. Loss of carcass weight during prolonged transportation is due to dehydration of the carcass tissues and also related to the mobilisation of depot fat and muscle glycogen (Connell, 1984; Lambooy et al., 1985). However glycogen is only a very small amount of the total body (less than 1%). From data of our experiments it appeared that at an environmental temperature of about 16°C weight loss and heat production are lowest. Therefore this will be in the optimal temperature range for minimal metabolic rate and weight loss.

Animals were last fed about 16 hours before entrance at the calorimeter. This means that at slaughter they were at least 60 hours without feed. Thus the overall thermogenic effect of feed in the intestinal tract can be neglected (Kidder and Mannen, 1978). Thus the animals have been below maintenance during most of the experimental period and it may be assumed that during the experiment the effect of the remaining feed will be very small. From the results of this experiment it was calculated how much fat was catabolized in the animals if it is assumed that only fat is oxidized for heat production. Thus the differences in fat mobilisation calculated will be a measure for the differences in heat production (Table 6).

Heat production at maintenance can be assumed normally at about 420 kJ/ $kg^{0.75}$/day (Holmes and Close, 1977). At feeding time (twice a day) heat production will increase with about 30% (Van der Hel et al., 1986). In the present experiment the pigs did not receive feed at all. It can be assumed that heat derived from feed in the intestinal tract is neglectible during our experiments. Therefore it can be calculated how much fat is mobilized from the body for heat production. It can be calculated that the animals will loose 824 - 944 g body fat during the 44 hr period of exposure. The metabolic rate was as an average above the maintenance requirement as normally assumed (A.R.C., 1981). The mean heat production was 551 kJ/kg$^{0.75}$day (at 16°C). The animals have produced this heat as a result of their maintenance and response to the environment. The heat production in our experiments increased during light with a maximum early in the evening and decreased during the darkness (Figures 1, 2 and 3).

Table 6. Calculated body fat loss (g/pig) and extra thermoregulatory heat production (ETH: kJ/kg$^{0.75}$/day) and difference in critical temperature (T_{cr}) (Verstegen and Van der Hel, 1974) of non-fed slaughterpigs during different thermal conditions. It is assumed that per °C colder about 8 kJ of heat is produced additionally.

Temperature (°C)	8		16		24		24 with showering	
Air velocity (m/s)	0.2	0.8	0.2	0.8	0.2	0.8	0.2	0.8
Body fat loss	874	944	824	885	865	921	876	912
ETH	37	96	0	29	36	57	31	58
ΔT_{cr} (°C)	-4.4	-11.5	0	-3.5				

Heat production values during environmental temperatures of 8 and 24°C (Table 2) tended to be higher than during 16°C. This extra heat production may be due to some extra activity at both high (24°C) and low (8°C) temperatures. From data derived here it appeared that 8°C is below thermoneutrality which agrees with data derived from literature by Holmes and Close (1977). In Table 6 it has also been calculated how much extra thermoregulatory heat (ETH) is produced at each condition if it is assumed that 16°C and an air velocity of 0.2 m/s are at thermoneutrality. At the same time, the pH in our experiments was clearly affected by environmental temperatures during two days before slaughtering (Table 3 and 5). This indicates that an environmental temperature of 16°C will be near the optimal temperature range for minimal stress. It should be noted that the pigs were kept extremely crowded in our experiments. The animal density, however, was similar to the conditions during transport of pigs to the slaughter-house (Lambooy et al., 1985). Therefore these non-fed pigs at 24°C were considered above their zone of thermoneutrality. At lower densities pigs of 110 kg could be expected to be in their thermoneutral zone at this temperature (Holmes and Close, 1977).

Effects of showering at 24°C

Pigs under normal housing conditons (with feeding) will drink 9 to 18 liter water daily per 100 kg body weight (Mount et al., 1971). The average water uptake during the different treatments here was low (4.6 liter/pig/44 hr). During showering the water uptake was minimized to 1.1 liter/pig/44 hr. In spite of the low water uptake the weight loss was not affected by this treatment (Table 3). A low water uptake combined with the use of nipples as showers was observed by Lambooy (1983) and Lambooy et al. (1985) during long distance transports of pigs. In addition it has been observed that showering during warm weather conditions decreased weight loss (Hails, 1978).

Showering at an environmental temperature of 24°C as applied here appeared not to affect the heat production, however, the activity related heat production was decreased as was the rigor mortis value after slaughter in this treatment group. Apparently stress was reduced and showering had provided the ability to evaporate additional water. This means that heat load of stressed pigs and the activity is reduced by showering during transport and lairage (Hails, 1978).

Effects of air velocity

The significant higher live weight loss of pigs subjected to the air velocity of 0.8 m/s may be related to the lower water consumption and the higher heat production, ETH and body fat loss compared to the lower air velocity of 0.2 m/s (Tables 5 and 6). Pigs are generally transported in lorries with fixed vents. The speed of the lorrie, which affects the air velocity in the compartment, and the weather conditions will change the conditions in the compartment. During transport the ventilation rate cannot be altered. Therefore a solution might be to use artificial ventilation systems (Van Putten and Lambooy, 1982). It should be noticed that the ventilation rate should be increased during daytime to match the increased heat production (Figure 3). This should also be applied at higher environmental temperatures.

From the results of these experiments it was calculated that the pigs at an air velocity of 0.2 m/s lost on average 56 g/pig less body fat than they lost at 0.8 m/s (Table 6). This is associated with the increased critical temperature. We estimated that 0.8 m/s will be similar to an increase of 3.5°C in lower critical temperature compared to 0.2 m/s provided that each °C will require about 8 kJ heat per °C per $kg^{0.75}$ (Table 6). This agrees with Verstegen and Van der Hel (1976) who observed that the critical temperature was increased with 1.9 °C when the air velocity increased from 0.2 to 0.45 m/s. It may be concluded that an environmental temperature of 16°C and an air velocity of 0.2 m/s may be near the optimum and should be efficient for minimal weight loss and better meat quality in slaughterpigs during their transport to the slaughter-house.

ACKNOWLEDGEMENTS

This research was financially supported by the E.C. Thanks are due to mr. K. van der Linden and N. van Voorst for excellent assistance and Drs. B. Engel for statistical analyses.

REFERENCES

Agricultural Research Committee (A.R.C.), 1981. The nutrient requirements of pigs. Commonwealth Agr. Bur.: 1-44.

Alvey, N., Galweyk, N. and Lane, P. 1982. An Introduction to Genstat. Academic Press. London.

Augustini, C., 1976. EKG- und Körper Temperatur Messungen Schweinen während der Mast und auf dem Transport. Die Fleischwirtschaft 56: 1133-1137.

Connell, J., 1984. International transport of farm animals intended for slaughter. Comm. Eur. Communities B-1040, Brussels.

Dantzer, R., 1970. Ètude des pertes de poids par des porcelets an cours de transports. Ann. Rech. Vét. 1:179-187.

Dantzer, R., 1982. Research on farm animal transport in France: a survey. In Moss, R., Transport of animals intended for breeding, production and slaughter. Curr. Topics Vet. Med. Anim. Sci., 18, Martinus Nijhoff, The Hague: 218-231.

Eikelenboom, G., 1972. Stress-susceptibility in swine and its relationship with energy metabolism in skeletal musculature. Ph. D. Thesis. Utrecht, 1972.

Hails, M.R., 1978. Transport stress in animals: a review. Anim. Regul. Stud. 1:289-343.

Hel, W. van der, Verstegen, M.W.A., Baltussen, W. and Brandsma, H. 1984. The effect of ambient temperature on diurnal rhythm in heat production and activity in pigs kept in groups. Int. J. Biometeor. 28: 303-315.

Hel, W. van der, Duijghuisen, R. and Verstegen, M.W.A. 1986. The effect of ambient temperature and activity on the daily variation in heat production of growing pigs kept in groups. Neth. J. Agric. Sc. 34: 173-184.

Holmes, C.W. and Close, W.H. 1977. The influence of clinical variables on energy metabolism and associated aspects of productivity in the pig. In: Haresign, Swan and Lewis (Ed.), Nutrition and the climate environment. Butterworth, London: 51-73.

Kidder, D.E. and Mannen, M.J. 1978. Digestion in the pig. Scientechnica, Bristol.

Lambooy, E., 1983. Watering pigs during 30 hour road transport through Europe. Die Fleischwirtschaft 63: 1456-1458.

Lambooy, E., Garssen, G.J., Walstra, P., Mateman, G. and Merkus, G.S.M. 1985. Transport by car for two days; some aspects of watering and loading density. Livestock Prod. Sci. 13: 289-299.

Lambooy, E., 1987. Road transport of pigs over a long distance. In prep.

Lockefeer, W.L.A., 1982. The logistics of animal transportation within Europe and into Europe from further a field. Proc. 2nd Eur. Conf. on the Protection of Farm Animals, May 25-26, 1982. Strasbourg: 42-49.

Logtestijn, J.G. van, Romme, A.M.T.C. and Eikelenboom, G. 1982. Losses caused by transport of slaughterpigs in The Netherlands. In Moss, R. Transport of animals intended for breeding, production and slaughter. Curr. Topics Vet. Med. Anim. Sci., 18. Martinus Nijhoff, The Hague: 105-114.

Markov, E., 1981. Studies on weight losses and death rate in pigs transported over a long distance. Meat Industry Bulletin 14: 5.

Mount, L.E., Holmes, C.W., Close, W.H., Morrison S.R. and Start, I.B. 1971. A note on the consumption of water by the growing pig at research environmental temperatures and levels of feeding. Anim. Prod. 13: 561-563.

Putten, G. van and Lambooy, E. 1982. The international transport of pigs. Proc. 2nd Eur. Conf. on the Protection of Farm Anim., May 25-26, 1982, Strasbourg: 92-108.

Sybesma, W. und Logtestijn, J.G. van, 1967. Rigor mortis und Fleischqualität. Die Fleischwirtschaft 4: 408-410.

Vajrabukka, C., Thwaites, C.J. and Farrell, D.J., 1986. The effect of duration of sprinkling on total moisture Losses from pigs. Int. J. Biometeor. 30: 185-188.

Verstegen, M.W.A. and Hel, W. van der, 1974. The effects of temperature and type of floor on metabolic rate and effective critical temperature in groups of growing pigs. Anim. Prod. 18: 1-11.

Verstegen, M.W.A. and Hel, W. van der, 1976. Energy balances in groups of pigs in relation to air velocity and ambient temperature. Energy Metabolism of Farm Animals (Ed. M. Vermorel) EAAP 19: 347-350.

Verstegen, M.W.A., Hel, W. van der, Brandsma, H.A., Henken, A.M. and Bransen, A.M., 1987. The Wageningen Respiration Unit for Animal Production Research: A description of the equipment and its possibilities In: Energy Metabolism of Farm Animals (M.W.A. Verstegen and A.M. Henken, eds.). Martinus Nijhoff Publishers, Dordrecht.

Wenk, C. and Es, A.J.H. van, 1976. Eine Methode zur Bestimmung des Energieaufwandes für die Körperliche Aktivität von wachsenden Küken. Schweiz. Landwirtsch. Monatst. 54: 232-236.

EFFECTS OF CLIMATIC CONDITIONS ON ENERGY METABOLISM AND PERFORMANCE OF CALVES

M. VERMOREL

ABSTRACT

The effects of cold are especially critical in newborn calves due to their high energy losses through the wet hair coat and the great variability in summit metabolism. Nevertheless, high-vitality calves can withstand low ambient temperatures, whereas premature or dystocial calves are unable to increase their heat production to compensate for their high heat loss and they become hypothermic.

During the rearing period, the energy expenditures and the lower critical temperature of calves depend on breed, physiological stage, feeding level and climatic factors (temperature, wind and rain). The performance of healthy calves kept outdoors in winter in moderate climatic conditions is not significantly different from that of calves reared in a heated barn or under shelter, except at low temperatures associated with very high humidity. The weight gain of suckler calves is reduced only at very low temperatures. Shelters with walls alone have a greater protective effect than a roof alone. Shelters with both walls and roof almost eliminate the depressive effect of climate.

Calves are more sensitive to heat than to cold, especially when their production level is high. The effect depends on ambient temperature, air humidity and velocity, and solar radiation. Adaptation to heat is long and incomplete. However, calves are able to withstand high temperature and humidity for several hours, without any significant reduction in performance, if they can recover at a lower temperature during the night.

The physiological responses of calves to cold and heat are briefly discussed in relation to energy metabolism, thermoregulation and performance, especially in the case of dystocial calves born in winter and rapidly growing calves in Mediteranean regions during the summer period.

INTRODUCTION

Adverse climatic conditions affect calves, especially during the first weeks of life, either by increasing heat losses or by inducing secondary effects on health. In both cases, growth rate or feed intake is affected. In the temperate zones, this is one of the reasons why calves are often reared in heated barns. But healthy, newborn calves have been placed outdoors in hutches more and more frequently without any obvious increase in mortality rate, even during winter. Nevertheless, the effects of cold exposure on calf performance are still controversial. Rapidly-growing veal calves are sometimes stressed by heat in summer. They reduce their food intake and weight gain. Even when the heat stress has disappeared, they do not grow normally.

Several studies were carried out to determine the lower and upper critical temperatures of calves. However, the experiments were conducted during short periods, often without previous adaptation. The energy expenditure of calves exposed to various climatic conditions was measured either by direct calorimetry or estimated from their heat production determined by indirect calorimetry. Furthermore, acclimatization may reduce the effects of adverse climatic conditions to some extent and, on the other hand, animal performance may be reduced in the thermoneutral zone without any increase in heat production (Young, 1981; Christopherson and Young, 1986). To state the optimum climatic conditions, the effects of the various climatic factors (temperature, humidity, rain, wind, solar radiation) on heat loss, heat production, and performance of calves have to be accurately known, as well as their acclimatization ability.

EFFECTS OF CLIMATIC CONDITIONS ON NEWBORN CALVES

Birth corresponds to a disruption in the body heat balance of the calf, which abruptly passes from an ambient temperature of 38.8°C in utero to ambient temperatures generally lower than 20°C and sometimes below 0°C when calving occurs outdoors in winter. The heat loss of the calf is proportional to the difference between skin and air or ground temperature, as the thermal insulation of the wet hair coat is very small. Thus, the newborn calf immediately has to face a tremendous heat loss while being in poor physiological condition caused by hypoxia that arises during parturition. This is one of the reasons why hypothermia often takes place during the first day of life and causes death or morbidity in weaker calves. The early mortality rate is, therefore, higher in winter

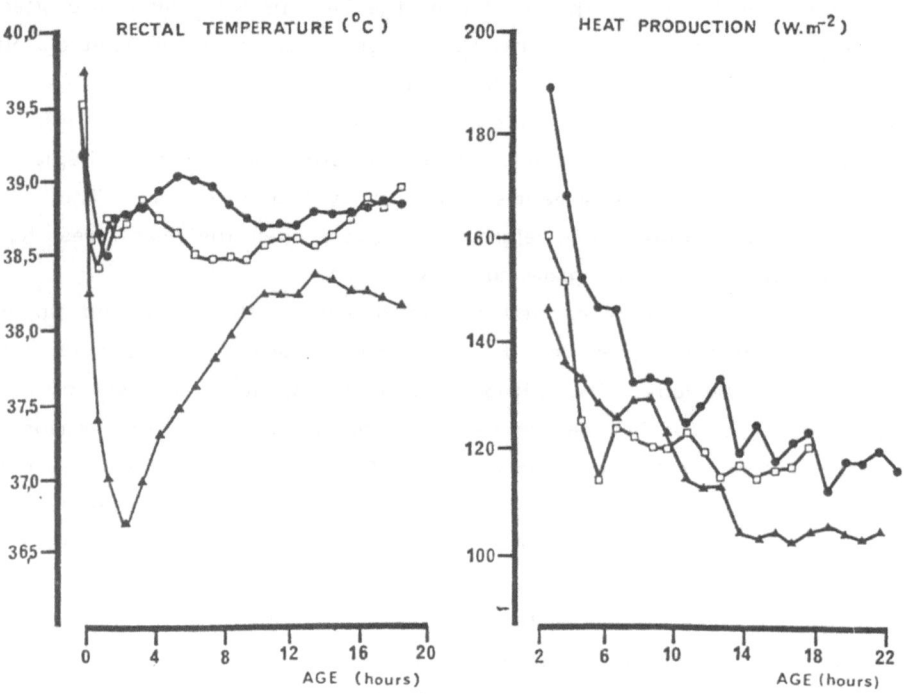

Figure 1. Variations in rectal temperature and heat production of Holstein × Friesian calves kept at 10°C during the first day of life as related to calving conditions (● normal; □ difficult; ▲ very difficult; Vermorel et al., 1987b).

than in summer (Speicher and Hepp, 1973; Fink, 1980) and during periods when large fluctuations in environmental climatic conditions occur (Martin et al., 1975) or very low ambient temperatures (Jordan et al., 1969) which can cause some weak calves to chill and die before standing up (Grenet et al., 1982).

As a matter of fact, the basal or resting heat production of newborn calves kept in a 38°C water bath (75-82 $W.m^{-2}$, Okamoto et al., 1986; Vermorel et al., 1987a) is close to that of the foetus (Comline and Silver, 1976). However, the newborn calf is able to enhance its heat production to compensate for the high heat loss when the ambient temperature drops. The heat production of 6 hour-old Ayrshire calves increased from 97 to 168 $W.m^{-2}$ between 25 and 5°C (Thompson and Clough, 1970). The summit metabolism of 2.5 to 15 hour-old Holstein calves exposed to cold in a water immersion system averages 275 $W.m^{-2}$, which is about 3.7 times the basal metabolic rate, and increases with birth weight (Okamoto et al., 1986). However, a very large individual variability (from 171 to 335 $W.m^{-2}$) was reported by Okamoto et al. (1986) and confirmed by Vermorel et a.l (1987a), which could explain the differences in cold resistance of newborn calves. Nevertheless, healthy calves can withstand very low temperatures (down to -20°C) in winter, in Central France or in Colorado, without an increase in early postnatal mortality, when they are placed outdoors in hutches with straw bedding, two to six hours after birth, and after the first colostrum meal.

The rectal temperature of newborn calves kept at an ambient temperature of 10°C decreases from 39.2 ± 0.4°C to 38.5 ± 0.4°C during the 90 minutes following birth, and stabilizes at 38.8°C five hours later (Vermorel et al., 1983; Fig. 1). The heat production of newborn calves kept at 20°C is at its highest 15 minutes after birth and remains so for three hours due to the evaporation of amniotic fluids, and then decreases slowly (Thompson and Clough, 1970). Furthermore, heat production is higher and decreases curvilinearly during the first day of life, from 189 ± 29 to 118 $W.m^{-2}$ between 1.5 and 22 hours in Holstein x Friesian calves held at 10°C. This is in relation to drying of the hair coat, which becomes more and more insulating (Vermorel et al., 1983; Fig. 1). Given the initial 0.7°C drop in rectal temperature, the heat loss can be estimated at 210 $W.m^{-2}$ during the first hours of life, which is below the

summit metabolism of most calves. Furthermore, heat production increased 13% after a 2 kg colostrum meal, 30 to 100% when the calves were

Table 1.   Variations in heat production $(W.m^{-2})$ of Holstein
× Friesian preruminant and ruminant calves at thermoneutrality as related to body weight and daily weight gain.

Type of	weight	daily weight gain (kg)				
calves	(kg)	0.5	0.8	1.0	1.2	1.5
Preruminant	50	108	120			
	70		106	115		
	100			115	125	
	150			112	120	130
	180			116	124	139
Ruminant	100		112	116	120	
	150		125	130	135	140
	200		133	137	142	150

struggling to get up, and 40% while they were standing during the first
hours of life (Vermorel et al., 1987a). Thus physical activity contributes
very efficiently to the required heat production of high-vitality newborn
calves.

By contrast, low-vitality calves, especially dystocial calves, are unable to withstand adverse climatic conditions. The rectal temperature of
Holstein × Friesian and Charolais dystocial calves kept at 10°C dropped
from 39.6 to 36.6 ± 1.3°C within two hours after birth, increased slowly
afterwards but still remained 0.8°C below that of eutocial calves
(Fig. 1). Their lower heat production was also significantly lower, 25%
and 10% on average, 2.5 and 15 hours after birth, respectively (Fig.
1). Their lower heat production resulted mainly from a severe acidosis,
a reduced mobilization of body reserves, and from lower plasma thyroid
hormone levels (Vermorel et al., 1987b). Similarly, premature calves

185

cannot face cold. Their rectal temperature drops linearly and they have
to be rewarmed. Thus, the heat production of a 26.5 kg calf born one
month before term was only half that of the high-vitality calves. It had
a slight acidosis, a low plasma triiodothyronin level, and a low physical
activity (Vermorel et al., 1983).

EFFECTS OF COLD DURING THE REARING PERIOD

Effects on energy expenditure

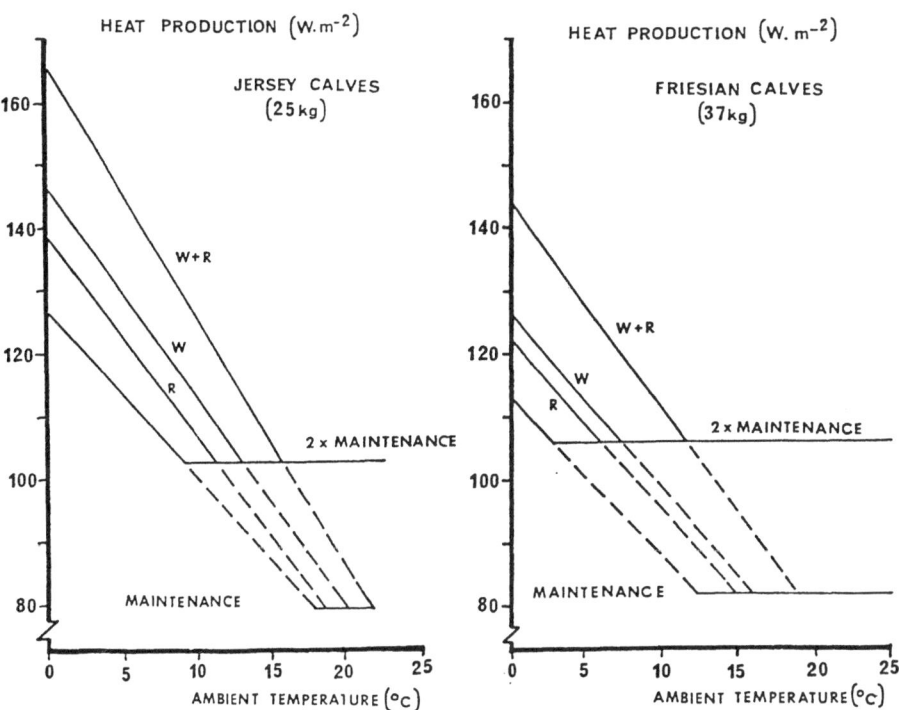

Figure 2. Variation in the calculated mean daily heat production $(W.m^{-2})$
of Jersey and Friesian calves in relation to ambient temperature, rain
(R), wind (W, 1.55 $m.s^{-1}$), wind and rain (W+R) and feeding level
(maintenance and twice maintenance), from Holmes and McLean, 1975.

On the whole, energy expenditure depends on climatic factors (ambient temperature, wind, rain, etc.) and on factors related to the animal itself (breed, sex, physiological stage and feeding level). Heat production increases with feeding level or growth rate and, in the thermoneutral zone, ranges from 105 to 130 $W.m^{-2}$ in preruminant calves and from 110 to 150 $W.m^{-2}$ in ruminant calves due to lower efficiencies of metabolizable energy utilization in the latter (Table 1).

A drop in ambient temperature from 20° to 3°C caused a 15% increase in heat production in 1 to 8 week-old calves fed 60% above maintenance. Their lower critical temperature (Lct) was close to 7°C (Holmes and McLean, 1975). However, there are large differences between breeds, as in the same conditions the rise in heat production of Jersey calves was 21%, their Lct being close to 12°C. These differences may result from the unfavourable surface body mass ratio of the Jersey calves, from their lighter hair coat (160 vs 350 $g.m^{-2}$) and from their higher whole body (tissue + haircoat) conductance (Holmes and McLean, 1975). Similarly, in the thermoneutral zone, heat production was 13% lower in Charolais than in Holstein x Friesian calves (Vermorel et al., 1983) and 4 to 12% lower in Hereford x Friesian calves than in pure Friesian calves, without a significant difference in whole body thermal conductance (Webster and Gordon, 1977; Webster et al., 1978).

In other respects, the lower critical temperature of young calves decreased during the 3 first weeks of life, from 13 to 8°C (Gonzales-Jimenez and Blaxter, 1962). This variation could result from a reduction in whole body thermal conductance and especially of tissue thermal conductance due to both greater amounts of subcutaneous adipose tissue, and more efficient peripheral vasoconstriction (Holmes, 1970).

Finally, the lower critical temperature depends on the feeding level: in young Friesian and Jersey calves it was estmated at 12° and 18°C respectively, for animals fed at maintenance and at 3° and 9°C respectively, for animals fed twice the maintenance level (Holmes and McLean, 1975; Fig. 2). The increase in heat production compensates for the higher heat loss at low temperatures (Van Es et al., 1969). Similarly, the Lct was close to 10°C in 1 to 8 week-old Friesian veal calves having a daily weight gain (DWG) of 0.8 kg and lower than 5°C in heavier calves (100-180 kg) having a DWG of 1.4 kg. However, the heat production of the latter was 20% higher and their tissue thermal insulation was probably bettter (Webster and Gordon, 1977; Webster et al., 1976).

Raising air velocity from 0.22 to 1.55 m.s^{-1} caused an increase in heat production of 10 day-old Friesian calves of 5, 18 and 23% at 12, 8 and 3°C, respectively. The increases amounted to 13, 28 and 36% at the same temperatures in the Jersey calves. These increases may result from a reduced thermal insulation at the higher air velocity. It caused a rise in Lct from 3 to 7°C and from 9 to 13°C in the Friesian and Jersey calves, respectively (Holmes and McLean, 1975).

Rain and wind (1.55 m/s) separately caused a 12 and 9% higher heat production, respectively, in young calves kept at 5°C. But the joint effect of both factors amounted to 28% and induced a 6 to 8°C increase in Lct (Holmes and McLean, 1975; Fig. 2). These effects could result from an increase in the whole body thermal conductance, especially that of the wet hair coat "stirred" by wind. The greater heat loss may be more pronounced in winter, especially at lower ambient temperature, and with stronger wind. Such climatic conditions can reduce the metabolizable energy available for production and, consequently, the performance of calves reared outdoors or sheltered in winter.

Effects on calf performance

In the Sixties, the recommended ambient temperature for young calves was above 20°C. However, Haartsen and Van Hellemond (1970) showed that neither health, growth rate or feed efficiency of veal calves was altered when the temeprature in the byre was gradually reduced from 25 to 20°C or from 15 to 10°C between the first and the tenth week of the rearing period. Recently Kunz and Montandon (1985) did not note significant differences in the performance between calves reared from birth to 100 days of age in a heated, ventilated, house (17°C) and calves reared in cold stalls where the climate was similar to the outside climate (4.5°C). However, a low ambient temperature (7°C) associated with a very high humidity (95% rh) was detrimental to the growth rate of calves (345 vs 403 g/d) between 4 and 42 days of age (Kelly et al., 1982).

In countries with a clement winter (3 to 5°C), the growth rate and feed efficiency of healthy suckler calves, reared outdoors in a sheltered paddock or an open-air paddock, were satisfactory and not significantly different from the performances of calves reared in a heated byre

(15°C). Furthermore, their health status was often better than that of calves reared in a heated environment (Larsen et al., 1980; Bruce, 1981; Quillet, 1982). However, when the veal calves were in a poor health condition, the weight gain of the animals reared in a shelter at 5°C was 10% lower than that of the calves reared in a heated house (Ladrat and Jousselin, 1971).

During the winter in the French mountains at altitudes of 800m (average temperature 1.7°C, 89 days of frost, 64 days of precipitations), the mortality rate (11.4%) and the weight gain (862 g/day) of suckler calves kept outdoors (180 animals during 4 winters) were not significantly different from those of calves kept under an enclosed shelter (Grenet et al., 1982). However, this does not mean that the energy expenditures of the calves were similar as there was no information on their energy gain. As a matter of fact, at the very harsh climatic conditions of Canada (from - 6 to - 23°C between November and March), a 50% mortality rate was found for suckler calves born and reared in an open yard. For those born in a loose-housing barn and placed outside 3 days after calving the value was 42%. By contrast, it was reduced to 8% when the animals were placed outdoors 7 days after calving. Furthermore, the weight gain of the calves reared outdoors was 13% lower than that of the calves reared indoors (Jordan et al., 1969).

Shelters are very effective for the protection of young calves. They create a microclimate, with a temperature that is 3 to 4°C above outside temperature, and they reduce air velocity and fluctuations of outside temperature. The effectiveness of several types of shelters was compared. In Ireland, calves reared at 15°C in an uncovered area with a solid wall on three sides with a height of 1.25m had lower weight gains (400 versus 460 g/day) than those reared in a covered area with a solid wall on three sides (Harte and Fallon, 1982). At moderate climatic conditions, shelters with walls alone had a greater protective effect than a roof alone, and shelters with both walls and roof almost eliminated the depressive effect of climate on the weight gain of suckler calves (Bruce, 1984). Insulating the walls and the roof of the shelter did not anymore improve calf performance (Williams et al., 1981).

Physiological responses to cold

The mechanisms of thermoregulation and adaptation to cold in cattle were reviewed by Webster (1974), Young (1981), Christopherson and Young (1986). Therefore, these aspects will not be excessively dealt with here. In short: under acute cold stress, the hypothalamic-pituitary-adrenal axis is involved, with increased sympathetic neural activity, and release of catecholamine and steroid hormones, which enhance the mobilization of energy substrates and thermogenesis.

During prolonged cold exposure, the catecholamine secretion remains high, thyroid hormone secretion is increased, tissue protein turnover and cell sodium-potassium transport are accelerated, resulting in increased resting metabolism and summit metabolism. Furthermore, in newborn calves, the change of brown adipose tissue into white adipose tissue is delayed. On the other hand, whole body thermal insulation is improved through adaptation of peripheral circulation and increased hair coat weight, resulting in a fall of the lower critical temperature. However dry matter digestibility decreases by 0.21 unit per degree drop in ambient temperature in ruminant calves due to a faster rate of passage. These phenomena could explain the difference between the observed performance of calves in winter and that expected from their theoretical lower critical temperature.

EFFECT OF HEAT DURING THE REARING PERIOD

Many studies have focussed on the effects of high temperature and humidity on performance and thermoregulation of growing and lactating cattle, mainly to determine the adaptation ability of European or American breeds to tropical climatic conditions. The studies were carried out mainly by the Missouri (H.H. Kibler, H.D. Johnson, A.C. Kamal) and the Scottish (J.D. Findlay, W. Bianca, J.A. McLean) teams.

On the basis of the reviews written since 1958, especially by Bianca (1965) and Fuquay (1981) it appears that bovines are more sensitive to heat than to cold, especially when their production level is high, as they must dissipate a great amount of heat. This is particularly true for veal calves and rapidly growing ruminant calves. The intensity of heat exchanges depends on the temperature gradient between skin and air, on air humidity and velocity, and on solar radiation.

Effects of ambient temperature

An acute increase in ambient temperature causes a rapid decrease in food intake to reduce the heat increment of feeding. The food intake of growing bulls kept in climatic chambers was reduced by 25% when the ambient temperature rose from 10 to 27°C (Johnson et al., 1958). The decrease in metabolizable energy above maintenance as available for production was twice as high. Similarly, a rise of ambient temperature from 17 to 34°C induced a 40% reduction in food intake and a 58% drop in weight gain (410 versus 970 g per day) of 6 month-old Friesian calves, while water consumption was doubled (Holmes et al., 1980).

Adaptation to heat is a long process (more than 4 months) and may remain incomplete, as reported by Bond and McDowell (1972). Thus, Holstein calves were reared at temperatures varying each day between 24 and 35°C during the first three months of life. Weight gain was reduced 28% during the first month, and 38% during the other two months. Fur-

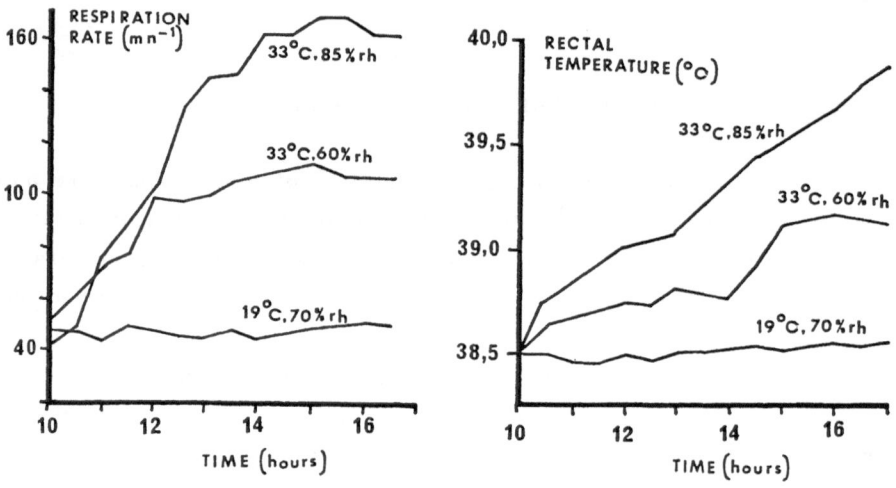

Figure 3. Effects of ambient temperature and relative humidity (rh) on rectal temperature and respiration rate of Friesian veal calves (Bouvier et al., 1974).

thermore, growth rate was still 30% lower than that of control calves during the two-month period following the heat stress (Randel and Russof, 1965). There was, hence, neither adaptation to heat nor compensatory growth. The depressive effects of high temperature continued during the whole period in Holstein or Friesian heifers, but were less severe in Jersey and Brown Swiss (Hancock and Payne, 1955; Johnson and Ragsdale, 1959).

Effects of high temperature and humidity

Veal calves are often exposed to high temperature and humidity in summer. This considerably reduces their weight gain and causes the farmers to send calves prematurely to the slaughterhouse. Their heat production ranges from 120 to 140 $W.m^{-2}$ (Table 1), while about 80% must be dissipated as latent heat (McLean and Calvert, 1972). The high relative humidity (rh) of the air reduces heat loss by sweating and induces panting.

An ambient temperature of 27°C with 95% rh or 30°C with 75% rh led to high respiration rates of 100-160 per minute in rapidly growing veal calves of 150 kg. They also showed an increased rectal temperature, which reached 42°C within 20 hours after the start of exposure (Van Es et al., 1969). These increases also depend on rh alone, being more pronounced at 85% than at 60% rh, both at 33°C (Bouvier et al., 1974; Fig. 3). Heat production increased 10% and 3% in 150 kg and 100 kg veal calves, respectively, due to this rise in rh. Latent heat loss doubled. However, rapidly growing veal calves can withstand high temperature and humidity for several hours, without significant decreases in food intake or growth rate if they can recover at a lower temperature during the night (Bouvier et al., 1974).

Effects of wind

Increasing air velocity favours sensible and latent heat losses by convection and sweat evaporation. Consequently, it reduces rectal temperature, heart rate and respiration rate of dairy cows kept above 27°C (Kibler and Brody, 1954) and improves milk production (Johnson, 1965). To our knowledge, there are no direct data on calves, but increasing air velocity may improve growth rate in heat stressed calves, as in growing bulls (Bond et al., 1957; Pontif et al., 1974).

Effects of solar radiation

Solar radiation contributes to heat stress in cattle when ambient temperature is high. Thus, exposure to simulated solar radiation (1060 $W.m^{-2}$) of 6 month-old Friesian calves in a calorimeter (32°C, 50% rh) induced higher levels of rectal temperature (41.45 versus 40.11°C) and respiration rate (124 versus 81 per minute) after seven hours (Holmes et al., 1980). Shade for feedlot steers improved their weight gain 8% and their feed efficiency 10% in Iowa (Self and Hoffmann, 1974) and their growth rate 15% in Louisiana (Pontif et al., 1974). It is, therefore, important to avoid "roof lighting" in calfbarns and to provide the calves reared outdoors with shade during hot periods.

Physiological responses to heat

These aspects were reviewed by Bianca (1965), Webster (1974) and Fuquay (1981) and will not be described in detail here. Reducing food intake is known to be the first means for mammals to lower their heat production rapidly. Acute heat stress induces a prolonged rise in catecholamine secretion, which stimulates sweating (Christison and Johnson, 1972). The rise in cortisol secretion could explain the higher protein catabolism as observed by Colditz and Kellaway (1972), the reduction of nitrogen retention and the increased protein requirement as noticed by several authors. Furthermore, heat stress causes reduced thyroid hormone secretion followed by decreased heat production. On the whole, prolonged heat stress induces a decrease in the secretion of all hormones involved in anabolism, a.o. thyroxin, growth hormone, insulin (Mitra et al., 1972), which can explain the reduction of animal performance, even when food intake is artificially maintained.

CONCLUSION

The effects of cold are critical to newborn calves due to their high heat loss through the wet hair coat and to incomplete vasoconstriction. Furthermore, there is a great individual variability in summit metabolism and in heat production between calves exposed to cold. The origins of lower thermogenesis are worth studying. Does it result from lesser amounts of brown adipose tissue, from lower plasma thyroid hormone levels, or from a poor functioning of thermoregulatory mechanisms, due to prolonged parturition?

As a matter of fact, a difficult parturition often causes hypoxia, metabolic acidosis, depression of the sympathetic nervous system after birth, reductions of hormone secretions, of shivering and physical activity, which can all restrict increases in thermogenesis. Hypoxia can be rapidly suppressed by oxygen-enriched air-breathing; hypothermia can also be limited by using, for example, a heating cover. However, there is an interesting field of research on the development of appropriate treatments to suppress acidosis and to stimulate energy metabolism of low-vitality calves, with great practical uses in husbandry.

Older, healthy calves can withstand the low ambient temperatures encountered in winter in the temperate zones, without any detrimental effect on health and growth rate, when shelters against wind and precipitations are available. Nonetheless, the first 2 or 3 weeks of life are a critical phase for calves exposed to temperatures around or lower than -10°C. In effect, cold acclimatization mechanisms are still poorly understood in young calves and so are their effects on energy expenditures.

By contrast, heavy veal calves are mainly affected by a combination of high temperature and humidity in summer in the temperate zones of Europe. However, they can tolerate heat for several hours, without any deleterious effect on food intake and weight gain if they are not exposed to solar radiation, provided that:

1. they be able to reduce their body temperature during the night in well ventilated housing;
2. they be fed very early in the morning and late in the evening to dissipate their postprandial heat production during the cooler period;
3. water be available all day long.

Heat stress may be more serious for rapidly growing calves in the Mediterranean regions because the ambient temperature remains high during the night. Calves reared outdoors, therefore, must have shade and water available and have the possibility to graze or be fed during the night. Also, there is a great individual variability in heat resistance of calves. Heat-tolerant animals, especially Holsteins, which are used for milk production in these regions, should be bred with this in mind.

REFERENCES

Bianca, W., 1965. Reviews of the progress of dairy science. Section A. Physiology. Cattle in a hot environment. J. Dairy Sci. 32: 291-345.

Bond, T.E., Kelly, C.F. and Ittner, N.N., 1957. Cooling beef cattle with fans. Agri. Eng. 38: 308-321.

Bond, J. and McDowell, R.E., 1972. Reproductive performance and physiological responses of beef females as affected by a prolonged high environmental temperature. J. Anim. Sci. 35: 820-829.

Bouvier, J.C., Espinosa-Moliner, J. and Vermorel, M., 1974. Influence de températures et d'hygrométries élevées pendant une partie de la journée sur la thermorégulation du Veau préruminant à l'engrais. Ann. Biol. anim. Bioch. Biophys. 14: 721-727.

Bruce, J.M., 1981. The effect of winter environment on the performance and health of autumn calving suckler cows. Suckler cow Meeting, Dublin, June 1981.

Bruce, J.M., 1984. Climate and the value of shelter for suckler cows and calves. Farm Building Progress 78: 21-25.

Christison, G.I. and Johnson, H.D., 1972. Cortisol turnover in heat stressed cows. J. Anim. Sci. 35: 1005-1010.

Christopherson, R.J. and Young, B.A., 1986. Effects of cold environments on domestic animals. In: O. Gudmundsson (Editor), Grazing Research at Northern Latitudes. Plenum publishing corporation: 247-257.

Colditz, P.J. and Kellaway, R.C., 1972. The effect of diet and heat stress on feed intake, growth and nitrogen metabolism in Friesian, Brahman x Friesian and Brahman heifers. Aust. J. agric. Res. 23: 717-725.

Comline, R.S. and Silver, M., 1976. Some aspects of foetal and utero-placental metabolism in cow with indwelling umbilical and uterine vascular catheters. J. Physiol. 260: 571-586.

Es, A.J.H. van, Nijkamp, H.J., Weerden, E.J. van and Hellemond, K.K. van, 1969. Energy, carbon and nitrogen balance with veal calves. In: K.L. Blaxter, J. Kielanowski and G. Thorbek (Editors), Energy Metabolism of Farm Animals. Oriel Press Ltd., Newcastle: 197-201.

Fink, T., 1980. Untersuchungen über den Einfluss von Aufstallungsart,

Stallklima und Management auf den Gesundheitszustand von Kälbern (Praxisstudie). Inaugural Dissertation, Tierartzliche Hochschule, Hannover: 120 pp.

Fuquay, J.W., 1981. Heat stress as it affects animal production. J. Anim. Sci. 52: 164-174.

Grenet, N., Melet, L., Chupin, J.M., Le Neindre, P. and Malterre, C., 1982. Performances de vaches allaitantes en plein air intégral. In: Actions du climat sur l'Animal au Pâturage. INRA, Paris: 31-44.

Gonzales-Jimenez, E. and Blaxter, K.L., 1962. The metabolism and thermal regulation of calves in the first month of life. Brit. J. Nutr. 16: 199-212.

Haartsen, P.I. and Hellemond, K.K. Van, 1970. De invloed van verschillende staltemperaturen op de groei, de voederconversie en de gezondheidstoestand van mestkalveren. Landbouwkundig Tijdschrift 82: 136-142.

Hancock, J. and Payne, W., 1955. The direct effect of tropical climate on the performance of European-type cattle. 1. Growth. Empire J. Exp. Agric. 23: 55-74.

Harte, F.J. and Fallon, R.J., 1982. Effect of various environments on calf performance. In: J.P. Signoret (Editor), Welfare and Husbandry of calves. Martinus Nijhoff Publishers: 196-208.

Holmes, C.W., 1970. Effects of temperature on body temperatures and sensible heat loss of Friesian and Jersey calves at 12 and 76 days of age. Anim. Prod. 12: 493-501.

Holmes, C.W., King, C.T. and Sauwa, P.E.L., 1980. Effects of exposure to a hot environment of Friesian and Brahman x Friesian cattle, with some measurements of the effects of exposure to radiant heat. Anim. Prod. 30: 1-11.

Holmes, C.W. and McLean, N.A., 1975. Effects of air temperature and air movement on the heat produced by young Friesian and Jersey calves, N.Z. Journal of Agric. Res. 18: 277-284.

Johnson, H.D., 1965. Response of animals to heat. Meteor. Monographs 6: 109-122.

Johnson, H.D. and Ragsdale, A.C., 1959. Environmental Physiology and Shelter Engineering, with Special Reference to Domestic Animals. LII. Effects of constant environmental temperatures of 50° and 80°F on the growth responses of Holstein, Brown Swiss and Jersey calves. Missoury, Agr. Exp. Sta. Res. Bull. 705.

Johnson, H.D., Ragsdale, A.C. and Yeck, R.G., 1958. Environmental Physiology and Shelter Engineering. XLIX. Effects of constant environmental temperatures of 50° and 80°F on the feed and water consumption of Brahman, Santa Gertrudis and Shorthorn calves during growth. Missoury Agr. Exp. Sta. Res. Bull. 683.

Jordan, W.A., Lister, E.E. and Commeau, J.E., 1969. Outdoor versus indoor wintering of fall calving beef cows and their calves. In: Can. J. Anim. Sci. 49: 127-129.

Kelly, T.G., Dodd, V.A., Ruane, D.J., Tuite, P.J., Fallon, R.J. and Dempster, J.F., 1982. Effects of moderate high levels of relative humidity at 15°C and 7°C on calf health and performance. Livestock Environment. II. Proc. 2nd Int. Livestock Environment Symp., April 20-23, 1982. Scheman Center, Iowa State University, Ames, Iowa: 392-399.

Kibler, H.H. and Brody, S., 1954. Environmental Physiology and Shelter Engineering, with Special Reference to Domestic Animals. XVII. Influence of wind on heat exchange and body temperature regulation in Jersey, Holstein, Brown Swiss and Brahman cattle. Missoury Agr. Exp. Sta. Res. Bull. 552.

Kunz, P. and Montandon, G., 1985. Vergleichende Untersuchungen zur Haltung von Kälbern im Warm- und Kaltstall während der ersten 100 Lebenstage. Betriebswirtschaft und Landtechnik, Switzerland, 26: 126 pp.

Ladrat, J. and Jousselin, W., 1971. Production du veau de boucherie en étable ouverte. C.R. Acad. Agric., June 16: 996-1007.

Larsen, H.J., Tenpas, G.H. and Cramer, C.O., 1980. Rearing dairy calves in warm and cold housing. J. Dairy Sci., 63, Suppl.: 72-73.

Martin, S.W., Schwabe, C.W. and Franti, C.E., 1975. Dairy calf mortality rate: influence of meteorologic factors on calf mortality rates in Tulare Country California. Am. J. Vet. Res. 36: 1105-1109.

McLean, S.A. and Calvert, D.T., 1972. Influence of air humidity on the partition of heat exchanges of cattle. J. Agric. Sci. 78: 303-307.

Mitra, R., Christison, G.I. and Johnson, H.D., 1972. Effect of prolonged thermal exposure on growth hormone secretion in cattle. J. Anim. Sci. 34: 776-779.

Okamoto, M., Robinson, J.B., Christopherson, R.J. and Young, B.A.,

1986. Summit metabolism of newborn calves with and without colostrum feeding. Can.J. Anim. Sci. 66: 937-944.

Pontif, J.E., Nipper, W.A., Loyacano, A.F. and Braud, H.J., 1974. Effects of windbreaks and roofs in winter and shades and fans in summer on feedlot performance of cattle in the South. In: Livestock Environment, affects Production, Reproduction, Health, Proc. Int. Livestock Environment Symp., A.S.A.E.: 305-309.

Quillet, J.P., 1982. Housing veal calves. Some results on housing systems. In: J.P. Signoret (Editor), Welfare and Husbandry of calves. Martinus Nijhoff Publishers: 185-195.

Randel, P.F. and Rusoff, L.L., 1965. Responses of Holstein calves raised on limited milk to simulated Louisiana summer climatic conditions. J. Dairy Sci. 179: 65-72.

Self, H.L. and Hoffman, M.P., 1974. Influence of environment on feeding parameters. In: Livestock Environment, affects Production, Reproduction, Health, Proc. Int. Livestock Environment Symp., A.S.A.E.: 281-287.

Speicher, J.A. and Hepp, R.E., 1973. Factors associated with calf mortality in Michigan dairy herds. J. Am. Vet. Med. Ass. 162: 459-465.

Thompson, G.E. and Clough, D.P., 1970. Temperature regulation in the New-Born Ox (Bos taurus). Biology of the Neonate 15: 19-25.

Vermorel, M., Dardillat, C., Vernet J., Saido and Demigne, C., 1983. Energy metabolism and thermoregulation in the newborn calf. Ann. Rech. Vet. 14: 382-389.

Vermorel, M., Vernet, J., Dardillat, C., Saido and Demigne, C., 1987a. Energy metabolism and thermoregulation in the newborn calf. 1. Variations during the first day of life and differences between breeds. Can. J. Anim. Sci. (in press).

Vermorel, M., Vernet, J., Dardillat, C., Saido, Demigne, C., and Davicco, M.J., 1987b. Influence of difficult parturition on energy metabolism of newborn calves. In: P.W. Moe, H.F. Tyrrell and P.J. Reynolds (Editors), Energy Metabolism of Farm Animals, E.A.A.P., Publ. No. 32. Rowman and Littlefield: 34-37.

Webster, A.J.F., 1974. The influence of climatic environment on metabolism in cattle. In: H. Swan and W.H. Broster (Editors), Principles of Cattle Production, Butterworths, London: 103-120.

Webster, A.J.F. and Gordon, J.G., 1977. Air temperature and heat

198

losses from calves in the first weeks of life. Anim. Prod. 24: 142.

Webster, A.J.F., Gordon, J.G. and McGregor, R., 1978. The cold tolerance of beef and dairy type calves in the first weeks of life. Anim. Prod. 26: 85-92.

Webster, A.J.F., Gordon, J.G. and Smith, J.S., 1976. Energy exchanges of veal calves in relation to body weight, food intake and air temperature. Anim. Prod. 23: 35-42.

Williams, P.E.V., Day, D., Raven, A.M. and McLean, J.A., 1981. The effect of climatic housing and level of nutrition on the performance of calves. Anim. Prod. 32: 133-141.

Young, B.A., 1981. Cold stress as it affects animal production. J. Anim. Sci. 52: 154-163.

CLIMATIC CONDITIONS AND ENERGY METABOLISM OF LAYING HENS

M. VAN KAMPEN

ABSTRACT

The effects of climatic conditions such as temperature, wind speed, humidity, light and air composition on metabolic and production rate are discussed. The physiological and production responses are at the same time depending on the factors: breed, age, body weight, activity, feathering, food intake, nutritional and temperature history.

The mass exponent in the metabolic body weight unit, which is used for intraspecific comparison, varies generally between 0.60 and 0.67.

There is a direct depressing effect of ambient temperature on the increase of food intake per centigrade at decreasing temperatures and on egg production at increasing temperatures.

For comparative and predictive purposes it is necessary to incorporate many factors in a "total effective ambient temperature".

INTRODUCTION

The values of energy metabolism in laying hens may vary within wide limits, due to factors such as age, body weight, body form, body posture, feather cover, group size, egg production, activity, alertness, food intake, light-dark periods, ambient temperature, wind speed and humidity.

It requires numerical measurements in order to obtain the contribution of the separate factors to the total metabolic rate. But even under stan-

dard conditions in chamber-calorimetry measurements not all the factors
are under control. Thus, the prediction of the energy need of hens under commercial conditions, of which not all the factors are quantified,
has its shortcomings (MacLeod, 1984).

METABOLIC BODY SIZE

Of the multiple factors which affect the metabolic rate it appears that
body size has a great influence. For interspecific comparison the metabolic rate should be expressed in a unit that makes the metabolic level of
an animal independent of its body mass.

Brody (1945) and Kleiber (1947) empirically found an allometric relation ($M = aW^b$) between basal metabolism (M) and body mass (W) with a
mass exponent (b) of 0.73-0.75 for adult, mainly domestic animals.

Table 1. The effect of body weight on heat production.

Body weight	HP/animal	HP/kg	HP/m^2	HP/kg$^{0.75}$
0.1 kg	89 kJ	890 kJ	4134 kJ	500 kJ
1 kg	500 kJ	500 kJ	5000 kJ	500 kJ
10 kg	2810 kJ	281 kJ	6050 kJ	500 kJ
100 kg	15810 kJ	158 kJ	7327 kJ	500 kJ
1000 kg	88915 kJ	89 kJ	8871 kJ	500 kJ

Table 1 shows that there is, in animals with the same metabolic level
or with equal amounts of heat produced per metabolic kg ($kg^{0.75}$), a
positive relationship between body mass and heat production expressed
per animal or square meter body surface area. However, expressing the
heat production per kg body weight results in a negative relationship.

Even under basal conditions (awake, in a post-absorptive state, no
reproductive or physical activity and in a thermoneutral environment)
the mass coefficient (a) or the metabolic level varies. Poczopko (1971)
concluded that there are at least four different metabolic levels in adult

homeotherms, e.g. eutherian mammals have a metabolic level of 293 kJ and non-passerine birds 335 kJ/kg$^{0.75}$·24h. Heusner (1985) stated that the 0.75 mass exponent is valid for interspecific comparison, but he calculated a value of 0.67 for intraspecific comparison in mainly domestic animals.

Hayssen and Lacy (1985) excluded the domestic animals in their study of taxonomic differences in the allometry of basal metabolic rate and body mass. They found a mass exponent of 0.70 for 248 eutherian species and an exponent of 0.66 within 8 eutherian orders, in which four metabolic levels were distinguishable.

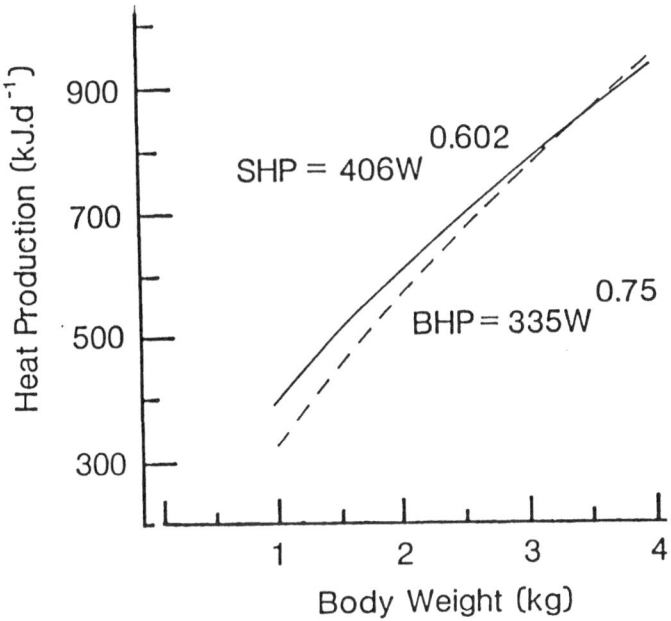

Figure 1. Relationships between body weight and starving or basal heat production of domestic fowl (———) and non-passerine birds (-----).

Prothero (1986) suggests that even in unicelullar organisms there will be different regressions with energy metabolism. In experiments with chicks, in which the exponent was treated as a variable, the b-value varied between 0.3 for mature birds (Brody, 1945) and 2.1 for young chickens (Freeman, 1964). Johnson and Farrell (1985) found a relationship between starvation heat production and body size of mature domestic fowl of SHP = 406 $w^{0.602}$. In their analysis they used data of layer-type and broilers of both sexes under thermal neutral conditions, independent of their productivity and age. Predicting the starving heat production for domestic fowl with the non-passerine basal value of 335 kJ/$kg^{0.75}$·24h results in lower values for small hens and are identical at a body weight of 3.66 kg (Figure 1).

The higher value for small fowl may be caused by the egg production or that they were not under thermal neutral conditions. On the other hand, the crossing of the lines suggest that heavy fowl are not producing eggs. Especially older hens may contain a lot of body fat, which acts mainly as inert mass and contributes to a lower b-value. There are no indications that the b-value will change when the metabolic rate is not measured under basal conditions (Taylor et al., 1978; Berman and Snapir, 1965), although a filled gastro-intestinal tract may increase the inert mass, up to 10% of body weight in hens, and lower the b-value. Due to many conflicting data concerning the thermoneutral zone in poultry it cannot be excluded that Johnson and Farrell (1985) included also data obtained at temperatures below the thermoneutral zone. Below the lower critical temperature heat production will be more and more caused and related to heat loss with decreasing ambient temperature. Since the rate of heat transfer below the lower critical temperature varies with body weight as $w^{0.50}$ (Kendeigh et al., 1977), the mass exponent of 0.67 under neutral conditions may be lowered to 0.50 under cold conditions. If a certain condition affects all the active metabolic body mass with the same intensity, then there is no reason for a change in the b-value. But it will change when the mass of metabolically active organs alters without an increase in total body weight (Koong et al., 1985) and may change during acclimation to a new condition.

MINIMUM HEAT PRODUCTION

The relation between ambient temperature and heat production is described by a linear equation (H, kJ·hen^{-1}·d^{-1}, = 1336 - 19.152 T, °C,; Marsden and Morris, 1980) in a limited temperature range or by parabolic equations (Tzschentke and Nichelmann, 1986) in an extended temperature range within the homeothermic temperature range. There is no clear zone in which heat production is minimal and constant. Besides that the mini-mum heat production, even under fasting conditions, is achieved at an ambient temperature above the thermoneutral zone where the body tem-perature is increased and the respiratory evaporative heat loss is acti-vated. Thus basal metabolic rate is not equivalent to minimum heat pro-duction.

In the thermoneutral zone heat production increases slowly with de-creasing ambient temperature. Below the lower critical temperature the heat production curve can be described by a linear function even over a considerable temperature range, but better with a curvilinear line (McArthur, 1981). This suggests that the insulation of the hen changes and only at very low T_a the line may be straight (Figure 2).

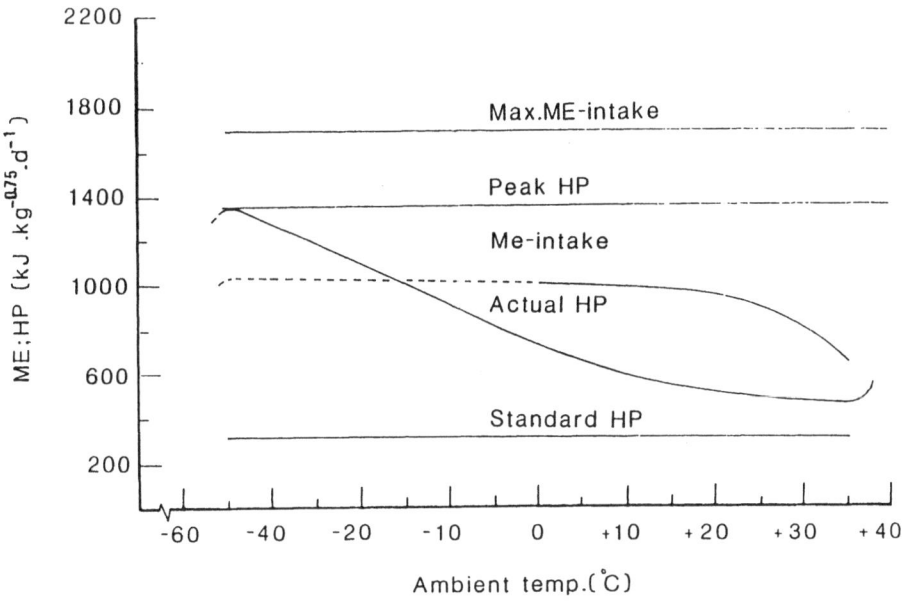

Figure 2. Metabolizable energy intake and heat production of laying hens in relation to ambient temperature.

The level of minimum heat production depends on factors such as breed, food intake, production rate, nutritional and temperature history, activity, age, season and time of the day.

BREED

There are several reports in which metabolic rate of light and medium-heavy strains are compared. In general the medium heavy strains have a lower metabolic rate, if it is expressed per kg body weight or metabolic kg. Possible causes are differences in activity, feathering, laying rate, unit body weight, ambient temperature and minimum metabolic rate.

Unit body weight

Farrell (1975) found a difference in fasting metabolic rate between Australorp and Leghorn hens of 39% when expressed per kg body weight. This percentage will be 24 and 17%, when expressed per $kg^{0.75}$ and $kg^{0.602}$, respectively.

Ambient temperature and feathering

In heavier or well-feathered hens the heat production curve (Figure 2) will be displaced to the left and the slope below the lower critical temperature less steep. Thus when there is no clear thermoneutral zone and a light and a heavy bird are compared at the same ambient temperature, the light bird will have a higher metabolic rate. This difference increases with decreasing ambient temperature. Comparing well- and poorly feathered birds at the same ambient temperature results in an identical effect. Well-feathered hens have more fat (Damme and Pirchner, 1984). That means that in poorly-feathered hens rate of heat transfer is extra high and also body volume body surface ratio is less favourable due to a lower body weight.

Laying rate

Laying hens have a 20% higher metabolic rate than non-laying hens (Balnave, 1974). But Leghorns even with a 50% difference in laying rate had the same starving heat production (Damme and Pirchner, 1984).

Activity

Especially White Leghorns seem to be more active than other strains (MacLeod, 1984).

NUTRITIONAL HISTORY

Heat production is linearly related with metabolizable energy intake (MacLeod, 1984). This increment of heat is with ad libitum feeding 15-30%, but depends on the diet. There are also long-term effects of food intake on metabolic rate. A long term energy intake reduction of 20% in cocks resulted in a starving heat production reduction of about 30% if expressed per $kg^{0.75}$.

Probably a reduction in metabolic rate takes place within days after underfeeding. For example in man metabolic rate was reduced by 9% after 1 week and 15% after 2 weeks of underfeeding. Overfeeding resulted in a rise of metabolic rate (Garrow, 1986).

Also the composition of the diet fed in the pre-experimental period may affect the heat production (Rufeger and Bottin, 1980). Many measurements of starving heat production were started after 24-30h of starvation. Lengthening this period resulted in a further decline of heat production (Van Kampen, 1974).

TEMPERATURE HISTORY

A metabolic rate measurement directly after a temperature change, for instance from a cold environment with a high food intake and metabolic rate to a higher temperature, may result in relatively high results (Freeman, 1984) until organs such as liver and intestinal tract are regressed (Koong et al., 1985). There are several reports describing immediate responses to temperature and wind speed changes. However, if this first line of defense operates only during a short period results of a measuring period of one or two days immediately after the change may be more a reflection of the previous temperature. The response will be affected by the magnitude of the stimulus (absolute and relative to time). The ultimate temperature effect on metabolic rate itself can be measured after the metabolic, insulative and behavioural modifications are complete. For many physiological variables the adjustments are achieved for at least 90% after 3 days, but takes sometimes weeks for the final adjustment.

It is also possible that a temporary cold stimulus early in life has an everlasting effect on morphological (Deaton et al., 1976) and physiological variables (Decuypere, 1979).

FOOD INTAKE

There are published several equations for the prediction of food requirements of layers at different ambient temperatures (Marsden and Morris, 1980; Polin, 1983).

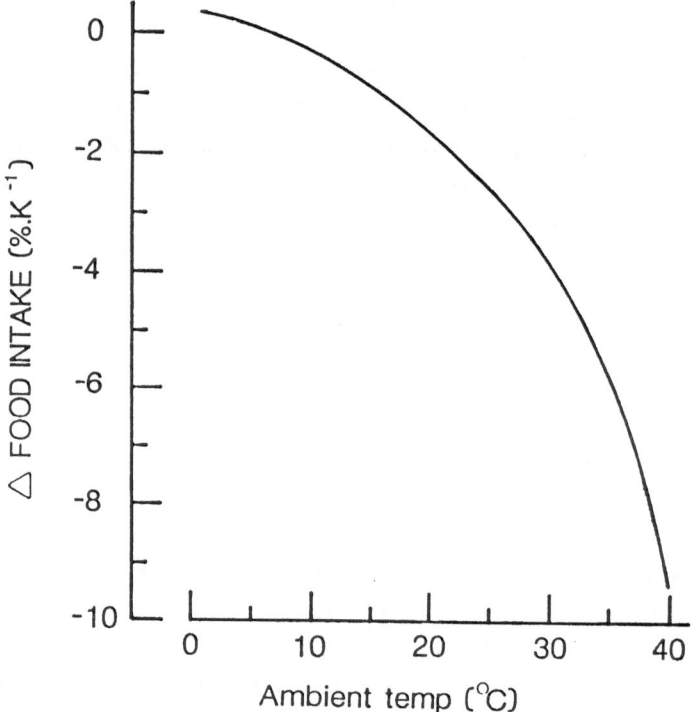

Figure 3. Food intake change at different ambient temperatures (data recalculated from Polin, 1983).

Figure 3 shows that there is no linear decrease in food intake with decreasing ambient temperature. A maximum food intake is reached at 5°C. There is hardly any change in food intake at low ambient temperatures while heat production changes linearly (Van Kampen, 1984). The reduction in the increase of food intake at low temperatures may not only be a result of the filling capacity but also a direct temperature effect. After changing the ambient temperature from 0 to 20°C food intake was increased with 25% during the first two days (Van Kampen, unpublished results). Also the amount of food intake of a low energy diet at 24°C may be higher than of a normal diet at 5°C (Polin, 1983).

It would be of interest to compare the size and volume of the gastro-intestinal tract of hens fed ad libitum at 0°C with hens fed a single meal in the afternoon at 20°C. If filling capacity is the limiting factor then concentrating energy in the diet by adding fat may be helpful.

The curve in Fig. 3 may shift depending on many environmental factors as mentioned with regard to heat production. There is a progressive decrease in food intake at higher ambient temperatures while heat production changes hardly or even increases. This results in a sharp decline of energy for production (Figure 2).

Egg weight and quality are reduced at high temperatures, but there are many conflicting data concerning laying rate. Several authors indicate that diets must be enriched with essential amino acids. Others indicate that the reduction in egg mass output is caused by a shortage of energy. However, increasing nutrient density and essential amino acid supply seems to have no clear effect on egg production (De Schutter and Morrison, 1986). Thus there seems to be also a temperature effect per se.

If energy flow is a limiting factor then adding fat to a diet does not improve energy intake. Only the low thermogenic effect of fat may reduce heat stress and improve energy intake and egg output.

Free choice feeding with separate calcium by shells (Picard, 1986) or free choice between a protein and an energy supplement (Scott and Balnave, 1986) seems to improve egg quality and production.

Starving hens for a few hours on a sudden "tropical" day may present severe heat stress. Feeding hens only late in the afternoon equals ad libitum food intake (MacLeod and Jewitt, 1984) and protein utilization may be even better (Rogers and Pesti, 1986). Whether this, combined with a lengthening of the photoperiod, improves performance of hens under heat stress needs further research.

RELATIVE HUMIDITY

Increasing water vapour pressure results in a lower evaporative heat loss. This heat loss is only a minor part of total heat loss at low ambient temperatures. A change in P_{water} has little effect on the heat balance up to 25°C. Above this temperature heat production does not decrease much, while the sensible heat loss decreases due to the lower temperature gradient, between surrounding and heat exchanging surface of the animal.

Evaporative heat loss increases above 25°C passively, because the water vapour capacity of air increases progressively with ambient temperature rise. It increases actively after a small rise in body temperature by raising respiratory frequency and minute volume. Evaporative water loss decreases with increasing ambient water vapour pressure by 0.7 mg of water for each gram of live-weight per hour per kilopascal (Richards, 1976).

To compensate for the depressed evaporative heat loss body temperature will increase as well as the temperature gradient with the environment, which results in a higher sensible heat loss. Increasing water vapour pressure, therefore, has a similar effect as raising ambient temperature.

The contribution of the humidity in the physiologically effective temperature was obtained by weighing the dry- and wet-bulb temperatures by 0.64 and 0.36 respectively (Egbunike, 1979). The weighing factor for the wet-bulb temperature is much less than for animals with active cutaneous evaporation by sweating. This is caused by the facts that the evaporation at the skin generally occurs at a lower temperature than in the respiratory tract and the water vapour capacity is higher at higher ambient temperatures. The water vapour pressure gradient at the higher ambient temperature will be higher. Thus a rise in ambient water vapour pressure has relatively less effect on animals with mainly active respiratory evaporative heat loss.

It is clear that the reported effects of wind speed on physiological variables such as a 12K displacement of thermoneutral temperature (Tzschentke and Nichelmann, 1986), requires also a wind scale index in the effective temperature. But as a matter of fact all the factors which shift a physiological variable with regard to ambient temperature have to be incorporated in a total effective ambient temperature index (Nichelmann et al., 1986).

DAILY RHYTHM

Minimum metabolic rate of larger, diurnally active, birds (> 500g) is 10 to 25% higher during the day, if measured in darkness, than during the night. However, there could not be found a significant difference between the day and night lower critical temperatures of non-passerine birds (Kendeigh et al., 1977). A lowering of body temperature during the night results in a decrease of the lower critical temperature, but a lowering of heat production in an increase. Only a part of the day-night change in metabolic rate can be explained by the body temperature change (Q_{10}-effect). Another possibility is a lower disturbance of the air boundary layer during the night by body movement, muscle tone and shivering (Van Kampen et al., 1979), resulting in a lower rate of heat transfer and a lower critical temperature.

If heat production is related to $W^{0.60}$ and heat loss to the surface area as $W^{0.67}$, then it should be expected that larger birds are less sensitive to higher ambient temperatures. However, the rate of heat transfer through the feathered skin varies with body weight as $W^{0.50}$ (Kendeigh et al., 1977). This means that heavier birds will show earlier signs of heat stress, such as elevated body temperature, elevated metabolic rate and reduction in egg production (Becker, 1983). When the rate of heat transfer varies as $W^{0.50}$ and metabolic rate as $W^{0.60}$, then the lower critical temperature should vary as $W^{-0.10}$. A 3 kg weight increase of a 1.5 kg hen will then shift the critical temperature 3.3°C downwards. It is not clear if the shift in upper critical temperature is of the same magnitude.

In fed hens daily variation in heat production may be more pronounced than in starving birds but that depends on the lighting and feeding regime. Other factors will be physical activity, length of starvation period and ambient temperature (Klandorf et al., 1981). MacLeod and Jewitt (1984) found that metabolic rate during the night was 35%, 15 or 7% lower than during day-time with ad libitum feeding, single meal in the morning or evening, respectively.

In starving birds metabolic rate decreases rapidly after the lights are off and increases already gradually before and with a temporary overshoot after the lights are on. The periods in the night with the lowest metabolic rate represent the minimum observed metabolic rate.

AGE

It is difficult to estimate an age effect on metabolic rate of laying hens due to factors such as difference in laying rate, feather cover and body weight. Fasting metabolic rate in older hens is often higher than in young ones, despite a lower laying rate. This is caused mainly by a weaker feather insulation (Damme and Pirchner, 1984). Often body weight of mature hens increases with age due to fat deposition. This fat is metabolically almost inert. Comparing heat production data per same metabolic body weight unit without correcting for body fat will result in lower values for the older hens.

SEASON

Cold acclimation stimulates aerobic metabolism in several gallinaceous species (Aulie, 1977). In domestic fowl the resting metabolic rate is increased during the winther (Arieli et al., 1979).

PHOTOPERIOD

Metabolic rate of birds tends to increase with lengthening photoperiods (Kendeigh et al., 1977) and intensity of illumination (Pohl, 1970), which is correlated with the number of body movements (Boshouwers and Nicaise, 1987).

ACTIVITY

Locomotor activity contributes 10 to 20% in total energy expenditure. Some head and neck movements raise heat production by 9%, while standing requires 16% more energy than sitting (Van Kampen, 1976). Physical activity involved with food intake may increase heat production temporarily by 37%. Walking speed is linearly related with heat production (Van Kampen, 1976). A peak heat production may be reached which is about twelve times higher than the resting metabolic rate (Brackenbury and Avery, 1980).

DISEASES

Extreme temperatures may induce diseases by weakening preventive mechanisms. To prevent low temperatures birds are often kept in insulated houses, in which sensible heat loss of the hens is used for raising the temperature by reducing ventilation. This reduction in ventilation results in an increase of carbon dioxide and ammonia concentrations, air humidity and moisture content of litter. Carbon dioxide concentration is a good indicator for the ventilation rate, but has no direct harmful effect on hens up to a concentration of 2.5% and even improves shell quality at high temperatures. High humidity increases ammonia production and decreases dust, which serves as a pathogen disseminator. When the ammonia concentration exceeds 20 ppm it increases the infection rate.

The infections cause often fever, resulting in an enhancement of the immune response, with body temperatures sometimes raised even above 44°C (Stauch, 1979). During fever body temperature is regulated at an elevated set-point and metabolic rate increases about 10% for each centigrade rise. In the same time food intake is impaired, through which the amount of energy available for production declines. The hens fall into a negative energy balance resulting in a decline of egg production and body weight. This loss may be reduced by raising ambient temperature a few centigrades.

CONCLUSIONS

It is clear that many factors contribute in the metabolic rate. For comparative studies a standardization of the involved factors is a necessity.

The small thermoneutral zone combined with the fact that basal metabolic rate is not identical with minimum metabolic rate complicates the standardization.

There is general interest for possible breed differences in minimum metabolic rate as an explanation for food intake and conversion variation (Damme et al., 1986) and sensitivity to high temperatures. Due to the variety of circumstances under which data are collected it is difficult to estimate whether a measured effect or difference exists and at which temperature, unless the factors get a weight in a total effective ambient temperature. However, an average effect of climate on energy metabolism is presented in Figure 2. This figure shows that the energy available for egg production (ME-intake minus Actual HP) is nearly constant over the temperature range 10-25°C.

In this temperature range egg weight decreases with about 0.1 g per centigrade and above 25°C up to 0.3 g per centigrade (Van Kampen, 1981). Rate of lay can be maintained up to 30°C or even increases with an egg per bird per year per centigrade (Emmans and Charles, 1977). Above 25°C the egg mass production cannot be kept constant due to the progressive decrease in egg size.

The maximum shivering and non-shivering thermogenesis is about 4-fold of standard heat production, which allows a homeothermic zone of -50°C to +35°C. Maximum metabolizable energy intake is even a 5-fold of standard heat production (Kirkwood, 1983). When it is possible to stimulate a hen to such an intake then even at -50°C there is energy left for production and at +25°C egg production may increase 3 times.

REFERENCES

Arieli, A., Meltzer, A. and Berman, A., 1979. Seasonal acclimatisation in the hen. Br. Poult. Sci. 20: 505-513.
Aulie, A., 1977. The effect of intermittent cold exposure on the ther-

moregulatory capacity of bantam chicks, Gallus domesticus. Comp. Biochem. Physiol. 56a: 545-549.

Balnave, D., 1974. Biological factors affecting energy expenditure. In: T.R. Morris and B.M. Freeman (Editors), Energy requirements of poultry. Br. Poult. Sci. Ltd., Edinburgh, pp. 25-46.

Becker, C., 1983. Die Bedeutung von Körpergewicht sowie Körpergrösse und Körperkondition für das produktive Adaptionsvermögen der Legehenne unter hoher Umgebungstemperatur. Thesis, Berlin, 154 pp.

Berman, A. and Snapir, N., 1965. The relation of fasting and resting metabolic rates to heat tolerance in the domestic fowl. Br. Poult. Sci. 6: 207-216.

Boshouwers, F.M.G. and Nicaise, E., 1987. Physical activity and energy expenditure of laying hens as affected by light intensity. Br. Poult. Sci. 28: 155-163.

Brackenbury, J.H. and Avery, P., 1980. Energy consumption and ventilatory mechanisms in the excercising fowl. Comp. Biochem. Physiol. 66a: 439-445.

Brody, S., 1945. Bioenergetics and growth. Reinhold Publishing Corporation, New York, 1023 pp.

Damme, K. and Pirchner, F., 1984. Genetic differences of feather-loss in layers and effects on production traits. Arch. Geflügelk. 48 (6): 215-222.

Damme, K., Pirchner, F., Willeke, H. and Eichinger, H., 1986. Fasting metabolic rate in hens. 2. Strain differences and heritability estimates. Po. Sci. 65: 616-620.

Deaton, J.W., May, J.D., Kubena, L.F. and Reece, F.N., 1976. Physiological changes associated with acclimation of broiler chickens to constant temperatures. Int. J. Biometeorol. 20: 333-336.

Decuypere, E., 1979. Effects of incubation temperature pattern on morphological, physiological and reproduction criteria in Rhode Island Red birds. Agric. (Heverlee), 27: 65-280.

Egbunike, G.N., 1979. The relative importance of dry- and wet-bulb temperatures in the thermorespiratory function in the chicken. Zentralblatt für Veterinärmedizin A26: 573-579.

Emmans, G.C. and Charles, D.R., 1977. Climatic environment and poultry feeding in practice. In: W. Haresign, H. Swan and D. Lewis

(Editors), Nutrition and the Climatic Environment. Butterworths, London, pp. 31-50.

Farrell, D.J., 1975. A comparison of the energy metabolism of two breeds of hens and their reciprocal cross using respiration calorimetry. Br. Poult. Sci. 16: 103-113.

Freeman, B.M., 1964. The effect of diet and breed upon the oxygen requirements of the chicken during the period of rapid growth. Br. Poult. Sci. 4: 169-178.

Freeman, B.M., 1984. Some responses of the domestic fowl to environmental temperature. Arch. exper. Vet. med. 38 (3): 392-398.

Garrow, J.S., 1986. Chronic effects of over- and under-nutrition on thermogenesis. Internat. J. Vit. Nutr. Res. 56: 201-204.

Hayssen, V. and Lacy, R.C. 1985. Basal metabolic rates in mammals: taxonomic differences in the allometry of BMR and body mass. Comp. Biochem. Physiol. 81A (4): 741-754.

Heusner, A.A., 1985. Body size and energy metabolism. Ann. Rev. Nutr., 5: 267-293.

Johnson, R.J. and Farrell, D.J., 1985. Relationship between starvation heat production and body size in the domestic fowl. Br. Poult. Sci. 26: 513-517.

Kampen, M. van, 1974. Physical factors affecting energy expenditure. In: T.R. Morris and B.M. Freeman (Editors), Energy requirements of poultry Br. Poult. Sci. Ltd., Edinburgh, pp. 47-59.

Kampen, M. van, 1976. Activity and energy expenditure in laying hens. J. agric. Sci. (Cambridge) 87: 81-88.

Kampen, M. van, Mitchell, B.W. and Siegel, H.S., 1979. Thermoneutral zone of chickens as determined by measuring heat production, respiration rate, and electromyographic and electroencephalographic activity in light and dark environments and changing ambient temperatures. J. agric. Sci. (Cambridge) 92: 219-226.

Kampen, M. van, 1981. Thermal influences on poultry. In: J.E. Clark (Editor), Environmental aspects of housing for animal production. Butterworths, London, pp. 131-147.

Kampen, M. van, 1984. Physiological responses of poultry to ambient temperature. Arch. exper. Vet. Med. 38 (3): 384-391.

Kendeigh, S.C., Dol'nik, V.R., Gavrilov, V.M., 1977. Avian energetics. In: J. Pinowski and S.C. Kendeigh (Editors), Granivorous birds in ecosystems. Cambridge Univ. Press, Cambridge, pp. 127-204.

Kirkwood, J.K., 1983. A limit to metabolisable energy intake in mammals and birds. Comp. Biochem. Physiol. 75A: 1-3.

Klandorf, H., Sharp, P.J. and MacLeod, M.G., 1981. The relationship between heat production and concentrations of plasma thyroid hormones in the domestic hen. Gen. comp. Endocr. 45: 513-520.

Kleiber, M., 1947. Body size and metabolic rate. Physiol. Rev. 27: 511-541.

Koong, L.J., Ferrell, C.L. and Nienaber, J.A., 1985. Assessment of interrelationships among levels of intake and production, organ size and fasting heat production in growing animals. J. Nutr. 115: 1383-1390.

MacLeod, M.G., 1984. Factors influencing the agreement between thermal physiology measurements and field performance in poultry. Arch. exper. Vet. Med. 38 (3): 399-410.

MacLeod, M.G. and Jewitt, T.R., 1984. Circadian variation in the heat production rate of the domestic fowl, Gallus domesticus: effects of limiting feeding to a single daily meal. Comp. Biochem. Physiol. 78a: 687-690.

Marsden, A. and Morris, T.R., 1980. Egg production at high temperatures. In: Intensive Animal Production in Developing Countries. Occasional publication no. 4. Brit. Soc. Anim. Prod., London.

McArthur, A.J., 1981. Thermal insulation and heat loss from animals. In: J.A. Clark (Editor), Environmental aspects of housing for animal production. Butterworth, London, pp. 37-60.

Nichelmann, M., Baranyiová, E., Goll, R. and Tzschentke, B., 1986. Influence of feather cover on heat balance in laying hens (Gallus domesticus). J. therm. Biol. 11 (2): 121-126.

Picard, M., 1986. Heat effects on the laying hen - protein nutrition and food intake. Zootechnica International 5: 64-67.

Poczopko, P., 1971. Metabolic levels in adult homeotherms. Acta Theriol. 16: 1-21.

Pohl, H., 1970. Zur Wirkung des Lichtes auf die circadiane Periodik des Stoffwechsels und der Aktivität beim Buchfinken (Fringilla coelebs L.). Z. vergl. Physiol. 66: 141-163.

Polin, D., 1983. The influence of environmental temperature on the feed intake of laying hens examined. Feedstuffs USA 55 (5): 21-22.

Prothero, J., 1986. Scaling of energy metabolism in unicellular organisms: a re-analysis. Comp. Biochem. Physiol. 83A (2): 243-248.

Richards, S.A., 1976. Evaporative water loss in domestic fowls and its partition in relation to ambient temperature. J. agric. Sci. (Cambridge) 87: 527-532.

Rogers, S.R. and Pesti, G.M., 1986. Comparison of protein utilization in morning vs. afternoon fed chicks. Po. Sci. 65 (suppl. 1): 113 (abstract).

Rufeger, H. and Bottin, U., 1980. Der Ruhe-Nüchtern-Sauerstoffverbrauch der Albinoratte und seine Abhängigkeit von der Körpermasse bei Ernährung mit proteinhaltiger und N-freier Kost. Tierphysiol. Tierernähr. u. Futtermittelkd. 43: 1-17.

Schutter, A.C. de and Morrison, W.D., 1986. The influence of nutrient density on performance of laying hens subjected to short term heat stress. Po. Sci. 65 (suppl. 1): 34 (abstract).

Scott, T.A. and Balnave, D., 1986. The influence of dietary self-selection on performance of young pullets under hot and cold environmental temperatures. Po. Sci. 65 (suppl. 1): 122 (abstract).

Stauch, R., 1979. Literaturübersicht über die Körpertemperaturen bei Vögeln. Thesis, München, 170 pp.

Taylor, C.R., Seeherman, H.J., Malory, G.M.O., Heglund, N.C. and Kamau, J.M.Z., 1978. Scaling maximum aerobic capacity (VO_{2max}) to body size in mammals. Fed. Proc. 37: 473.

Tzschentke, B and Nichelmann, M., 1986. The influence of wind speed on heat production in laying hybrids (Gallus domesticus) of different ages at various relative humidities. J. therm. Biol. 11 (2): 109-113.

CLIMATIC ENVIRONMENT AND ENERGY METABOLISM IN BROILERS

C.W. SCHEELE, W. VAN DER HEL, M.W.A. VERSTEGEN AND
A.M.HENKEN

ABSTRACT

Effects of climatic environment on performance, energy metabolism, energy retention and protein to fat ratio in deposited energy of growing broilers are discussed. The thermal requirement of chickens is related to both environmental and internal factors.

Quantitatively, physical activity accounts for the highest waste of dietary energy. It is released as heat to the environment. Muscular activity in combination with active behaviour is one of the processes involved in normal circadian rhythms of metabolism. Excesses of excitement and physical activity may be avoided by an environment adapted to natural rhythms in metabolism and activity.

Attention is paid to application of ambient temperatures below or partly below the zone of thermoneutrality combined with high energy diets containing sufficient amino acids to stimulate feed intake and growth after three weeks of age. The limited significance of the zone of thermoneutrality for fast growing birds is discussed.

Experimental results indicate that the heat production which seems to be related with protein deposition cannot be solely considered as the direct costs of protein synthesis. The higher levels of heat production in accompaniment with a higher rate of protein deposition may be more related to a simultaneous enhanced level of physical activity.

It is important to know to what extent environmental conditions affect activity. This knowledge can be used to define optimal conditions for broiler production. The fast development of modern techniques will facilitate application of such knowledge.

INTRODUCTION

Haet production is one of the quantitatively most important result of metabolic processes. The climatic environment has a profound effect on metabolic activity and on production characteristics of animals. Heat production is primarily affected by feed intake and by ambient temperature (Ta). Environmental factors include also humidity, light (length of day and intensity), partial pressures of oxygen and carbondioxide, ammonia concentration, wind velocity, density of population and housing system. Emmans (1986) showed that a few simple ideas about growth, body composition and feed intake can be combined to make useful predictions of animal performance. It should be noticed that the collection of available knowledge about important effects of the environment on performance and on metabolism can be used to complete that model for growth.

The following deals with the effects of several environmental factors on performance criteria, body composition and energy metabolism of broilers. First the effects of humidity, light regimens, air composition, population density and housing system are discussed. The main attention of this review article however is attributed to the relation between ambient temperature, physical activity and protein accretion.

Published results in literature and results of own experiments are reviewed to find essential factors which can be used to predict maintenance requirements of chickens and deposition of fat and protein.

Especially the importance of ambient temperature in relation to maximal weight gain and protein synthesis is discussed. Maintenance requirements (MEm) of animals are defined as the heat which is produced at EB = 0 within the zone of thermoneutrality with minimal activity. EB = 0 means that no energy is stored in the body nor is broken down.

In some studies on energy metabolism in growing animals the concept of maintenance is used in a broader sense, maintenance requirements then are calculated as the total heat production which is not directly related to deposition of protein and fat. The heat produced due to physical activity now is part of the maintenance requirement. This is called herein: "total maintenance requirements (MEmt). In those studies it is assumed that the direct energy costs of protein and fat are constant and independent of the level of deposition of protein and fat.

HUMIDITY

Romijn and Lokhorst (1961) observed that total heat loss of adult chickens was not influenced by relative humidity (RH) of the environment (23.8°C).

Barrott and Pringle (1949) found no differences in growth of chickens from 1 to 18 days of age at a RH from 35 to 75 percent while maintaining brooding temperature. Prince et al. (1965) showed that differences in RH from 52 to 90% at two temperatures (12.6 and 23.8°C) had no effect on feed consumption or on weight gain of male chickens from four to eight weeks of age. Milligan and Winn (1964) did not find any effect of high or low humidities on performance criteria of broilers between 35 and 70 days of age. Thus at moderate temperatures below 30°C performances and heat loss of broilers is not importantly dependent on RH.

Studies at ambient temperatures above 30°C revealed that the amount of respired moisture in the air or the latent heat production increases rapidly to values of 80 and 90% of total heat production (Farrell and Swain, 1977; Reece and Lott, 1982).

Veerkamp (1986) suggests that if latent heat production is an important part of the total heat production then the total heat content of the air including the moisture may be more appropriate to predict heat loss of birds to the environment than just ambient temperature. Total heat content of the ambient air can be expressed as the enthalpy.

LIGHT REGIMENS

Studies of Vogt et al. (1980) and Van Es (1981) showed that light regimens and light intensities may have an important effect on physical activity of chickens and thus on total maintenance requirements (MEmt).

Wenk (1980) found a variation of 100-200 $kJ/W^{0.75}$ per day in metabolizable energy requirements for MEmt, due to differences in physical activity of chicken. This was 20-40% of their total maintenance requirements. In the dark poultry has a low physical activity and a low heat production.

Barott and Pringle (1951) reported that the number and length of the light and dark periods were critical for chicks up to 32 days of age. They advocated that young chicks require about 1 hour light for eating, after that about 3 hours of darkness should be provided.

Over the past 30 years there has been a dramatic increase in broiler growth rate and a significant decrease in the feed to gain ratio. Chickens in studies of Barott and Pringle (1946) exhibited body weights at 8 weeks of age of 590 grams. Modern broilers at this age have weights of 2500 grams. Moreover higher dietary energy levels and nutrient contents are used. Light schedules proposed by Barott and Pringle may therefore be no longer optimal for modern broiler husbandry.

Results of recent studies on intermittent lighting with the aim to improve feed efficiency are not identical (Malone et al., 1980; Deaton et al., 1980; Cherry et al., 1980; Haye et al., 1984; Simons and Haye, 1985). Long dark periods led to low feed intake and reduced growth. Van Es (1981) suggested that frequent shorter dark periods might give a reduction in total physical activity and consequently in MEmt. If light and dark periods would be attuned more accurately to changes in metabolic rate, a reduction in total physical activity and a higher efficiency of dietary energy utilization for production may be attained. Changes in metabolism can be considered as a result of feed intake, digestion, anabolism, circadian rhythms in activity and sleep behaviour. Each of these parts can have its own different requirement for an optimal environment. More research is needed to establish the response of each part of metabolism to changes in ambient temperature combined with different light schedules.

If broilers are grown in continuous light with periods of different light intensities the response is similar to that found for conditions of light and dark alterations (Beane et al., 1965). This illustrates the importance of light intensity in broiler houses. It has been found that intensities as high as 130 lux are detrimental to growth (Barott and Pringle, 1951). Little is known of the lower limit of the light intensity necessary for effects on growth. A detailed review on effects of light intensity on performance of broilers has been published by Siegel (1977).

PARTIAL PRESSURES OF AIR COMPONENTS

The effects of a low partial pressure of oxygen on energy metabolism are not clear. However diseases as high altitude disease resulting in right ventricular hypertrophy, cholangiohepatites and ascites are well known (Cueva et al., 1974; Sillau et al., 1980; Julian and Wilson, 1986).

Atland (1961) reported that chickens have a much lower altitude tolerance than other small warm-blooded animals. They determined the highest altitude tolerated without deaths and the lowest altitude at which there were no survivors. The low hypoxic resistance of the chicken could be related to relatively low hematocrit values, a low oxygen carrying capacity, and an inefficient oxygenation of the fowl lung. This means that broilers will be nearer their summit metabolism than other animals at similar levels of feed intake.

The higher incidence of ascites in broiler chickens coincides with a genetic and nutritional improvement in feed efficiency and rate of growth accompanied with a shortage on oxygen supply (Van Blerk, 1985; Julian and Wilson, 1986).

The increased incidence of ascites in the cold season in countries with a moderate climate (Van Blerk, 1985; Buys and Barnes, 1981) may be related to the increased oxygen requirement.

The poor ventilation in cold weather with relative high concentrations of carbondioxide and ammonia could have an extra initiating effect on metabolic disorders. Reece and Lott (1980) found that exposure of chickens from 0 to 4 weeks of age to 1.2% CO_2 in the air resulted in a negative effect on growth. Lower contents did not effect performances. The concentration of different air components is controlled by the ventilation in poultry houses. Charles et al. (1981b) investigated the effect of different ventilation rates at 20°C after brooding on the performance of broilers. They advocated a minimum ventilation rate for broilers of 1.5 × $10^{-4} m^3/s$ per $W^{0.75}$. The first limiting factor governing ventilation rate is often the ammonia concentration in the air which should be not higher than 25 ppm (Quarles and Kling, 1974). Therefore the minimum ventilation rate may be higher after about 4 to 5 weeks of age. Use of this relatively low ventilation rate will permit the attainment and the application of optimum temperatures at low heating costs. From studies of Charles et al. (1981b) it appeared that the heating cost increased exponentially as ventilation rate was increased.

POPULATION DENSITY

Chickens kept at low environmental temperatures may decrease their rate of heat loss by huddling together. Kleiber and Winchester (1933) found that at an environmental temperature of 14°C 3 week old chicks produced 15% less heat per hour when allowed to huddle than when kept separated. This huddling effect can be considered as one of the components of physical thermoregulation.

The contribution of huddling to thermoregulation is dependent on group size, the number of chickens per square meter and on the behaviour of the birds such as active movement or inactivity when clustered together.

The temperature of the micro-climate is thus not the same as the ambient temperature as the latter means the average temperature in an experimental room or in a poultry house.

HOUSING SYSTEM

From experiments with pigs it is demonstrated that the housing system can have an important effect on thermoregulation (Verstegen and Van der Hel, 1974; Verstegen et al., 1977).

Analoguous to these results Morrison and McMillan (1985 and 1986) found that chicks of 8 days of age on litter at the same Ta (18°C) as those on wire, were really in an effective environmental temperature (Te) 3.2°C warmer (21.2°C). The difference between Te and Ta certainly will decrease when Ta increases and with an increased age of the birds. More research is needed to find effective environmental temperatures of broilers in relation to housing at the various ambient temperatures used in practice.

AMBIENT TEMPERATURE

Many experiments have been carried out to study the effect of ambient temperature on performance and energy metabolism of chickens.

Newly hatched domestic chicks are poor thermoregulators with a relatively narrow zone of thermal neutrality (Romijn, 1954; Wekstein and Zolman, 1970).

Young chicks in the first week after hatching have some poikilothermic characteristics: their body temperature is dependent upon Ta. Romijn (1954) demonstrated that within this first week the birds exhibit a rapid development of homeothermy. The thermoregulating capacity is relatively mature at 2 to 3 weeks of age according to Osbaldiston (1968) and Freeman (1976).

Muscular activity appears to be closely related to regulation of homeothermy. It is dependent on a fast degrading of available free energy as stored in adenosinetriphosphate (ATP) into heat energy. Increased muscle tone appeared to be the only mode of producing heat for thermogenesis in fasting birds after about 2 weeks of age (Wekstein and Zolman, 1971; Freeman, 1976). Muscular activity is especially related to acting intensively on the environment in search for feed and water and in surviving within the group.

During the first 14 days of age the bird is vulnerable with regard to its environment. The resting oxygen consumption of fully fed young birds was found to rise substantially during the first 14 days of age from 1.64 ml/g/h to 2.52 ml/g/h at 33°C and from 2.40 to 2.70 ml/g/h at 29°C and then to remain at a relatively constant level (Freeman, 1963). In this early stage of development the metabolic rate increases at a faster rate than body weight (Denbow and Kuenzel, 1978).

Ten Have and Scheele (1980) found a linear relationship between heat production and ambient temperatures ranging from 25 to 37°C with broilers from 0-7 days of age. From 7-14 days of age the chicks started to regulate their heat production independently of environmental temperatures in the range from 28°C to 34°C.

In this early period the recommended ambient temperature lies above 26°C. A lower temperature during the first 14 days after hatching can lead to imbalances. It may lead to diseases (ascites) and death (Van Blerk, 1985; Buys and Barnes, 1981).

Deaton et al. (1976) measured physiological changes in young broilers from 0-4 weeks of age exposed to 24 and 32°C. At 24°C a significant increased heart-body weight ratio, liver-body weight ratio and an elevated hematocrit level was observed. A higher hematocrit can result in a higher viscosity of the blood which is related to a lowered circulation rate (Gilbert, 1963) and so to diseases (ascites).

Ota (1967) noted that hot air brooding may be started at 30°C with 60% relative humidity for chicks averaging more than 35 g. Smaller chicks averaging 30 g would require a higher brooding temperature (34°C, 60% RH). After 2 to 3 days the Ta can be reduced daily. In reviewing a great number of experimental results, Charles (1986) recommended a gradually reduced Ta until 21°C at an age of about 17 to 21 days of age in draught free houses provided with good dry bedding. For systems employing radiant spot brooders in houses free from draught and cold surfaces, a Ta as low as 20 to 25°C might be adequate even at the start, but only if there is an easy access to feeders and drinkers. In Table 1 schedules for Ta are given as are used in practice for broilers from 0-5 weeks of age.

Table 1. Ambient temperature schedules for broilers used in practice (CADP, 1986).

Age (weeks)	Ta (°C)
0 - 1	33 - 29
1 - 2	29 - 26
2 - 3	26 - 23
3 - 4	23 - 20
4 - 5 and continued	20 - 18

A rather striking element in these recommendations is the low value of Ta for chickens after approximately 2 weeks of age.

These temperatures are definitely below the zone of thermoneutrality, as published in the literature, even when an increased effective temperature by a good litter bedding and by huddling is taken into account.

THE THERMONEUTRAL ZONE

Within the thermoneutral zone the fasting and resting animal has a minimal rate of heat production, called the basal metabolic rate (Kleiber, 1975). In this zone the heat production is independent of environmental temperature.

For fed birds the definition of the thermoneutral zone has been extended to the range of ambient temperatures in which the heat exchange is minimal and constant at a given feeding level.

This thermoneutral zone may also be defined as that zone which lies between two climatic extremes, i.e. the lower critical temperature higher than in the zone of thermoneutrality. In the literature no complete unanimity exists about the range of temperatures being the zone of thermoneutrality.

Some reports indicate only a point of flexure or a very narrow zone. In Table 2 the values are given for the Lct found by different authors in different years for different strains of chickens. Also within a strain differences can be found related to feeding level and time of the day.

Van Es et al. (1973) explained that each combination of feeding, production and activity level had its own thermoneutral zone.

In Table 2 it is also shown that light and time of the day are important factors in regulating heat production of the birds.

The different values for the Lct of chickens shown in Table 2 indicate that at three to four weeks of age when thermoregulating capacity is approximately mature the Lct is still higher than 23°C. Especially at day time the birds exhibit an increased thermal demand.

Nevertheless in practice the ability of the broilers at that age to cope with a relatively wide range of temperatures is utilized by keeping them at a Ta below the Lct. Besides saving heating costs it is found that also better growth rates can be obtained.

As at this low Ta feeding costs are increased it is important to know more about the effect of a Ta below the Lct on metabolism and perfor-

Table 2. Values for the Lct found for B (broilers) and P (pullets) in °C. A comparison of findings.

References		Age in weeks						
		1w	2w	3w	4w	5w	6w	7w
Barott and Pringle (1949)	-	35	35					
Romijn (1950)	B	35	35					
Freeman (1963)	P	35	31-35	30-33	26-31			
Van Kampen et al. (1978)	P							
in light								32
in dark								27.5
Ten Have and Scheele (1980)	B			27-29				
Henken et al. (1982a)								
fed ad libitum: at day						29.9		
at night						23.1 - 23.2		
fed restricted: at day						27.0 - 30.5		
at night						25.6 - 27.6		
Meltzer et al. (1982)	P	33	31	29	27.5	26	24.5	23
Meltzer (1983)	B	32	29.5	27.5	26.0	24.5	23.5	23
Misson (1982)	P	28						
Nichelman et al. (1983)	P		26.7	26.5			25.4	

mance of broilers.

CIRCADIAN RHYTHMS IN HEAT PRODUCTION

The differences in heat production between day and night as reported by Henken et al. (1982a) can not simply be attributed to the effect of light. Extra thermoregulatory heat production of restricted fed pullets below Lct was higher in the first 4.5 hours of the night than in the second 4.5 hours at darkness. This is in agreement with other experiments in which differences in heat production were independent of temperature and light but which were related to time of the day.

Barott et al. (1938) found that chickens exhibit a very definite rhythm in oxygen consumption during the 24 hour period of a day. At an age of one week the chickens revealed an amplitude in metabolic rate of about 12% of the lowest level during the daily period. An increasing age was accompanied by a decreased amplitude (5.7% at 14 weeks of age). During the experiments the birds were starved and the temperature was held constantly above the Lct, the birds were limited in their movement and kept in the dark.

More insight in the circadian variation in heat exchange was obtained by Van der Hel et al. (1984) in experiments with growing pigs. The animals were housed in groups in large calorimeters (Verstegen et al., 1987). During these experiments the physcial activity of the pigs was measured by registrating their movement with a method based on the doppler effect using ultra sound.

Van der Hel and coworkers (1984) observed a large difference in heat production between night and day. The amplitude of the circadian rhythm was 50% of the lowest heat production. From their measurements the authors calculated correlation coefficients and concluded that 60% of the variations in circadian heat production was associated with physical activity. Most interesting was also that the variation in metabolic rate of the pigs within a day was not diminished at temperatures below the Lct.

This points out that in certain periods within a day animals appear to have other priorities than saving energy and voluntary produce more heat than is demanded by the environment even at a low Ta. These priorities are closely related to the time of feed intake and to the duration of digestion of the food. It is obvious that an increased physical activity leads to an increased total maintenance requirement. This could be detrimental for an efficient utilization of dietary energy. The question arises to which extent a circadian variation in metabolic rate and activity is a prerequisite for a high production level and, if so, to what extent a periodically increased total maintenance requirement (MEmt) should be stimulated by light and/or by ambient temperature.

METABOLIC RATE, BODY TEMPERATURE AND MUSCULAR ACTIVITY

According to the law of Van 't Hoff the speed of reaction in metabolic processes, and thus metabolic rate, is dependent on body temperature.

Lamoreux and Hutt (1939) found the variability in body temperature of chickens to be affected by diurnal variation in metabolism.

Baldwin and Kendeigh (cited by Wilson, 1948) observed that metabolic activity in the muscles including muscle tone was one of the important factors causing variations in body temperature.

Blokhuis (1983) elicited the significance of sleep for poultry explaining that during fast wave sleep neck muscle tonus and heart rate of poultry is decreased compared with awakeness. Similar changes in muscle tone and heart rate can be obtained by changing Ta of poultry. Chaffee et al. (1963), Wekstein and Zolman (1970) and Freeman (1976) showed that a decrease in metabolic rate and in heat production at an increasing Ta was correlated with a decrease in electromyographic (EMG) activity.

Van Kampen et al. (1979) found that the EMG amplitude of chickens declined with increasing Ta up to 27.5°C in the light and 22.3°C in the dark. It was higher in the light than in the dark.

Van Kampen et al. (1978) found that heart rate of chickens was positively correlated with metabolic rate measured as oxygen consumption. Heart rate was lower during dark periods than in light periods and decreased with increasing ambient temperatures.

These experimental results demonstrate that changes in metabolic rate can be affected both by changes in environmental and internal factors.

This means that an increased metabolic rate and a higher level of heat production below the Lct cannot simply be translated in a higher requirement of the bird for heat from the environment. On the contrary if a higher environmental temperature is offered in periods that apparently a minimal heat loss is not desired, this can have a negative effect on feed intake and thus on production level.

A high turn-over of energy is required to produce the relative large amounts of free energy stored in ATP (the driving force) which are necessary for fast speed reaction (Wilkie, 1960). These reactions occur to ensure muscular activity. The used energy has to be released as heat to the environment. This is in agreement with experimental results with growing chickens. Van Kampen et al. (1979) found that the LCT of growing chickens in the light was 5°C higher than in the dark. The birds were kept at the same feeding level with no feed intake during measurements of heat production. These results indicate that there might be no zone of thermoneutrality during light periods (day time) as the chickens did not produce a minimal level of heat in the light.

Also interesting in these experiments was that an increase in body temperature at an increasing Ta coincided with a decrease in metabolic rate and also in heat production related to the lower muscular activity. This reflects the important role of muscular activity in regulating metabolic rate associated with changed Ta.

Experiments with growing pigs (Verstegen et al., 1986) also showed that during day time the Lct was higher than in the night when metabolic rate was decreased, although differences in metabolic rate did not always result in different critical temperatures. It appears that especially during light periods the direct effect of Ta on heat production is reduced and the effect of internal factors catalyzing metabolism are more important.

THE EFFECT OF FEED INTAKE ON THE Lct

At increasing levels of feed intake and a higher heat increment of feeding a decreased Lct can be expected. Relationships between metabolic rate, ambient temperature and the calorigenic effect of food in homeotherms have been discussed by Kleiber (1975) and Mount (1979). They indicated that at temperatures below the Lct all the heat produced will be used for thermoregulation.

Thus below the Lct heat production becomes independent of the amount of food consumed. It means also that the heat increment of feeding is discernable only at temperatures above the Lct when the heat production at a given feeding level is minimal. However in experiments of Kleiber and Dougherty (1934) it was shown that the difference in metabolic rate between fed and starved chickens increased further as Ta decreased more below the Lct. This was also found in experiments with pigs exposed to cold conditions during several weeks (Close, 1981).

Similar results were obtained in experiments with growing chickens by O'Neill et al. (1971), O'Neill and Jackson (1974). Farrell and Swain (1977) and by MacLeod et al. (1979) with 3 week old turkey pullets.

Both chickens and turkeys were exposed to a wide range of ambient temperatures. At low temperatures it was shown that heat dissipated from heat increment of the diet was used in an uneconomical way and not only for keeping the body temperature constant as compared to thermoneutral conditions.

Misson (1982) demonstrated that both ad libitum fed and starved one week old chickens had the same Lct of 28°C independent of feeding level. Paradoxically it appeared that the fed birds had no advantages in the cold by having a lower Lct. Misson found that the cold chicks fed ad libitum regulated their body temperatures 2°C above those of starved birds, therefore fed birds do not exhibit the same energy saving mechanisms as starved birds.

Ten Have and Scheele (1980) measured the heat production of young broilers, 7-14 days of age, exposed to different temperatures ranging from 22°C to 34°C and fed different diets. The diets contained 27.7, 43.6 and 58.8 grams of protein per kg dry matter respectively. Due to energy costs of uric acid formation and excretion, high protein diets will have a much higher heat increment compared to the low protein diets. More heat was produced with high protein diets than with low protein diets. The difference was 100 kJ per kg body weight per day. The same difference however was found at high and low temperatures. This similarity in heat increment was not expected.

The significance of these results is that in high productive growing broilers the stimulus to gain body weight may have a higher priority than saving energy even at low temperatures.

This could also mean that the birds utilize the cooler environment for a voluntary higher heat production to maximize processes involved in a high production rate.

One of these processes could be feed intake behaviour demanding a high level of production of free energy (or ATP). Internal processes like an enhanced level of circulation, intestinal mobility, peristalsis and secretion of enzymes are related to this feed intake behaviour.

A high rate of muscular activity is generally considered as a waste of energy. Wilkie (1960), however, illustrated that in producing animals it appears to be important to have periods of a high energy turn-over. This is needed for fast speed reactions as occur during feed intake behaviour, and this leads inevitably to a high level, both of waste and of achievement.

For economical reasons it is important to know to what extent physical activity can be reduced without negative effects on production rate.

From a physiological point of view it is also understood that periods of inactivity and rest may be important in the daily cycle of metabolism. Blokhuis (1983) mentioned three important themes associated with sleep, rest and inactivity in poultry:

1. Saving energy at times when activity is not desired;
2. Processes governing growth and tissue restoration;
3. The role of sleep in the functioning of the central nervous system.

Meddis (1975) concluded that it is advantageous for an animal to schedule its behaviour so that activities are concentrated at those times in a day that permit most economical performance. Such an advantage may have caused the development of rest-activity cycles during evolutionary history. This is why activity has such a clear association with feed intake. This also suggest that environmental factors, i.e. a high T_a, which have a reducing effect on metabolic rate and activity should be offered only at times when feed intake is finished.

REGULATION OF FEED INTAKE

Modern fast growing broilers must consume large quantities of feed in order to attain maximal growth rate. Control of feed intake has been ascribed to dietary composition (chemostatic regulation) as well as to environmental factors including conditions of body temperature regulation (thermostatic regulation). The chemostatic control is associated with chemical components in the blood or in other body components (glucose level, insuline, pH). The most important factor in the chemostatic feed intake control seems to be the metabolizable energy (ME) level of the diet. At a constant Ta birds tend to eat a constant amount of ME per day (Hill and Dansky, 1954; Olson et al., 1972). A chemostatic regulation is also partly related to genetical factors determining weight gain and partly to physiological limitations in energy turn-over (Osbaldiston, 1968; Hurwitz et al., 1980).

Strains of birds with increased growth rates may be nearer physiological limitations than low producing strains. Feed intake controlled by a maximum rate of energy turn-over is not necessarily different from a thermostatic regulation of feed intake. Both regulations will be related to the capacity to convert chemical feed energy in heat energy, as heat energy is inevitable coupled to synthesis of body tissues. The thermostatic mechanism of feed intake control results in an inverse relationship between feed or energy intake and environmental temperature.

The importance of a thermostatic regulation of feed intake has been reported by Prince et al. (1965), Olson et al. (1972), Ahmad et al. (1974), Kirchgessner et al. (1978) and Cerniglia et al. (1983).

Thermostatic control of energy intake means that feed intake partly determines body temperature. Body temperature will rise when feed intake and thus heat production rises.

If the peak heat production during feed intake cannot easily be released to the environment further feed intake and thus growth can be inhibited.

A reduced energy intake at a higher Ta leads simultaneously to a reduced intake of other nutrients. This is expected to be a direct reason for reduction of growth.

Thus, lowering the total maintenance requirements for energy (MEmt) by a higher Ta and supplying diets with higher contents of essential nutrients and so maintaining a high level of nutrient intake, could lead to a continuous high level of growth at lower costs of feed energy. However, Cowan and Michie (1978) revealed that diets with increased concentrations of protein did not reduce the depression in growth rate of broilers after three weeks of age reared at 26°C compared with those reared at a Ta below the Lct. In experiments by Charles et al. (1981a) an extra supply of both energy and protein was offered to broilers at different ambient temperatures ranging from 15 to 27°C. The depression of weight gain at increased temperatures could not be prevented by adjusting overall nutrient concentrations. They mentioned the distinct contrast between the absence in broilers of an interaction between temperature and nutrition with the "well documented" interaction in laying hens. According to these authors it seems that in broilers the effects of temperature are specific and direct and not through nutrition. Also the findings of Adams et al. (1962), Adams and Rogler (1968) and Kubena et al. (1972) clearly illustrate that the growth rate depression of broilers which occurs at a T_a above about 23°C is not simply a reflection of inadequate nutrition.

The results of the experiments indicate that birds require a cool environment at least during a part of the day.

Interesting results were obtained in experiments with growing chickens which were allowed to choose their own environmental temperature, or as it is called, their zone of comfort.

Haller and Sunde (1973) concluded from their experiments that group housed chicks from 0-4 weeks of age preferred a warm environment (within the zone of thermoneutrality) in the dark and in periods of rest. The cooler zones, during awakeness, were used for feed and water intake.

Morrison and McMillan (1986) found that even at very low temperatures young chickens voluntarily preferred to be some time without supplementary heat. At 4°C they did not choose supplementary heat for 27 percent of the time. They even deliberately choose to be cool, even cold, at certain times. At these times the zone of comfort was apparently extended over the Lct below the zone of thermoneutrality.

REGULATION OF METABOLISM

According to Kleiber (1975) the metabolic rate of animals is regulated by two controlling systems: the nervous system, which mediates rapid changes in metabolic rate and the endoctrine system which controls slower and longer lasting changes.

Metabolic rate and also animal growth is regulated by complex hormonal interactions. Many hormones are known to be involved: thyroid hormones, growth hormones, gonadal steroids, adrenal corticoids, so-matomedins and insuline.

Discussing all these hormones is beyond the scope of this paper. However, some attention is given to thyroid hormones as thyroid activity is related to growth, metabolic rate and physical activity.

In fed animals it has been established that decreased environmental temperatures are accompanied with an increased thyroid activity (Ring, 1939; Dempsey and Atwood, 1943; Huston and Carmon, 1962; Freeman, 1970). At low temperatures the output of thyrotropin from the anterior pituitary increases which results in a high thyroid output. It was found by Leung et al. (1984) that thyrotropin releasing hormone (TRH) stimulates body weight gain and increases thyroid hormones and growth hormones in cockerels.

Brody (1945) detailed what was known of the effect of thyroid hormone production on growth and metabolism in animals at that time. He reported that after thyroidectomy all anabolic processes are retarded in different species including the rate of digestion, peristalsis, circulation rate, egg production and also muscular activity. Thyroid hormones also accelerate the metabolism of nerve tissue and affect corresponding neuromuscular activity, cortical alpha rhythms and thresholds to light.

The reports of Leung et al. (1984) and Brody (1945) may illustrate the possible link between activity patterns and growth rate as both are affected by TRH.

According to Brody (1945) there is a physiologically optimal thyroid hormone level. Above this level the catabolic effect overbalances the anabolic effect. Singh et al. (1968) showed that thyroxine administration in small doses improved growth of normal chickens. Higher doses of thyroxine depressed growth and accelerated catabolic processes. Leung et al. (1986) found that administering growth hormone to chickens resulted in positive effects when low levels were injected. They found no effect at high levels.

These reports strongly suggest an optimal level for thyroid and growth hormones.

Thus, if ambient temperature affects secretion of these hormones, only moderate changes in environmental temperature could have beneficial effects.

Keshavarz and Fuller (1980) observed a profound effect of T_a on thyroid size of broilers. The largest thyroid size was found in birds held in the cold (12.8°C and lower) and the smallest in the birds held at the highest temperature (23.9-35°C).

Thyroid sizes found between 12.8°C and 23.9°C were considered as normal. These authors demonstrated that widely fluctuating temperatures (amplitudes of more than 5°C) resulted in significantly smaller thyroid sizes compared with constant temperatures.

Pethes et al. (1979) observed a reduced plasma growth hormone con-centration response to TRH in cold (10°C) adapted ducks. This would be consistent with elevated thyroid hormone secretion (Scanes and Lauterio, 1984) and illuminates the relation between Ta and hormone secretion.

Little is known about the effects of low temperatures and of duration of exposure to temperatures below the Lct on hormonal activity. If phy-sical activity in some periods within a day is essential for growth, it is important to know what the minimal length of these periods is. Reduc-ing physical activity may decrease the MEmt. High total maintenance re-quirements will result in a low energetic efficiency of food energy for growth. Wenk and Van Es (1980) demonstrated that in broilers 25% of the MEmt could be attributed to physical activity. Boshouwers and Nicaise (1985) measured physical activity of laying hens. They demon-strated that at day time 25% of the total heat production is related to physical activity. No information is available on minimal levels of physical activity and on the effect of activity on development of muscles.

Total maintenance requirements for energy in growing chickens are approximately 600 kJ per $W^{0.75}$ (W*) per day. It is assumed that 25% of these 600 kJ (= 150 kJ) can be attributed to activity. If it is possible to reduce this to 15% then this would result in a sparing effect of 60 kJ metabolizable energy per W* per day. Thus, with an average body weight of 1 kg (W* = 1) during a growth period of 7 weeks about 3 MJ ME will be saved (= 49 x 60). This corresponds to 240 grams of food, which is about 6% of the total amount of food consumed per bird in 7 weeks. This reflects the importance of research for synchronizing the environment with optimal rhythms in metabolic rate in broilers.

GROWTH OF BROILERS AS AFFECTED BY Ta

A considerable amount of information has been published about the effect of environmental temperature on growth, feed intake and feed conversion of broilers.

A beneficial effect of ambient temperatures below 22°C, which is distinctly lower than the average value of Lct, on growth rate is reported by many authors (Barott and Pringle, 1950; Adams et al., 1962; Huston, 1965; Prince et al., 1965; Adams and Rogler, 1968; March and Biely, 1972; Olson et al., 1972; McNaughton et al., 1978; Cowan and Michie, 1978; Hurwitz et al., 1980; Charles et al., 1981a; McNaughton and Reece, 1982; Reece and McNaughton, 1982; and Cerniglia et al., 1983). The agreement between them is convincing. In most of the publications feed efficiency was highest at temperatures above 22°C indicating that the effective temperature was indeed below the Lct.

The results of McNaughton and Reece (1982) indicate that maximum performance of 23 to 48 day old broilers may be attained at 15.6°C environment by feeding them a high energy diet (14.1 MJ/kg).

Reece and McNaughton (1982) demonstrated that at a Ta of 18.3°C body weight increased linearly up to 3.1% when the energy level of the diet was changed from 13.3 to 13.9 MJ/kg. The broilers fed 13.9 MJ ME/kg and reared in a 15.6°C environment had equivalent feed efficiencies as those fed 13.6 MJ/kg and reared in a 21.1°C environment. At 26.7°C the body weight gain was lower. No significant differences were found between energy levels in 7 week body weights.

Adams and Rogler (1968) determined production characteristics of two different groups of growing chicks from one flock. The groups were composed at 4 weeks of age and differed in growth rate. All chicks of both groups gained faster at 21°C than at 29°C. The depression in growth rate at 29°C was greater for the fast growing chicks which also had the highest feed to gain ratio. If it is assumed that the digestibility of the diet and the energy costs for growth were similar for slow and fast growing birds, their results suggest that the fast growing birds had the highest total maintenance requirements (MEmt). They also found the response to T_a in feed to gain ratio dependent on the energy level of the diets. At high energy levels the best feed conversion was found at 21°C. At lower energy levels the best feed conversion was found at

29°C.

Especially fast growing birds with high density diets show an improved production rate at temperatures below the Lct after three weeks
of age.

Holsheimer (1985) stipulated the difference in growth rate between
fast growing male and female broilers and the higher requirements for
essential amino acids of males compared with females. McNaughton et al.
(1978) reported that in a cool environment higher lysine levels were required than in a warm environment to maximize weight of 4 week old
cockerels. Thus, fast growing male chickens fed diets with high contents
of energy and essential amino acids presumably will profit more from a
cool environment than female broilers.

It should be noticed that broilers kept in individual wire cages have a
higher thermal requirement than is provided at 22°C (Guill and Washburn, 1972).

CYCLIC TEMPERATURE REGIMENS

Several experiments were done with cyclic temperature regimens. In
some experiments a high temperature (more than 30°C) was used as the
upper limit of the temperature cycle simulating warm days in summer
time.

Deaton et al. (1984) used 24 hours linear temperature cycles ranging
from 35°C to either 26.7 or 21.1°C in experiments with broilers after
three weeks of age. Results showed that decreasing the lower limit of
the temperature cycle from 26.7 to 21.1°C significantly increased body
weight at 48 days of age.

Male and female chicks, grown in a 15.6 to 35.6°C diurnal cycled
temperature environment, had a higher body weight gain, a better feed
conversion and a lower mortality rate than those grown in a 24 to 35.5°C
environment (Griffin and Vardaman, 1970). In experiments with a moderate Ta (below 25°C) as highest temperature it appeared that widely
fluctuating cycles (amplitudes of more than 5°C) had detrimental effects
on production characteristics.

In experiments of Siegel and Drury (1970) temperatures were cycled
with amplitudes of 5.5, 11.1 and 16.6°C per 12 hours. The latter two

amplitudes caused a significant reduction in growth compared to more narrow cycling (± 5.5°C) or non cycling. Harris et al. (1974) found that feed consumption and gain were enhanced by a diurnal cycle of 18.3 to 23.9°C as compared with a diurnal cycle of 23.9 to 35°C.

Olson et al. (1972) obtained the highest growth rate and feed efficiency with an amplitude of ± 5.5°C at a highest Ta of 24°C compared with cycles with similar amplitudes but lying completely within the thermoneutral zone.

Although these results are difficult to compare it appears that amplitudes of cycles with a maximum Ta near the Lct should not exceed 5°C.

AMBIENT TEMPERATURE AND DEPOSITION OF PROTEIN AND FAT

In some studies on energy metabolism the MEmt is negatively correlated with deposition of fat.

Van der Wal et al. (1976) examined the energy metabolism of different lines of mice. Lean mice which retained only a little more protein than the obese ones, showed a strikingly higher maintenance requirement than the obese mice. This was associated with a higher level of physical activity.

Reviewing a great number of articles about thyroidal influence on body growth, King and May (1984) reported that changes in fat and water content in the body are the most obvious alterations of thyroid manipulations.

Noteworthy is the observation by Koger and Turner (cited by Brody, 1945) that the total energy retained in growth accelerated mice (thyroid administration) was similar to that in control mice. However, the thyroid dosed mice had a higher water and protein content and a lower fat content. Zorn (cited by Brody, 1945) concluded that feed utilization in partial thyroidectomized pigs was shifted from muscle and skeletal growth to fattening.

Johnson and Crownover (1976) found in experiments with growing chickens that a higher rate of protein synthesis was correlated with a high maintenance requirement. Berschauer et al. (1980) calculated from his results with growing pigs that higher estimates of the partial energetic efficiency for deposition of protein were associated with higher values for maintenance. Scheele et al. (1987a) found in experiments with broilers that energy costs for protein deposition were higher than those for deposition of fat. The difference was such that it could only be explained by a higher total maintenance requirement in birds exhibiting the highest protein retention. Adams and Rogler (1968) found fast growing chickens to have a higher feed to gain ratio than birds with a lower growth rate.

Assuming similar energy costs per gram of body weight gain for both groups of birds, the total maintenance requirements (MEmt) of fast growing birds must have been higher than those of slow growing chickens. The difference in growth and thus in protein deposition was more pronounced at a low temperature (21°C) than at a high temperature (29°C). The stimulation of growth by a low temperature was greater for the fast growing chicks than for slow growing birds.

Olson et al. (1972) examined the effect of three different circadian temperature cycles on growth and body composition of growing chickens from 2 to 4 weeks of age. Carcass dry matter, fat and energy were increased with an increasing temperature but protein content was decreased. At the lowest temperature cycle with a mean Ta below the Lct the growth rate was markedly improved compared with temperature cycles with a higher mean Ta. The result was an important increase in protein yield obtained at the lowest mean Ta.

Kubena et al. (1972) also noted a significant decrease in carcass ether extract with a concommitant increase in moisture content and a trend toward increasing protein contents in broilers at a lower Ta.

Conflicting with these reports are findings of Farrell and Swain (1977) and Henken et al. (1982b). In their experiments no increased protein accretion was found at lower temperatures below the Lct. They concluded that protein accretion was apparently independent of the temperature at which the chickens were grown. Henken et al. (1982b) pointed at the significance of dietary protein with respect to the presence or absence of temperature effects on protein gain. They suggested that effects of temperature on protein gain may be more easily found at lower dietary protein contents. The differences in experimental results concerning protein deposition in relation to Ta might also be due to the different temperature regimens applied. Siegel and Drury (1970) and Harris et al. (1974) demonstrated that fluctuating low temperatures can have detrimental effects on growth and thus on protein deposition. Also constant low temperatures can have a negative effect on growth and on protein accretion.

Therefore a study was conducted to examine further the effects of different ambient temperatures just below the thermoneutral zone on performances and energy metabolism of growing male and female broilers from 3 to 4 weeks of age (Scheele et al., 1987c).

EXPERIMENTS WITH BROILERS BELOW THE Lct

General

Three experiments were performed using three environmental controlled chambers in each. In all three experiments the same treatments were applied. Each time the treatments were allocated to another chamber. In each experiment each chamber contained 144 birds (6 replicates of 6 males and 6 females separately, fed ad libitum or restricted).

Temperature regimen

During the experimental period (21-28 days of age) the Ta of 24°C at the start was gradually reduced to 15, 17 and 19°C respectively for the three different chambers. This resulted in average temperatures of 18.8, 20.0 and 21.2°C respectively.

Diets

A diet based on maize, soybean-oil meal and animal fat was administered to all birds from 0-28 days of age. A diet with a high energy level and a high protein content was chosen. The average value (as fed) of the dietary ME was 12.96 MJ per kg. The calculated dietary protein, lysine and methionine + cystine contents were 22.0, 1.20 and 0.90% respectively. Restricted fed birds were given each day about 90% of the estimated ad libitum ME intake of broilers of this age kept under conditions comparable with the highest experimental Ta treatment.

Statistics

Effects of broods, chambers, temperature regimens, sex and level of feed intake on production characteristics, ME intake (MEi), energy retention (RE) and heat production (HP) were examined by analysis of variance. According to these results multiple regression equations were calculated to predict RE and HP from a range of factors. For each sex and feeding level the RE and HP were related to energy intake (MEi), metabolic body weight (W*), the retained protein energy as part of the total retained energy (FREp) and temperature regimen (Ta).

Results

In Table 3 the mean values of all data for growth and feed conversion are given per Ta, sex and feeding level.

Table 3. Mean values for weight gain and feed conversion of broilers between 21-28 days of age of females (F) and males (M), fed ad libitum (L) or restricted (R).

Mean Ta	Weight gain (g per bird/day)				Feed conversion			
(°C)	F	M	R	L	F	M	R	L
18.8	33.6	39.8	25.3	48.1	2.35	2.32	2.61	2.05
20.0	33.6	39.5	25.7	47.4	2.32	2.31	2.57	2.06
21.2	34.2	40.2	26.9	47.5	2.24	2.21	2.45	1.99

A significant ($p < 0.05$) interaction between Ta and feeding level was found for body weight gain and feed conversion. Restricted fed birds gained less and also less efficiently at a decreasing temperature. The data of ad libitum fed birds revealed a small positive effect of a lowered Ta on weight gain. The effect of Ta on feed conversion is much smaller in ad libitum fed broilers than in restricted fed ones.

In Table 4 the mean values for ad libitum feed and energy intake are shown.

A lower temperature leads to a higher feed and ME intake. A decrease in Ta of 1°C results in an increase in consumption of 1.3 gram or 17.5 kJ per bird per day for females and 2.1 gram or 27.0 kJ per bird per day for males. Per unit of metabolic weight the consumption of MEi changed per degree difference in Ta with 22 kJ per day, averaged over

Table 4. Feed and ME intake (MEi) between 21-28 days of age of female (F) and male (M) broilers, fed ad libitum at different temperatures.

Mean Ta (°C)	Feed intake (g per bird/day)		MEi (kJ per bird/day)		MEi (kJ/W*/day)
	F	M	F	M	F + M
18.8	90.8	106.7	1177	1383	1660
20.0	90.1	104.5	1168	1354	1630
21.2	87.6	101.7	1135	1318	1607

Table 5. Heat production between 21-28 days of age of female (F) and male (M) broilers at different ambient temperatures.

| Mean Ta (°C) | Heat production (kJ per bird/day) | | | |
| | fed ad libitum | | fed restricted | |
	F	M	F	M
18.8	722	909	581	717
20.0	692	846	573	691
21.2	673	824	551	689

females and males. In Table 5 the effect of Ta on heat production is shown. At ad libitum feeding male chickens are more affected by a lowered Ta than females.

The energy retention was plotted against energy consumption for each sex within each temperature regimen. The following model was used:
$RE/W* = aMEi/W* + b$.
The $RE/W*$ was fixed at 500 kJ/W*/day and MEi/W* calculated. Heat production was calculated as $MEi/W* - RE/W*$.

A highly significant ($p < 0.001$) difference between males and females was obtained for both MEi and HP (Table 6). It is generally known that males attain a higher protein-fat ratio than females. However, it seems unlikely that all extra heat production in males compared with females can be attributed to differences in energy costs between deposition of protein and deposition of fat (Scheele et al., 1987a).

The results suggest that total maintenance requirements, including physical activity, are higher in males than in females. Sex and feeding level were shown to be the most important factors. Because of highly

Table 6. ME intake (MEi) and heat production (HP) between 21-28 days of age of female (F) and male (M) broilers at RE = 500 kJ per W* per day at different temperatures.

Mean Ta	MEi (kJ/W*/day)		HP (kJ/W*/day)	
(°C)	F	M	F	M
18.8	1442	1558	942	1058
20.0	1402	1515	902	1015
21.2	1377	1508	877	1008

significant ($p < 0.01$) interaction between these two factors, the effects of Ta on energy intake, total deposited energy and energy deposition in protein and fat are given per sex per level of feed intake separately (Table 7).

In restricted fed birds the highest energy retention is found at the highest Ta. In ad libitum fed birds the peak in energy deposition is at the intermediate Ta. A decreased deposition of protein was found at the highest temperature. These experiments show that relatively small dif- ferences in Ta can have a significant effect not only on fat deposition but also on protein deposition.

As all three temperatures are below the Lct it is shown that a higher production of protein can be obtained at slightly lowered temperatures.

It was demonstrated by Scheele et al. (1987a) that a higher protein deposition in chickens was accompanied with high energy costs or with an increased total maintenance requirement (MEmt).

The same phenomenon was observed in experiments with growing pigs by Van der Honing et al. (1987). Scheele et al. (1987b) found that the clearly higher total maintenance requirements (MEmt) of growing rabbits kept at temperatures just below the Lct compared with rabbits kept at a higher Ta for the greatest part could be attributed to the higher level of protein deposition below Lct.

In Table 6 it was shown that at the same level of energy retention

Table 7. Energy metabolism characteristics (kJ/bird/day) in female and male broilers, fed restricted or ad libitum.

Sex	Feeding level	Mean Ta (°C)	MEi	RE	REp	REf	FREp fraction	W* 4th week
females	restr.	18.8	790	209	127	82	0.60	0.68
		20.0	789	216	125	91	0.58	0.67
		21.2	790	239	99	140	0.41	0.67
females	ad lib.	18.8	918	201	146	55	0.73	0.73
		20.0	918	227	139	88	0.61	0.73
		21.2	918	229	124	105	0.54	0.73
males	restr.	18.8	1174	452	191	261	0.42	0.73
		20.0	1174	472	199	273	0.42	0.73
		21.2	1136	463	163	300	0.35	0.72
males	ad lib.	18.8	1401	492	229	263	0.47	0.80
		20.0	1338	492	219	273	0.45	0.80
		21.2	1305	481	191	290	0.40	0.79

the heat production of male broilers is considerably higher than that of female broilers. It was suggested that this higher heat production was caused by an increased MEmt.

Calculations were carried out separately for females fed ad libitum or restricted and for males fed ad libitum or restricted. The basic model used was: RE or HP = A + aW* + b MEi. The constant A accounts for energy turn-over in broilers, which may not be directly related to W* or MEi. As MEi was constant in restricted fed birds the model for these birds was reduced to RE or HP = A + aW* (the contributions of MEi will be included in the constant A).

In ad libitum fed birds the factors MEi and W* were highly correlated. The factors A + W* + MEi should here be considered as one explanatory bloc explaining part of the variation in RE or HP. The calculations were intended to find out whether the residual variation could be significantly reduced by adding Ta, the fraction of deposited protein energy in the total deposited energy (FREp) and/or interactions to the basic model. The final model used was: RE or HP = Explanatory bloc + cTa + dFREP + interactions between covariates.

The complete regression formulas are published by Scheele et al. (1987c). In Table 8, 9 and 10 the effects of Ta, FREp and interactions between Ta and FREp and with other factors on heat production (HP) and on energy retention respectively are shown.

In Table 8 the highly significant effects of the fraction of protein energy in deposited energy on both heat production and energy retention are shown for restricted fed birds. These effects occur in both sexes. The similarity in the regression coefficients between females and males is striking. The factor Ta could not provide any further significant explanation. As all birds consumed the same amount of ME at all temperature treatments the differences in HP and RE occurring between the temperature treatments can be ascribed to the factor FREp.

It can be shown that if the fraction of retained protein energy is increased from 0.50 to 0.60 heat production per bird per day will increase by approximately $0.1 \times 221 = 22.1$ kJ. This is equal to 2.5% of the ME intake and 10% of the retained energy.

If the regression coefficient of FREp in the equation explaining HP represents the real difference in energy costs between synthesis of protein and synthesis of fat then it also can be calculated that the deposition of 1 kJ of protein energy requires the double amount of ME compared with deposition 1 kJ of fat. This difference in direct energy costs between protein and fat synthesis seems to be too large. However, also

246

Table 8. Regression coefficients, with P values of the regressors
FREp and Ta, in explaining HP and RE of female (F) and male (M)
broilers at restricted feeding (21-28 days of age).

Dependent variables	Sex	Regressors	Regression coefficient	P values
HP	F	FREp	215.6	0.000
	M	FREp	225.8	0.000
	F	Ta	-	-
	M	Ta	-	-
RE	F	FREp	- 1.0	0.000
	M	FREp	- 1.0	0.000
	F	Ta	-	-
	M	Ta	-	-

other, simultaneous, changes may occur which affect heat production,
e.g. changes in physical activity. These simultaneous changes may better explain the important effect of the factor FREp in the equations than
the direct energy costs of protein and fat synthesis.

Table 9 shows the effects of FREp and its interactions with Ta and
Ta^2 on heat production of ad libitum fed birds. Effects on energy retention (RE) are shown in Table 10. In Table 11 the effect of a changed
FREp on heat production is calculated for three different temperatures
within the experimental range.

The values given in Table 11 demonstrate that the meaning of highly
significant effects of Ta on the relation between FREp and heat production is not clear and is different for males and females. However both
equations produced comparable results showing that an increase of 0.1 in
FREp is related to an increase in heat production of approximately 80 kJ
per bird per day. In restricted fed birds the same change in FREp resulted in 20 kJ per bird per day. Assuming that the energy costs of fat
deposition in restricted fed broilers are not higher than in broilers fed

ad libitum, this may suggest that the energy costs of protein synthesis in ad libitum fed birds is four times higher than in restricted fed birds. This, however, is not likely.

These results support the hypothesis that heat loss which is related with deposition of protein determined at various conditions of feeding and environment cannot solely be considered as the direct costs of pro-

Table 9. Regression coefficients with P values of the regressors FREp, Ta and their interactions in explaining HP of female (F) and male (M) broilers at ad libitum feeding (21-28 days of age).

Dependent Variable HP	Sex	Regressors	Regression coefficient	P values
	F	FREp	$-$ 136740	0.000
	M	FREp	10636	0.018
	F	FREp.Ta	13884	0.001
	M	FREP.Ta	$-$ 1018	0.023
	F	FREp.Ta^2	$-$ 349.8	0.001
	M	FREp.Ta^2	26.4	0.019
	F	Ta.W*	$-$ 8976	0.000
	M	Ta.W*	-	-
	F	Ta^2.W*	226	0.000
	M	Ta^2.W*	-	-

tein synthesis. This experiment may show that within the range of changes in protein-fat ratio in deposited energy, especially in ad libitum fed birds, other heat producing processes which are related to FREp must have attributed to the effect of FREp.

This is in agreement with the concept of the important role of internal regulation systems in metabolism which affect both protein deposition and total maintenance requirements (MEmt) simultaneously and which may be stimulated by changes in Ta.

The effect of FREp on RE is much higher in ad libitum fed birds than in birds fed restrictedly. The important effect of FREp on RE found here is in agreement with results found by Scheele et al. (1987a) in ad libitum fed birds. This points out that in ad libitum fed birds with an increased ratio of protein to fat in deposited energy simultaneously there is an important increase in energy released as heat to the environment.

The ad libitum fed birds demonstrated a complicated but highly significant relationship between HP or RE and the explanatory variables and

Table 10. Regression coefficients with P values of the regressors FREp, Ta and their interactions in explaining RE of female (F) and male (M) broilers at ad libitum feeding (21-28 days of age).

Dependent Variable RE	Sex	Regressors	Regression coefficient	P values
	F	FREp	- 1.052	0.000
	M	FREp	- 0.995	0.000
	F	FREp.Ta	- 240.3	0.000
	M	FREP.Ta	- 118.1	0.000
	F	FREp.Ta^2	10.22	0.000
	M	FREp.Ta^2	3.826	0.000
	F	Ta	-	-
	M	Ta	11782	0.000
	F	Ta^2	-	-
	M	Ta^2	- 294.5	0.000
	F	Ta.W*	-	-
	M	Ta.W*	577.2	0.000
	F	Ta.MEi	0.8742	0.000
	M	Ta.MEi	5.35	0.000
	F	Ta^2.MEi	- 0.0235	0.000
	M	Ta^2.MEi	- 0.1262	0.000

interactions between these variables due to the variation in feed intake between birds.

As it is generally known that MEi of ad libitum fed birds is closely related to ambient temperature it could be expected that effects of Ta on

HP and on RE were incorporated in the effect of MEi. However, as is shown in Table 9 and 10, the factor Ta must at least partly be acting independently of MEi. Interactions of Ta with other factors affecting HP and RE exist.

Table 11. The effect of increase in FREp with 0.1 on heat production of ad libitum broilers at different mean temperatures. Heat production was estimated directly (HP) and by taking the difference between MEi and estimated RE (MEi-RE).

Mean Ta (°C)	The increase in HP (kJ per bird/day)	
	HP	MEi-RE
Females		
19	78	88
20	102	72
21	56	54
Males		
19	83	86
20	84	83
21	90	79

CONCLUSION AND COMMENTS

Experimental results found in literature and of own experiments show that the significance of a thermoneutral zone for fast growing birds after three weeks of age is limited.

It seems to be possible to use changes in environmental conditions, including ambient temperatures below the Lct, as stimulators in anabolism, protein accretion and growth in young broilers. Experimental results in literature and of own experiments suggest that a higher rate of growth and of protein deposition in broilers coincide with a higher level of physical activity during at least some periods within a day. The experimental results show that protein deposition is increased at a decreased Ta below the Lct. It is also shown that the energy costs accompanying this extra protein accretion are considerable.

Physical activity will reduce the energetic efficiency for production. It is therefore important to know to what extent ambient temperature and other environmental conditions, e.g. light, affect activity. This knowledge can then be used to define optimal conditions for broiler production. The fast development of modern techniques to control the environmental conditions in broilerhouses will facilitate application of such knowledge.

REFERENCES

Adams, R.L., Andrews, F.N., Gardiner, E.E., Fontaine, W.E. and Carrick, C.W., 1962. The effect of environmental temperature on the growth and nutritional requirements of the chick. Poult. Sci. 41: 588-594.

Adams, R.L. and Rogler, J.C., 1968. The effects of environmental temperature on the protein requirements and response to energy in slow and fast growing chicks. Poult. Sci. 47: 579-586.

Ahmad, M.M., Mather, F.B. and Gleaves, E.W., 1974. Feed intake response to changes in environmental temperature and dietary energy in roosters. Poult. Sci. 53: 1043-1052.

Atland, P.D., 1961. Altitude tolerance of chickens and pigeons. J. Appl. Physiol. 16: 141-143.

Barott, H.G. and Pringle, E.M., 1946. Energy and gaseous metabolism of the chicken from hatch to maturity as affected by temperature. J. Nutr. 31: 35-50.

Barott, H.G., Fritz, J.C., Pringle, E.M. and Titus, H.W., 1938. Heat production and gaseous metabolism of young male chickens. J. Nutr. 15: 145-167.

Barott, H.G. and Pringle, E.M., 1949. The effect of temperature and humidity of environment during the first 18 days after hatch. J. Nutr. 37: 153-161.

Barott, H.G. and Pringle, E.M., 1950. The effect of temperature of environment during the period from 18 to 32 days. J. Nutr. 41: 25-30.

Barott, H.G. and Pringle, E.M., 1951. Effect of environment on growth and feed and water consumption of chickens. IV. The effect of light on early growth. J. Nutr. 45: 265-274.

Beane, W.L., Siegel, P.B. and Siegel, H.S., 1965. Light environment as a factor in growth and feed efficiency of meat-type chickens. Poult. Sci. 44: 1009-102.

Berschauer, F., Gaus, G., Ehrensvärd, U. and Menke, K.H., 1980. Protein und Energie Verwertung sowie Blutharnstoff Konzentration bei Schweinen der Rassen Piétrain, Deutsches Landrasse und Deutsches Edelschwein. Pap. p 5/6.14, 31th. EAAP Meet., München, September, 1980.

Blerk, S. van, 1985. Changing disease picture for broilers. Poultry International, Dec.: 18-24.

Blokhuis, H.J., 1983. The relevance of sleep in poultry. W.P.S.A. Journal 39: 33-37.

Boshouwers, F.M.G. and Nicaise, E., 1985. Automatic gravimetric calorimeter with simultaneous recording of physical activity for poultry. Br. Poult. Sci. 26: 531-541.

Brody, S., 1945. Bioenergetics and growth. Reinhold Publ. Corp., New York, NY.

Buys, S.B. and Barnes, P., 1981. Ascites in broilers. Veterinary Record 108: 266-272.

C.A.D.P., 1986. Handboek voor de Pluimveehouderij. Consulentschap in Algemene Dienst voor de Pluimveehouderij "Het Spelderholt", Beekbergen, The Netherlands: 366 pp.

Cerniglia, G.J., Herbert, J.A. and Watts, A.B., 1983. The effect of constant ambient temperature and ration on the performance of sexed broilers. Poult. Sci. 62: 746-754.

Chaffee, R.R.J., Maghew, W.W., Drebin, M. and Cassuto, Y., 1963. Studies on thermogenesis in cold-acclimated birds. Canadian Journal of Biochemistry and Physiology 41: 2215-2220.

Charles, D.R., 1986. Temperature for broilers. W.P.S.A. Journal 42: 249-258.

Charles, D.R., Groom, C.M. and Bray, T.S., 1981a. The effects of temperature on broilers: interactions between temperature and feeding regime. Br. Poulty. Sci. 22: 475-481.

Charles, D.R., Scragg, R.H. and Binstead, J.A., 1981b. The effects of temperature on broilers: Ventilation rates for the application of temperature control. Br. Poult. Sci. 22: 493-498.

Cherry, J.A., Beane W.L. and Weaver, W.D. jr., 1980. Continuous versus intermittent photoperiod under low intensity illumination. Poult. Sci. 59: 1550-1551.

Close, W.H., 1981. The climatic requirement of the pig. In: J.A. Clark (editor), Environmental aspects of housing for animal production. Butterworths, London: 148-161.

Cowan, P.J. and Michie, W., 1978. Environmental temperature and broiler performance: the use of diets containing increased amounts of protein. Br. Poult. Sci. 19: 601-605.

Cueva, S., Sillau, H., Valuenzuela, A. and Ploog, H., 1974. High altitude indiced pulmonary hypertension and right heart failure in broiler chickens. Res. Vet. Sci. 16: 370-374.

Deaton, J.W., May, J.D., Kubena, L.F. and Reece, F.N., 1976. Physiological changes associated with acclimation of broiler chickens to constant temperatures. Int. J. Biometerol. 20: 333-336.

Deaton, J.W., Reece, F.N. and McNaughton, J.L., 1980. Effect of differing intermittent lighting regimes on broiler feed conversion. Poult. Sci. 59: 1342-1344.

Deaton, J.W., Reece, F.N. and Lott, .B.D., 1984. Effect of differing temperature cycles on broiler performance. Poult. Sci. 63: 612-615.

Dempsey, E.W. and Astwood, E.B., 1943. Determination of the rate of thyroid hormone secretion at various environmental temperatures. Endocrinology 32: 509-518.

Denbow, D.M. and Kuenzel, W.J., 1978. Gaseous metabolism of leghorns and broilers during early growth: Resting metabolic rate. Poult. Sci. 57: 1417-1422.

Emmans, G.C., 1986. Growth body composition and feed intake. A paper presented at the 28th British Poultry Breeders Round Table, September 1986.

Es, A.J.H. van, Aggelen, D. van, Nijkamp, H.J., Vogt, J.E. and Scheele, C.W., 1973. Thermoneutral zone of laying hens kept in batteries. Zeitschrift für Tierphysiologie, Tierernährung und Futtermittelkunde 32: 121-129.

Es, A.J.H. van, 1981. Poultry production in relation to energy utiliza-

tion and environment. In: C.W. Scheele and C.H. Veerkamp (editors)
World poultry production: where and how? Spelderholt Institute for
Poultry Research, Beekbergen, The Netherlands.

Farrell, D.J. and Swain, S., 1977. Effects of temperature treatments on
the energy and nitrogen metabolism of fed chickens. Br. Poult. Sci.
18: 735-748.

Freeman, B.M., 1963. Gaseous metabolism of the domestic chicken. IV.
The effect of temperature on the resting metabolism of the fowl during the first month of life. Br. Poult. Sci. 4: 275-278.

Freeman, B.M., 1970. Thermoregulatory mechanisms of the neonate
fowl. Comp. Biochem. Physiol. 33: 219-230.

Freeman, B.M., 1976. Thermoregulation in the young fowl (Gallus
domesticus). Comp. Biochem. Physiol. 54a: 141-144.

Gilbert, A.B., 1963. The effect of estrogen and thyroxine on blood
volume of the domestic cock. J. Endocrinol 26: 41-46.

Griffin, J.G. and Vardaman, T.H., 1970. Diurnal cyclic versus daily
constant temperatures for broiler performance. Poult. Sci. 49:
387-392.

Guill, R.A. and Washburn, K.W., 1972. Effects of temperature on feed
efficiency of broilers in individual cages. Poult. Sci. 51: 1229-1232.

Haller, R.W. and Sunde, M.L., 1973. effects of heat and light crossed
gradients room testing on the growth and performance of broiler type
chicks. Poult. Sci. 52: 74-80.

Harris, G.C. jr. Dodgen, W.H. and Nelson, G.S., 1974. Effects of
diurnal cyclic growing temperatures on broiler performance. Poult.
Sci. 53: 2204-2208.

Have, H.G.M. ten and Scheele, C.W., 1980. Maintenance requirements
of young chicks in relation to ambient temperature and feed composition. In: C.A. Kan and P.C.M. Simons (editors). Proceedings of the
2nd European symposium on poultry nutrition: 143-145. Spelderholt
Institute for Poultry Research, Beekbergen, The Netherlands.

Haye, U., Simons, P.C.M., Teunis, G.P. and Voorst, A. van, 1984.
Influence of housing design and lighting regime on the occurrence of
twisted legs in broilers. COVP Mededeling no. 416, Het Spelderholt,
Beekbergen, The Netherlands.

Hel, W. van der, Verstegen, M.W.A., Baltussen, W. and Brandsma,
H.A., 1984. The effect of ambient temperature on diurnal rhythm in

heat production and activity in pigs kept in groups. Int. J. Biomet. 28: 303-315.

Henken, A.M., Verstegen, M.W.A., Hel, W. van der and Knol E.F., 1982a. Effect of environmental temperature on heat production in pullets in relation to feeding level. In: A Ekern and F. Sundstøl (editors). Energy metabolism of farm animals, EAAP publication no. 29: 282-286.

Henken, A.M., Groote Schaarsberg, A.M.J. and Hel, W. van der, 1982b. The effect of environmental temperature on immune response and metabolism of the young chicken. IV. Effect of environmental temperature on some aspects of energy and protein metabolism. Poult. Sci. 62: 59-67.

Hill, F.W. and Dansky, L.M., 1954. Studies on the energy requirements of chickens. I. The effect of dietary energy level on growth and feed consumption. Poult. Sci. 33: 112-119.

Holsheimer, J.P., 1985. Technische resultaten van modelmatig voedings-onderzoek bij slachtkuikens. In: Primaire sector, Centrum voor On-derzoek en Voorlichting voor de Pluimveehouderij "Het Spelderholt", Beekbergen, The Netherlands.

Honing, Y. van der, Jongbloed, A.W., Boonzaaier, S. and Es, A.J.H. van, 1987. Sources of variation in the efficiency of utilization of me-tabolizable energy by growing pigs from compound feeds, differing in type and origin of carbohydrates and fat content. In: P.W. Moe, H.F. Tyrrel and P.J. Reynolds (editor) Energy metabolism of farm animals, EAAP publication no. 32: 264-268.

Hurwitz, S., Weiselberg, M., Eisner, U., Bartov, I., Riesenfeld, G., Sharvit, M., Niv, A. and Bornstein, S., 1980. The energy require-ments and performance of growing chickens and turkeys as affected by environmemtal temperature. Poult. Sci. 59: 2290-2299.

Huston, T.M., 1965. The influence of different environmental tempera-tures on immature fowl. Poultry Sci. 44: 1032-1036.

Huston, T.M. and Carmon, J.L., 1962. The influence of high environ-mental temperature on thyroid size of domestic fowl. Poult. Sci. 41: 175-179.

Johnson, D.E. and Crownover, J.C., 1976. Maintenances energy re-quirements of lean versus obese growing chicks at equal age and body energy. In: M. Vermorel (editor). Energy Metabolism of Farm Animals, EAAP Publ. no. 19: 121-124.

Julian, R.J. and Wilson, J.B., 1986. Right ventricular failure as a cause of ascites in broiler and roaster chickens. In: G.H.A. Borst (editor) Proc. IVth International Symposium of Veterinary Laboratory Diagnosticians, Amsterdam: 608-611.

Kampen, M. van, Mitchell, B.W. and Siegel, H.S., 1978. Influence of sudden temperature changes on oxygen consumption and heart rate in chickens in light and dark environments. J. Agric. Sci. Camb. 90: 605-609.

Kampen, M. van, Mitchell, B.S. and Siegel, H.S., 1979. Thermoneutral zone of chickens as determined by measuring heat production respiration rate and electromyographic and electroencephalographic activity in light and dark environments and changing ambient temperatures. J. Agric. Sci. Camb. 92: 219-226.

Keshavarz, K. and Fuller, H.L., 1980. The influence of widely fluctuating temperatures on heat production and energetic efficiency of broilers. Poult. Sci. 59: 2121-2158.

King, D.B. and May, J.D., 1984. Thyroidal influence on body growth. The Journal of Experimental Zoology 232: 453-460.

Kirchgessner, M., Gerum, J. and Roth-Maier, A., 1978. Körperzusammensetzung und Nährstoffansatz 3-5 Wochen alter Broiler bei unterschiedlicher Energie-und Eiweissversorgung. Archiv. für Geflügelk. 42: 6-69.

Kleiber, M., 1975. The fire of life, 2nd edn. Wiley, New York.

Kleiber, M. and Dougherty, J.E., 1934. The influence of environmental temperature on the utilization of food energy in baby chicks. J. Gen. Physiol. 17: 701-726.

Kleiber, M. and Winchester, C., 1933. Temperature regulation in the body of chicks. Proc. Soc. Exptl. Biol. Med. 31: 158-159.

Kubena, L.F., Lott, B.D., Deaton, J.W., Reece, F.N. and May, J.D., 1972. Body composition of chicks as influenced by environmental temperature and selected dietary factors. Poult. Sci. 51: 517-522.

Lamoreux, W.F. and Hutt, F.B., 1939. Variability of body temperature in the normal chick. Poult. Sci. 18: 70-75.

Leung, F.C., Taylor, J.E. and Iderstine, A. van, 1984. Thyrotropin-releasing hormone stimulates body weight gain and increases thyroid hormones and growth hormone in plasma of cockerels. Endocrinology 115: 736.

Leung, F.C., Taylor, J.E., Wien, S. and Iderstine, A. van, 1986. Purified chicken growth hormone (GH) and a human pancreatic GH-releasing hormone increase body weight gain in chickens. Endocrionology 118: 1961-1965.

MacLeod, M.G., Tullet, S.G. and Jewitt, T.R., 1979. Effects of ambient temperature on the heat production of growing turkeys. In: L.E. Mount (Editor) Energy metabolism of farm animals, EAAP publication no. 26: 257-361.

Malone, G.W., Chaloupka, G.W., Walpole, E.W. and Littlefield, L.H., 1980. The effect of dietary energy and light treatment on broiler performance. Poult. Sci. 59: 576-581.

March, B.E. and Biely, J., 1972. The effect of energy supplied from the diet and from environment heat on the response of chicks to different levels of dietary lysine. Poult. Sci. 51: 665-668.

McNaughton, J.L. and Reece, F.N., 1982. Dietary energy requirements of broilers reared in low and moderate environmental temperatures. 1. Adjusting dietary energy to compensate for abnormal environmental temperatures. Poult. Sci. 61: 1879-1884.

McNaughton, J.L., May, J.D., Reece F.N. and Deaton, J.W., 1978. Lysine requirements of broilers as influenced by environmental temperatures. Poult. Sci. 57: 57-65.

Meddis, R., 1975. On the function of sleep. Animal Behaviour 23: 676-682.

Meltzer, A., 1983. Thermoneutral zone and resting metabolic rate of and broilers. Br. Poult. Sci. 24: 471-476.

Meltzer, A., Goodman, G. and Fistul, J., 1982. Thermoneutral zone and resting metabolic rate of growing White Leghorn type chicks (Gallus domesticus). Br. Poult. Sci. 23: 383-391.

Milligan, J.L. and Winn, P.N., 1964. The influences of temperature and humidity on broiler performance in environmental chambers. Poult. Sci. 43: 817-824.

Misson, B.H., 1982. The thermoregulatory responses of fed and starved 1-week old chickens (Gallus domesticus). J. Therm. Biol. 7: 189-192.

Morrison, W.D. and McMillan, I., 1985. Operant control of the thermal environment by chicks. Poult. Sci. 64: 1656-1660.

Morrison, W.D. and McMillan, I., 1986. Response of chicks to various environmental temperatures. Poult. Sci. 65: 881-883.

Mount, L.E., 1979. Adaptation to thermal environment: Man and his productive animals. Arnold, London.

Nichelmann, M., Hewald, B. and Grune, B., 1983. Thermoregulatorische Wärmeproduktion bei Legehybriden - Beziehungen zwischen Lebensälter und Wärmeproduktion. Arch. Exper. Vet. Med., Leipzig 37: 341-352.

Olson, D.W., Sunde, M.L. and Bird, H.R., 1972. The effect of temperature on ME determination and utilization by the growing chick. Poult. Sci. 51: 1915-1922.

O'Neill, S.J.B., Balnave, D. and Jackson, N., 1971. The influence of feathering and environmental temperature on the heat production and efficiency of utilization of metabolizable energy by the mature cockerel. J. Agric. Sci. Camb. 77: 293-305.

O'Neill, S.J.B. and Jackson, N., 1974. The heat production of hens and cockerels maintained for an extended period of time at a constant environmental temperature of 23°C. J. Agric. Camb. 82: 549-552.

Osbaldiston, C.W., 1968. The effect of air temperature on food consumption of chickens. Br. Vet. J. 124: 110-115.

Ota, H., 1967. The physical control of environment for growing and laying birds. In: T.C. Carter (editor). Environmental Control in Poultry Production: 3-14. Edinburgh, Oliver and Boyd.

Pethes, G., Scanes, C.G. and Rudas, P., 1979. Effect of synthetic thyrotropin releasing hormone on the circulating growth hormone concentration in cold and heat stressed ducks. Acta Vet. Acad. Sci. Hung. 27: 175-177.

Prince, R.P., Whitaker, J.H., Matterson, L.D. and Luginbuhl, R.E., 1965. Response of chickens to temperature and relative humidity environments. Poult. Sci. 44: 73-77.

Quarles, C.L. and King, H.F., 1974. Evaluation of ammonia and infectious bronchitis vaccination stress on broiler performance and carcass quality. Poult. Sci. 53: 1592-1596.

Reece, F.N. and Lott, B.D., 1980. Effect of carbon dioxide on broiler chicken performance. Poult. Sci. 59: 2400-2402.

Reece, F.N. and Lott, B.D., 1982. The effect of environmental temperature on sensible and latent heat production of broiler chickens.

Poult. Sci. 61: 1590-1593.

Reece, F.N. and McNaughton, J.L., 1982. Effects of dietary nutrient density on broiler performance at low and moderate environmental temperatures. Poult. Sci. 61: 2208-2211.

Ring, G.C., 1939. Thyroid stimulation by cold. Amer. J. Physiol. 125: 244-250.

Romijn, C., 1950. Stofwisselingsonderzoek bij de kip proeven met Noord-Hollandse Blauwen. Invloed van verschillende factoren op de calorieproduktie. T. Diergeneesk. 75: 719-746.

Romijn, C., 1954. Development of heat regulation in chicks. 10th World's Poultry Congress, Section Papers: 181-184.

Romijn, C. and Lokhorst, W., 1961. Climate and poultry. Heat regulation in the fowl. T. Diergeneesk. 86: 153-172.

Scanes, C.G. and Lauterio, T.J., 1984. Growth hormone: Its physiology and control. Journ. of experimental zoology 232: 443-452.

Scheele, C.W., Es, A.J.H. van and Have, H.G.M. ten, 1987a. The efficiency of energy utilization by broiler chickens in relation to dietary composition. In: P.W. Moe, H.F. Tyrrell and P.J. Reynolds (editors). Energy metabolism of farm animals, EAAP publication no. 32: 232-326.

Scheele, C.W., Broek, A. van and Hendricks, F.A., 1987b. Maintenance requirements and energy utilization of growing rabbits at different environmental temperatures. In: P.W. Moe, H.F. Tyrrel and P.J. Reynolds (editors). Energy metabolism of farm animals, EAAP publication no. 32: 206-210.

Scheele, C.W., Hel, W. van der, Ehlhardt, D.A. and Schagen, P.J.W. van, 1987c. Energy metabolism and deposition of protein in broilers kept at temperatures below the thermoneutral zone (in preparation).

Siegel, H.S. and Drury, L.N., 1970. Broiler growth in diurnally cycling temperature environments. Poult. Sci. 49: 238-244.

Siegel, H.S., 1977. Effects of temperature and light on growth. In: K.N. Boorman and B.J. Wilson (editors). Growth and poultry meat production: 187-226.

Sillau, A.H., Cueva, S. and Morales, P., 1980. Pulmonary arterial hypertension in male and female chickens at 3300 M. Pfluegers Arch. 386: 269-275.

Simons, P.C.M. and Haye, U., 1985. Intermittent lighting has a positive effect on twisted legs. Poultry 1: 34-37.

Singh, A., Reineke, E.P. and Ringer, R.K., 1968. Influence of thyroid status of the chick on growth and metabolism, with observations on several parameters of thyroid function. Poult. Sci. 47: 212-218.

Veerkamp, C.H., 1986. Good handling gives better yield. Poultry 2: 30-33.

Verstegen, M.W.A. and Hel, W. van der, 1974. The effects of temperature and type of floor on metabolic rate and effective temperature in groups of growing pigs. Anim. Prod. 18: 1-11.

Verstegen, M.W.A., Hel, W. van der and Willems, G.J.M., 1977. Growth depression and food requirement of fattening pigs at low environmental temperatures when housed either on concrete slats or straw. Anim. Prod. 24: 253-259.

Verstegen, M.W.A., Hel, W. van der, Duyghuisen, R. and Geers, R., 1986. Diurnal variations in the thermal demand of growing pigs. J. Therm. Biol. 11: 131-135.

Verstegen, M.W.A., Hel, W. van der, Brandsma, H.A., Henken, A.M. and Bransen, A.M., 1987. The Wageningen respiration unit for animal production research: a description of the equipment and its possibilities. In: M.W.A. Verstegen and A.M. Henken (editors). Energy metabolism of Farm animals with special reference to effects of housing, stress and disease. Martinus Nijhoff publishers, Dordrecht.

Vogt, J.E., Bouwma, K., Weerd, J. van, Boekholt, H.A., Adrichem, P.W.M. van, and Es, A.J.H. van, 1980. The effect of different light regimens on feed intake, mobility, weight gain and energy and nitrogen retention of growing chickens. In: L.E. Mount (editor). Energy metabolism of farm animals, EAAP publication no. 26: 455-459.

Wal, H. van der, Verstegen, M.W.A., and Hel, W. van der, 1976. Protein and fat deposition in selected lines of mice in relation to feed intake. In: M. Vermorel (editor). Energy metabolism of farm animals, EAAP publication no. 19: 125-129.

Wekstein, D.R. and Zolman, J.F., 1970. Homeothermic development of young scaleless chicks. Br. Poultr. Sci. 11: 399-402.

Wekstein, D.R. and Zolman, J.F., 1971. Cold stress regulation in young chickens. Poult. Sci. 50: 56-61.

Wenk, C., 1980. Zur Verwertung der Energie beim wachsenden, monogastrischen, landwirtschaftlichen Nutztier. Habilitationsschrift ETH Zürich: 107 pp.

Wenk, C. and Es, A.J.H. van, 1980. Untersuchungen über den Stoff-
und Energiewechsel wachsender Küken unter besondere Berücksichti-
gung der Aktivität. Zeitsschrift für Tierphysiologie, Tierernährung
und Futtermittelkunde 43: 241-254.

Wilkie, D.R., 1960. Thermodynamics and the interpretation of biological
heat measurements. Progress in Biophysics. 10: 260-298.

Wilson, W.O., 1948. Some effects of increasing environmental temperature on pullets. Poult. Sci. 27: 813-817.

HEAT TOLERANCE OF ONE-DAY OLD CHICKENS WITH SPECIAL REF-
ERENCE TO CONDITIONS DURING AIRTRANSPORT[1]

A.M. HENKEN, W. VAN DER HEL, A. HOOGERBRUGGE AND
C.W. SCHEELE

ABSTRACT

To optimize airtransport quality for one-day old chickens a research
was carried out, which consisted of two parts.

In the first part (24 trials with 128 chickens each) it was determined
how heat production, weight loss, water loss and yolk loss are related to
the dry and wet bulb temperature (T_d and T_w respectively). Heat production increased with T_d (28.8 to 39.5°C) and T_w (22.0 to 32.7°C).
The effective temperature (T_{eff}) could be calculated as $0.81 * T_d + 0.19$
$* T_w$. Losses in weight and water increased at higher T_d. However, T_w
had an opposite effect. Yolk loss decreased at higher T_{eff}. Mortality occurred at a T_{eff} above about 37°C. Based on performances obtained afterwards during a recovering period of 14 days at normal conditions, it
was shown that the T_d during the first two days should preferably remain below about 37°C (T_{eff} about 35°C).

In the second part (4 trials with about 16.000 chickens each) it was
determined which temperatures normally can be expected inside the boxes
loaded on a pallet for export. It was also determined whether box type
and loading configuration are important in this respect. The results
showed that the temperatures inside the boxes loaded according to rou-

---

[1] This research was performed at the department of Animal Husbandry,
Agricultural University, Wageningen, in coöperation with: 1. Institute
for Poultry Research, Het Spelderholt, Beekbergen; 2. Department of
Veterinary Physiology, University of Utrecht; 3. KLM Royal Dutch
Airlines, Amstelveen.

M. W. A. Verstegen and A. M. Henken (eds.), Energy Metabolism in Farm Animals.
ISBN 0–89838–974–7, © 1987, Martinus Nijhoff Publishers, Dordrecht.

tine procedures are 8 to 14°C higher than that of the environment wherein the pallet is placed. The box temperatures depended not only on the
outside temperature but also on the type of box and method of spacing.
Boxes with two compartments for about 20 chickens each and vertical
spacing all around each stack gave the lowest increase in box temperature, the smallest variation in temperature between boxes, the lowest
weight loss and the least mortality as compared to other box types and
methods of spacing.

INTRODUCTION

In 1985 the total number of one-day old chickens exported form the
Netherlands amounted to 79×10^6, representing a value of DFL 123 million (Van Leeuwen, 1986). About 86% of the chickens was exported to
countries outside the European Community. These had to be transported
by air. Mean flying time and shipment value are increasing due to export
to more remote countries and of more (grand-)parent stock. Therefore
cargo handling and transport conditions for birds are attracting more
and more attention. Especially cargo compartment temperature was noted
as an essential factor (Müller, 1985). However, the temperature in the
chick boxes depends not only on the cargo compartment temperature, but
probably also on the loading configuration of the boxes on the pallets
(Hoogerbrugge and Ormel, 1982). These latter authors also noted a lack
of knowledge in this field, considering it a main reason for a lot of acci-
dents. Therefore a research program was developed, which consisted of
two parts: 1. effects of thermal conditions on some physiological parameters; 2. effects of box type and loading configuration on the difference between ambient conditions in cargo compartment and box.

MATERIALS AND METHODS

Part 1.   Effects of thermal conditions

A series of 24 trials was performed, using 128 unsexed one-day old
Hypeco chickens (meat type) in each. The trials started immediately after arrival of the chickens at the experimental facilities. At that time the

chickens were 2 to 12 hrs old and weighed about 46 g. Two identical climate-respiration chambers of 1.8 m³ each were used (Bransen and Kneepkens, 1982; Verstegen et al., 1987). Four cardboard 2-compartment chickboxes of 0.46 × 0.23 × 0.13 m were placed next to each other in each chamber on the perforated bottom. Each box contained 30 chickens, divided over the two compartments. The remaining eight chickens served as a sample for analysis of initial body composition. To simulate transport conditions no water and feed were supplied. Light was on continuously to allow visual observation. The climatic conditions applied in each trial are presented in Table 1. The dry bulb temperature (T_d) ranged from 28.8 to 39.5°C, while the relative humidity (RH) varied between 30 and 87% (wet bulb temperature T_w: 22.0 to 32.7°C). In order to enable air flowing from around the boxes into the boxes the lid was removed and replaced by wire netting. The airstream in the chambers is downwards from above. The conditions measured just above the wire net-

Table 1.   Climatic conditions tested in Part 1 (T_d = dry bulb temperature, T_w = wet bulb temperature, RH = relative humidity).

Chamber 1				Chamber 2			
Trial no.	T_d °C	T_w °C	RH %	Trial no.	T_d °C	T_w °C	RH %
1	30.3	22.6	52	2	32.9	26.1	60
3	33.7	22.0	36	4	34.7	23.2	38
5	34.9	22.2	34	6	35.9	22.2	30
7	28.8	23.3	64	8	33.7	24.2	46
9	32.6	28.3	72	10	31.1	29.1	86
11	34.8	29.9	70	12	36.5	29.2	58
13	36.1	31.1	70	14	33.6	30.1	78
15	38.7	29.8	52	16	31.9	28.6	79
17	34.5	25.5	48	18	37.0	27.8	49
19	38.2	31.1	61	20	39.5	31.9	59
21	38.5	32.7	67	22	34.3	28.7	66
23	34.2	32.2	87	24	38.5	26.6	40

ting are therefore considered to be similar to those encountered by the chickens. Each trial lasted 46 hrs, divided over two subsequent periods of 23 hrs each. Heat production, based on continuous measurements of O_2-consumption and CO_2-production during each period, was determined according to Romijn and Lokhorst (1961). Mean body weight was determined at the start and also at the end of each period. At the end of each trial a sample of 8 chickens was taken for body composition determination. This sample and the initial sample were analyzed for yolk percentage and water percentage of yolk and remainder separately. The results of these analyses were used to determine the origin (water and/or dry matter) of the weight losses observed, separated over yolk and the rest of the body.

The data on heat production of the first period and on total weight and water losses were used to fit a sigmoid curve via BMDPAR (Dixon et al., 1983). Preliminary analysis showed that such curves could be fitted at least as good as multiple regression functions. The other data (yolk loss and the difference in heat production between the first and second period) were subjected to multiple regression analysis via BMDP1R (Dixon et al., 1983).

Part 2.   Effect of box type and loading configuration

A series of 4 trials (no. 25 to 28) was performed, using about 16.000 male one-day old Euribrid chickens (layer type) in each. The trials started immediately after arrival of the chickens, which were 2 to 12 hrs old at that time and weighed about 40 g. One large climate-respiration chamber (80 m³) was used (Verstegen et al., 1987). To obtain a good simulation of practical conditions the following measures were taken: 1. Only box types commonly used in practice were chosen; 2. Boxes were stacked on an aluminium pallet (l × w = 3,18× 2.24 m) normally used in practice. About 200 cardboard 4-compartment boxes (about 16.000 one-day old chickens) can be placed on such a pallet, normally in 4 rows containing 6 stacks with 8 layers (Figure 1); 3. Adjustment of air flow in the chamber to a similar level as measured during flight in the lower forward cargo compartment of a Boeing 747 (air flow levels were measured in the spacings between boxes on the pallet); 4. Asking professional people to build the stack; 5. Observation of routine procedures and performing measurements during normally scheduled flights with

Table 2. Objectives and thermal conditions in the trials of Part 2 .

Trial no.	Thermal conditions T_d (°C)	T_w (°C)	Objectives
25	17.8-31.9	12.2-27.1	to determine the temperature profile on a pallet, as measured within boxes stacked according to routine procedures for air-transport at KLM.
26	20.0-28.0	12.2-19.0	to determine the effect of box type on the temperature profile. Four different cardboard box types were used: A. 0.458 x 0.462 x 0.11 m (4 compartments) B. 0.475 x 0.240 x 0.13 m (2 compartments) C. 0.465 x 0.435 x 0.12 m (4 compartments) D. 0.450 x 0.445 x 0.12 m (4 compartments)
27	20.0-28.0	12.2-19.0	to determine the effect of method of spacing on the temperature profile. Four spacings were tested: I spacing horizontally every other two layers. II spacing vertically around each stack of 8 boxes. III spacing vertically only parallel to the air flow. IV as III but with a forced air stream by building a plastic funnel between air inlet wall and the surface of the pallet at that side.
28	17.8-31.9	12.2-27.1	to test whether use of the best box types and spacing can ameliorate the conditions inside the boxes when compared to those measured in trial 25.

one-day old chickens.

The objectives of each trial of Part 2 and treatments applied are presented in Table 2. At about 30 locations around and in the boxes the dry bulb temperature was measured. Those locations were evenly distributed across layers and rows. Also the wet bulb temperature was measured at some locations. Before and after each trial mean body weight per box was determined. At the end also the number of deads was monitored per box. Each trial lasted about 48 hrs.

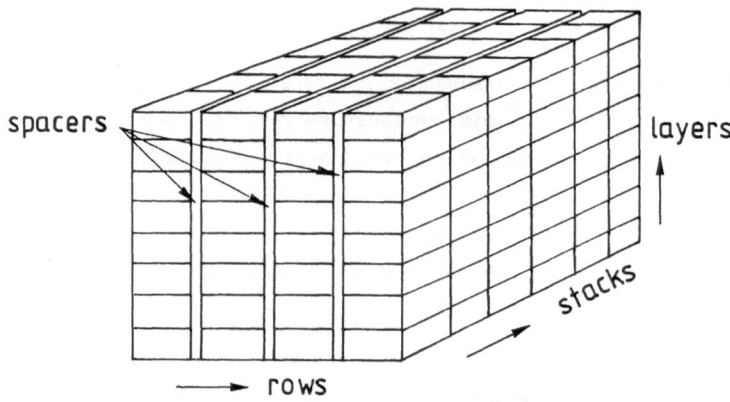

Figure 1. The normal loading configuration of boxes on a pallet.

Trial 25

The loading configuration of the pallet is shown in Figure 2. Type D boxes (Table 2) were used, which were stacked in 4 rows of 6 x 8 (l x h) boxes (198 boxes, 15840 chickens). One row was stacked 9 high. Between rows wooden spacers of 0.085 m were used. The temperature regime applied in the chamber is presented in Table 3. This regime is based on measurements during a flight with chickens from Amsterdam to Kuala Lumpur. The last (11th) period simulated the period between landing at and further transport by car from the airport in Malaysia. Between the 10th and 11th period mean chicken weight per box was determined.

Trial 26

The loading configuration of the pallet is shown in Figure 3 (192

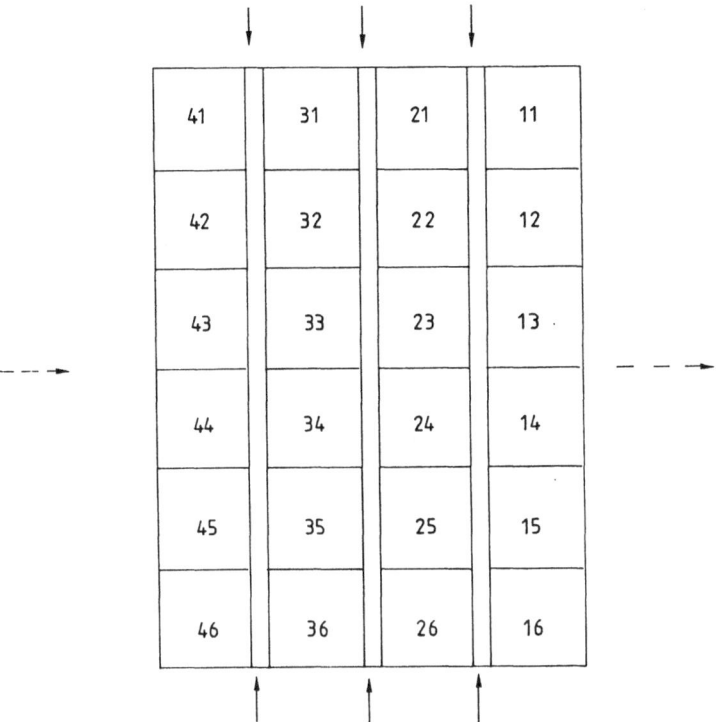

Figure 2.   Ground-plan of loading configuration in trial 25. One layer is
shown. Eight layers were used. Spacing (0.085 m) is indicated by →, direction of main air flow by --→ and box positions by numbers.

boxes,   15360 chickens). The   thermal   conditions applied   are shown   in
Table 4. Moderate conditions were chosen to reduce mortality rates. A
high mortality rate in some or a specific box type could affect the thermal measurements and thus confound the comparison between box types.

Each box type (A to D: Table 2) occupied one quarter of the pallet.
Every eleven hours box types were exchanged between quarters. This
was repeated four times (periods 1 to 3, 4 to 6, 7 to 9 and 10 to 12 respectively). Thus, each type was tested in each quarter.

Trial 27

Two hundred type D boxes were used containing 80 chickens each
(Table 2). The thermal conditions applied were identical to those in trial
26 (Table 4). During each time 11 hrs (periods 1 to 3, 4 to 6, 7 to 9
and 10 to 12 respectively) one of the four methods of spacing was tested

Table 3. Thermal conditions applied in the chamber in trial 25 .

Period no.	Length hrs	Td °C	T_w °C	RH %
1	3	17.8	12.2	56
2	3	19.5	13.6	50
3	4	23.7	15.5	41
4	4	28.5	18.0	35
5	4	25.0	16.2	42
6	3	25.1	16.1	39
7	4	19.7	14.1	52
8	4	27.2	17.2	36
9	4	31.3	20.0	34
10	4	22.9	16.1	50
11	6	31.9	27.1	69

(Table 2). Therefore the whole pallet was restacked four times.

Trial 28

In the trials 26 and 27 a moderate temperature regime was tested, while in trial 25 a more severe regime was used with higher temperatures (Tables 3 and 4). In trial 28 the same temperature regime as in trial 25 was adopted. Based on the results of trial 26 and 27, it was decided to test in trial 28 two box types (B and D: Table 2) with one method of spacing (II: Table 2). The loading configuration of the pallet is presented in Figure 4. As shown the type B boxes were placed two by two this time to save space. Both box types were stacked 10 high. Again the chickens were also weighed between period 10 and 11, as well as before and after the trial. It was tested whether these two box types combined with vertical spacing all around each stack could ameliorate the conditions inside the boxes at severe outside conditions. The results of trial 25 were looked upon as basic values, which had to be improved. As was noticed earlier, in trial 25 only routine procedures were used.

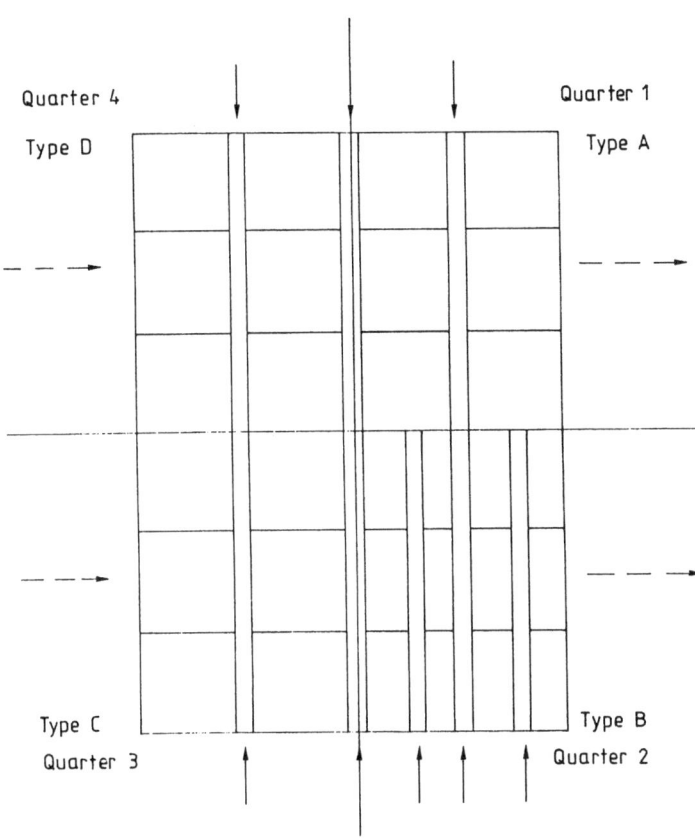

Figure 3. Ground-plan of loading configuration in trial 26
(spacing, →; air flow, --→).

Table 4. Thermal conditions applied in the chamber in trial 26 and 27.

Period	Length hrs	Td^* °C
1, 4, 7, 10	2	20
2, 5, 8, 11	4	28
3, 6, 9, 12	5	24

* RH = 40-44%

270

General

The data obtained in the four trials of Part 2 thus refer to: climatic conditions around and in pallet; weight loss; mortality. Measurements were
performed per box. Therefore, a profile of parameter values could be
made over the whole pallet.

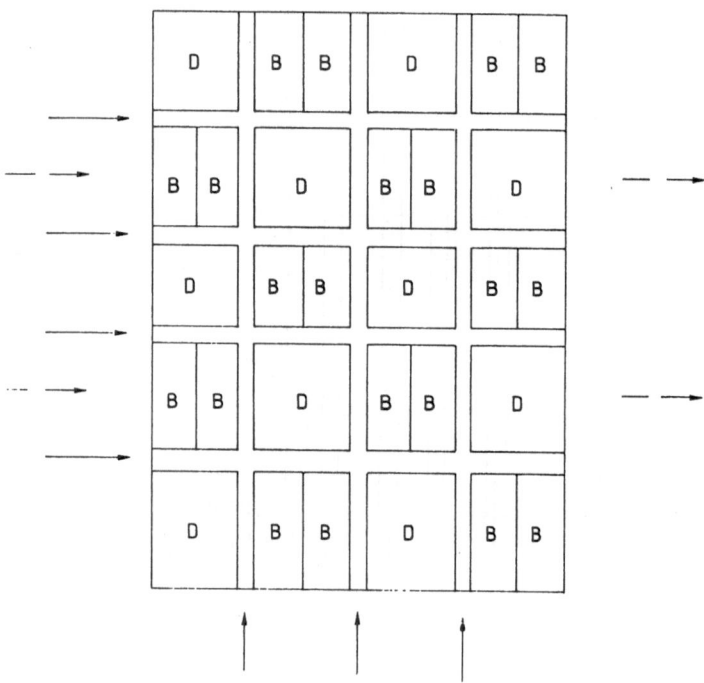

Figure 4.   Ground-plan of loading configuration in trial 28 (spacing, →;
air flow, --→; box type, B and D).

RESULTS AND DISCUSSION

Part 1.   Effects of thermal conditions

Weight loss and water loss

The results of the Part 1 trials are presented in Table 5.
Weight loss and water loss increased when the dry bulb temperature was
raised from 28.8 tot 39.5°C. From the data in Table 5 the following equations were derived:

Table 5. Losses in weight and water as percentages of initial body weight (μ ± SD, 46.18 ± 0.93 g, n=24) and heat production in relation to T_d and T_w in Part 1.

T_d	T_w	Whole body loss		Yolk loss*	Remainder loss	Heat production kJ. animal^{-1}.hr^{-1}	
		Weight	Water	Weight	Water	1e period	2nd period
°C	°C	%	%	%	%		
28.8	23.3	12.15	8.25	7.06	5.86	1.17	1.15
30.3	22.6	12.63	9.35	6.21	6.02	1.21	1.11
31.1	29.1	10.53	8.05	6.33	4.89	1.21	1.14
31.9	28.6	10.98	7.63	5.96	4.60	1.20	1.08
32.6	28.3	10.91	8.18	6.92	3.89	1.19	1.13
32.9	26.1	11.86	8.73	6.05	5.55	1.15	1.11
33.6	30.1	10.41	6.09	6.52	2.20	1.23	1.10
33.7	22.0	12.56	8.44	6.70	5.42	1.14	1.07
33.7	24.2	12.15	9.16	6.54	6.89	1.20	1.09
34.2	32.2	12.82	9.15	6.04	8.40	1.28	1.16
34.3	28.7	11.10	8.26	6.38	5.30	1.26	1.17
34.5	25.5	15.49	12.12	3.76	7.24	1.29	1.18
34.7	23.2	12.40	7.51	5.70	6.30	1.17	1.11
34.8	29.9	15.72	11.03	3.68	6.99	1.30	1.16
34.9	22.2	18.63	15.55	5.74	13.32	1.26	1.12
35.9	22.2	16.53	13.94	6.53	11.12	1.30	1.17
36.1	31.1	18.44	13.71	5.05	10.55	1.36	1.11
36.5	29.2	16.52	12.26	6.37	7.01	1.30	1.16
37.0	27.8	19.02	15.23	5.28	10.36	1.33	1.22
38.2	31.1	20.34	16.51	3.77	14.40	1.33	1.05
38.5	32.7	20.53	17.44	4.52	14.38	-	-
38.5	26.6	22.21	17.96	4.71	15.90	1.42	1.12
38.7	29.8	16.81	13.68	3.23	10.09	1.35	0.89
39.5	31.9	20.06	16.53	3.99	12.68	1.41	0.98

* Mean initial yolk weight = 4.40 ± 0.91 g (dry matter content is about 50%)

Weight loss of the whole body:

$$y\ (\%) = \frac{10.18}{1+2.81 * e^{-10.18 * 0.10 * ((T_d-29) - 0.06 (T_w-22))}} + 10\ (R^2=0.82)$$

Water loss of the whole body:

$$y\ (\%) = \frac{9.86}{1+285.14 * e^{-9.86 * 0.10 * ((T_d-29) - 0.10 (T_w-22))}} + 7(R^2=0.81)$$

Table 6. Weight loss, water loss and heat production at four chosen combinations of T_d and T_w as calculated with the derived formulae.

T_d	T_w	RH	Whole body		Heat production
			Weight loss	Water loss	
°C	°C	%	%	%	kJ.an^{-1}.hr^{-1}
32	24	52	18.854	7.518	1.167
32	28	74	18.846	7.355	1.183
36	24	38	20.154	14.307	1.293
36	28	55	20.147	13.495	1.322

As an example these two equations were solved for four chosen combinations of T_d and T_w (Table 6). As can be seen T_d and T_w have opposite effects on weight loss and water loss. Apparently the chickens cope with higher T_d by evaporating water. At a higher T_w however this possibility to loose heat by evaporation is decreased. Yolk loss decreased at higher T_d (yolk loss (%) = - 0.002*T_d*T_w - 0.61 * 10^{-4}* T_d^3 + 9.66 (R² = 0.52)). Thus, the weight loss of the whole body is strongly related to water loss of the remainder (body without yolk). The same phenomenon can be seen in the results of Macleod (1982) and Van der Hel et al. (1987).

Heat production

During the first period more heat was produced at a higher T_d. The following relation between the first period heat production on one hand

and T_d and T_w on the other hand gave the best fit:

Heat production during the first period ($kJ.an^{-1}.hr^{-1}$):

$$y = \frac{0.212}{1 + 170.90 * e^{-0.212* (3.71*(T_d-29) + 0.87*(T_w-22))}} + 1.15 \ (R^2=0.78)$$

This equation was solved for the same four combinations of T_d and T_w as for weight loss and water loss (Table 6). Not only a higher T_d, but also a higher T_w resulted in a higher heat production. The relative contribution of T_w in comparison with that of T_d is only about 0.19. This is much less than the 0.36 for 10-month old laying hens reported by Egbunike (1979). The latter value was based on respiratory rate and rectal temperature as measured during daytime. The dry and wet bulb temperatures applied in the Part 1 trials were used to calculate the effective temperature according to both estimates (T_{eff} = 0.81 * T_d + 0.19 * T_w and T_{eff} = 0.64 * T_d + 0.36 * T_w respectively). Estimates of the effective temperature using the latter equation were on average 1.25°C lower than those estimated with the former equation. The difference in relative contribution of T_d and T_w between the two equations may be due to the difference in age. One-day old chickens act as heterotherms outside the zone of minimal metabolism (Mission, 1976), while the thermoregulatory capacity develops rapidly after hatching and is relatively mature at 2 to 3 weeks of age (Osbaldiston, 1968; Wekstein and Zolman, 1970; Freeman, 1976). This development may induce an increase in dependence on T_w relative to that on T_d, especially at higher T_d where older birds try to maintain their body temperature constant.

Heat production during the second 23 hr period was lower than that during the first period (Table 5). It showed to be scattered around that lower level independent of T_d and T_w except at very high T_d where it fell to values below 1.0 $kJ.an^{-1}.hr^{-1}$. Thus, the difference in heat production between the first and second period increased more or less exponentially with T_d and T_w (H_1-H_2 = 2.18 - 0.10 * T_d + 0.75 * 10^{-6} * T_d^4 + 0.20 * 10^{-6} * T_w^4 + 0.20 * T_d/T_w (R^2 = 0.84)). The chickens were not able to cope anymore with the higher temperatures on the second day and became very weak. Visual observation showed some animals to be in coma and/or dying (at T_d > 38.2°C).

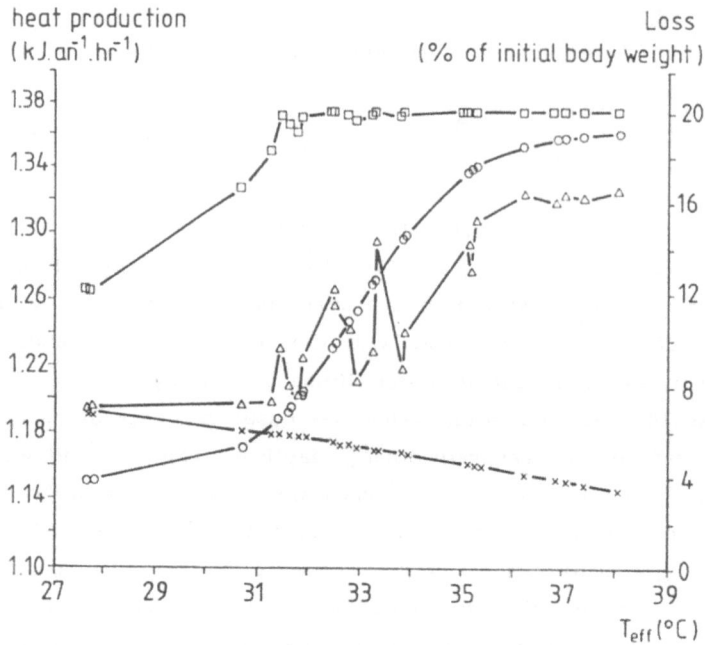

heat production
$(kJ.an^{-1}.hr^{-1})$

Loss
(% of initial body weight)

□ weight loss
△ water loss
× yolk loss
○ heat production

Figure 5. Heat production, weight loss, water loss and yolk loss as affected by $T_{eff}(= 0.81 * T_d + 0.19 * T_w)$.

General

The relation between weight loss, water loss, yolk loss and heat production with T_{eff} $(0.81 * T_d + 0.19 * T_w)$ is shown in Figure 5. It can be concluded that the chickens cope with heat by evaporating water. However, a limit seems to exist in the total amount of water available for this purpose. They apparently succumb when about 17% of their initial body weight has been evaporated (and about 3% oxidated). A higher T_w increases the thermal demand of the environment upon the chicken, because heat production raises. At the same time, however, it decreases the possibility for the chickens to cope with that demand, because less water can be evaporated. This resulted in mortality above about T_{eff} = 37°C.

Table 7. Mean performances of chickens, which were subjected to high temperatures during the first two days, but given the opportunity to re-cover at normal conditions for 14 days afterwards. The controls were kept at normal conditions from day 0 onwards and had free access to feed and water.

Parameters	Controls	Experimental groups			
	T_d=±30°C	T_d (°C) between:			
	RH=±40%	28-32	32-35	35-37	37-40
Body weight (g) at:					
day 0	45.8	46.3	45.9	46.6	46.2
day 2	61.9	41.1	39.7	39.0	36.9
day 14	313.5	-	-	-	-
day 16	-	323.1	315.3	306.2	253.8
Body weight change (g) between:					
day 0 and day 2	+16.1	-5.2	-6.4	-7.6	-9.3
day 0 and day 14	267.7	-	-	-	-
day 2 and day 16	-	282.0	275.6	267.2	216.9
Feed intake (g) between:					
day 0 and day 14	386.0	-	-	-	-
day 2 and day 16	-	395.4	384.3	366.2	292.0
Mortality (%) between:					
day 0 and day 14	0.7	-	-	-	-
day 2 and day 16	-	2.5	5.2	5.7	50.0

After the trials of Part 1 the chickens were given the opportunity to recover at normal conditions (T_D = about 30°C, RH = about 40%) with feed and water ad libitum. Groups, which were held at the higher tem-

peratures during the first 48 hrs, still had a higher mortality rate afterwards. They consumed less feed and weighed much less at two weeks of
age than groups held at lower temperatures or controls (Table 7).
Chickens kept at temperatures above T_d = 37°C (or T_{eff} = about 35°C)
during the first two days gained less weight than controls (216.9 vs
267.7 g between day 2 to 16 and day 0 to 14 respectively). This indicates that weight loss during transport (or water loss) may be a more
appropriate parameter to qualify transport conditions than only mortality
rate, especially in broilers.

Part 2.   Effects of box type and loading configuration.

Data obtained in a pallet loaded according to routine procedures
(trial 25)

The results of the temperature (T_d) measurements are shown in Figure 6, 7 and 8 for boxes on layer 2, 4 and 7 respectively. The box numbers refer to the positions as indicated in Figure 2. The temperature of
the chamber itself, as measured outside the pallet, is drawn in black. To
facilitate interpretation a dotted line is shown at 37°C. This temperature
is considered to be near the upper critical temperature (Freeman, 1963;
Misson, 1976; Macleod, 1982; Van der Hel et al., 1987). The mean temperature at layer 2, 4 and 7 was 31.2, 32.5 and 33.7°C respectively.
Within a layer the temperature in central boxes is on average 10°C
higher than that in peripheral boxes. The greatest difference measured
between chamber and box temperature was about 14°C. Thus, a large
difference may exist depending on the position of the specific box on the
pallet. Due to the large increase in box temperature above the chamber
level, the relative humidity remained low. It actually was always lower
than that of the chamber itself (< 48%).

The losses in body weight and mortality rate are given in Figure 9,
10 and 11 per row, layer and stack (across rows) respectively. Losses
in body weight were highest in central boxes in layers just below the
top. The same is true with respect to mortality rate. Relatively the most
weight was lost in the last period (11: Table 3) at the high temperature
(about 0.26 g.hr^{-1} in comparison to about 0.15 g. hr^{-1} during the first
10 periods). Overall the chickens lost 7.2 g, which is about 18.2% of
their initial body weight. Overall mortality was about 16.3%.

Figure 6. Temperatures (dry bulb) as measured in boxes
at layer 2 in trial 25.

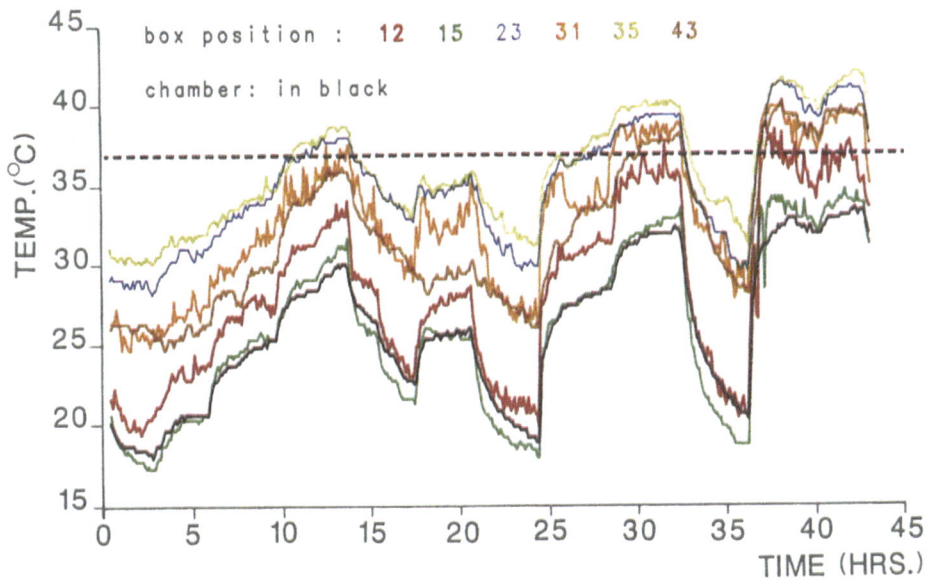

Figure 7. Temperatures (dry bulb) as measured in boxes
at layer 4 in trial 25.

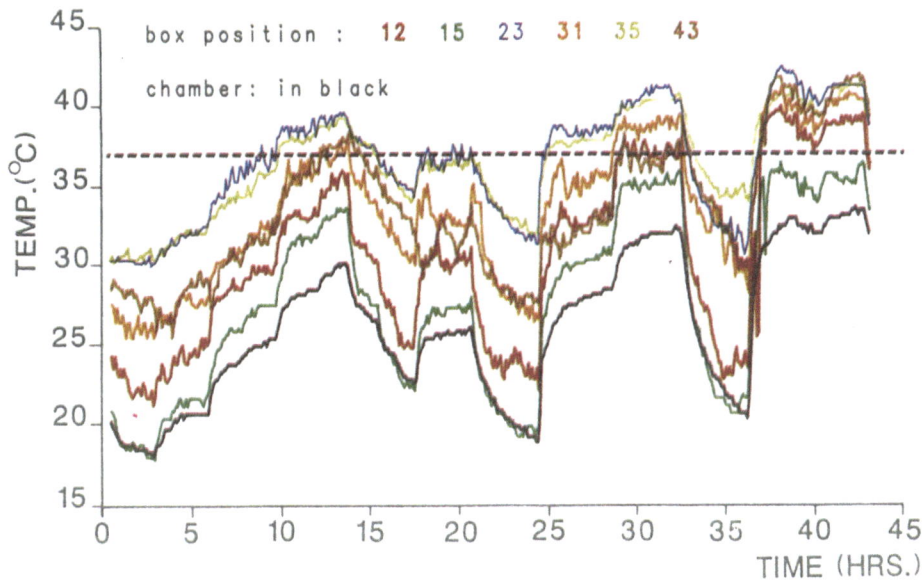

Figure 8. Temperatures (dry bulb) as measured in boxes
at layer 7 in trial 25.

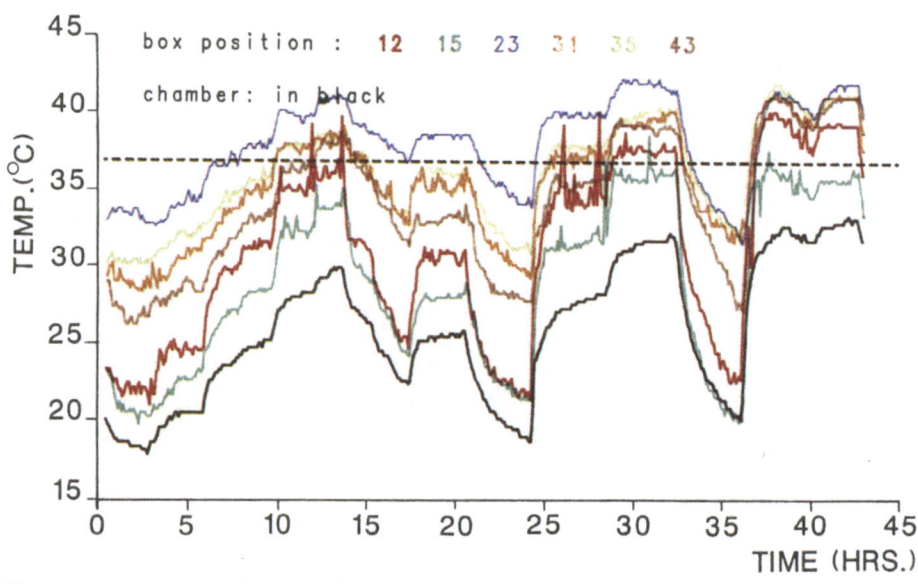

Effects of box type (trial 26)

The results of the measurements on box temperature are presented in Table 8. At both layers the lowest temperature was measured in type B

Table 8. Mean temperature in the tested box types as measured at layer 2 and 6 (trial 26)

Box type	Mean temperature (°C)		ΔT
	Layer 2	Layer 6	(6-2)
A	30.3	33.5	3.2
B	29.5	31.7	2.2
C	30.6	32.8	2.2
D	30.7	32.8	2.1

boxes. The difference in temperature between boxes at layer 2 and 6 was highest for type A boxes. Type C and D gave similar results. When only the measurements at 28°C (periods 2, 5, 8 and 11: Table 4) are taken into account, then the layer difference in temperature in box type D was smaller than in box type C (1.9 vs 2.2°C).

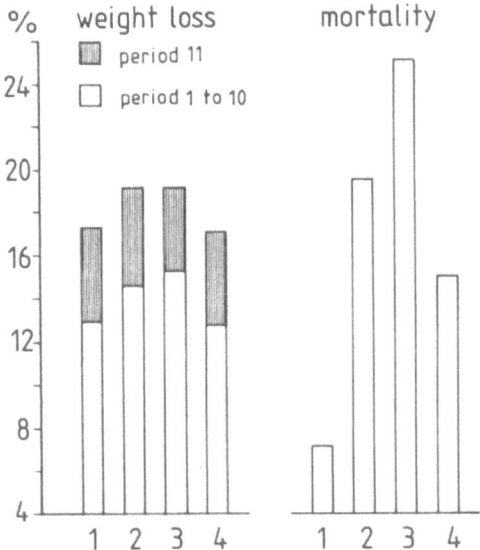

Figure 9. Weight loss and mortality per row as percentage of initial weight and number respectively (trial 25).

The data on weight loss and mortality are presented in Table 9. It can be concluded that, apart from chamber temperature, also the type of box is important with respect to weight loss and mortality. Type B boxes (2-compartment type, half the size of the other types) gave the lowest weight loss and mortality.

<u>Effects of spacing method (trial 27)</u>

The highest temperatures measured are presented in Table 10. Spacing around each stack of 8 boxes gave the smallest difference between chamber and box temperature. The difference between spacing vertically only parallel to the air flow with and without a forced air stream was only 0.5°C (39.6 vs 40.1°C). The results clearly indicate that vertical spacing all around each stack is the best method.

Data on weight loss and mortality are available only for the total of about 44 hrs of this trial. On average about 14.6% weight was lost. Sixteen chickens died (0.1%).

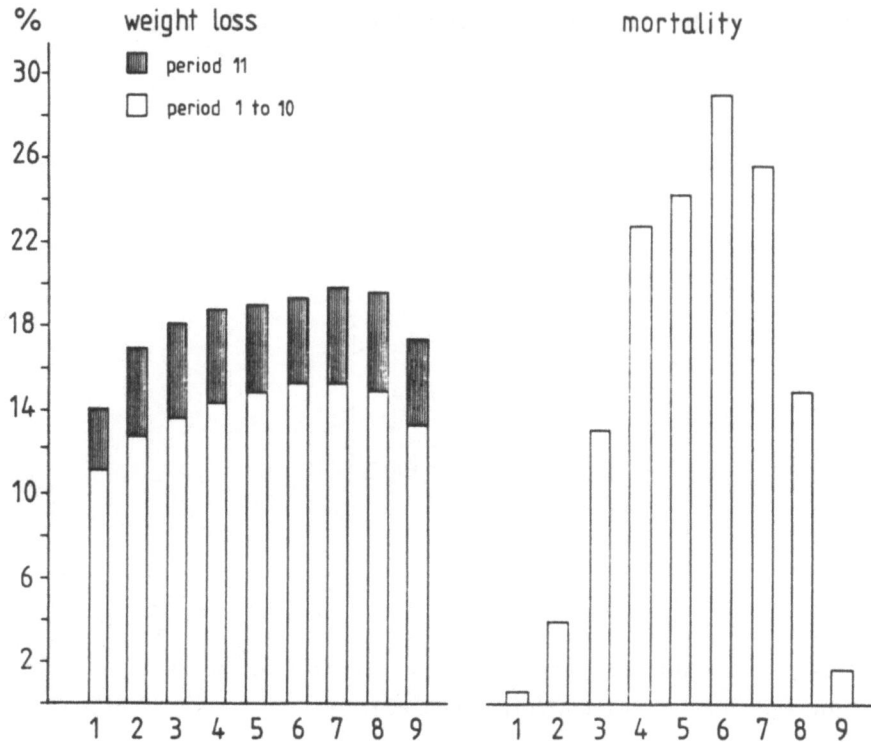

Figure 10. Weight loss and mortality per layer as percentage of initial weight and numer respectively (trial 25).

Table 9. Weight loss and mortality per box type (trial 26).

Layer no.	Weight loss (%)				Mortality (%)			
	A	B	C	D	A	B	C	D
1	13.2	13.5	12.7	14.2	0	0	0	0
2	16.4	13.8	14.6	16.7	0	0	0	0
3	17.7	14.1	16.1	17.1	0.21	0.21	0	0.21
4	19.6	15.5	17.5	18.4	2.08	0	0	0
5	19.9	16.2	18.2	19.3	0.83	0	1.04	0.42
6	21.3	16.6	18.3	19.2	6.88	0.21	3.96	0.62
7	20.7	16.9	19.5	19.9	16.04	0	3.33	1.88
8	18.3	16.6	16.2	17.9	0	0	0.21	0
mean	18.38	15.40	16.68	17.82	3.26	0.05	1.07	0.39

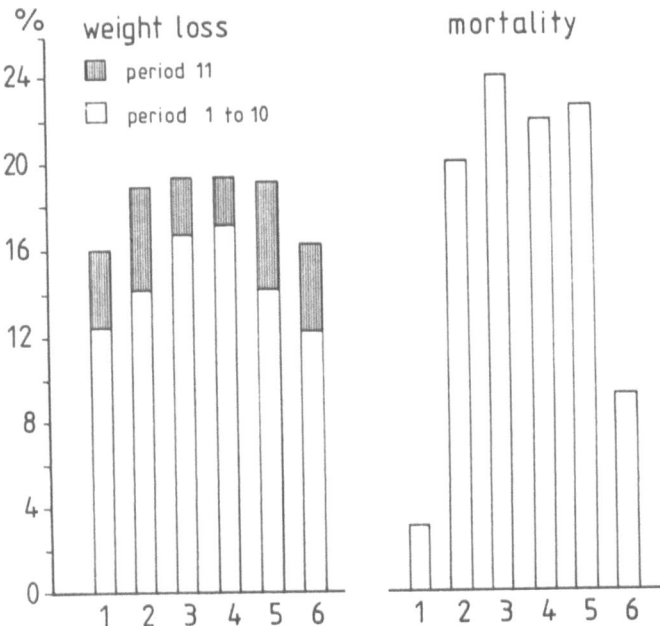

Figure 11. Weight loss and mortality per stack as percentage of initial weight and number respectively (trial 25).

Table 10. Highest temperature observed at four different methods of spacing (trial 27).

Type of spacing	Highest temperature (°C)
I spacing horizontally every other two layers	40.4
II spacing vertically around each stack	38.2
III spacing vertically parallel to air flow	40.1
IV as in III, but with forced air	39.6

Data obtained in a pallet loaded according to improved procedures (trial 28)

The results of the temperature measurements are shown in Figure 12, 13, 14 and 15. The highest temperature measured in type B boxes was 41.9°C, that in type D boxes 40.7°C. Apparently small 2-compartment boxes loose their advantages when stacked two by two to obtain the same

Figure 12. Temperatures (dry bulb) as measured in type B boxes at layer 4 in trial 28.

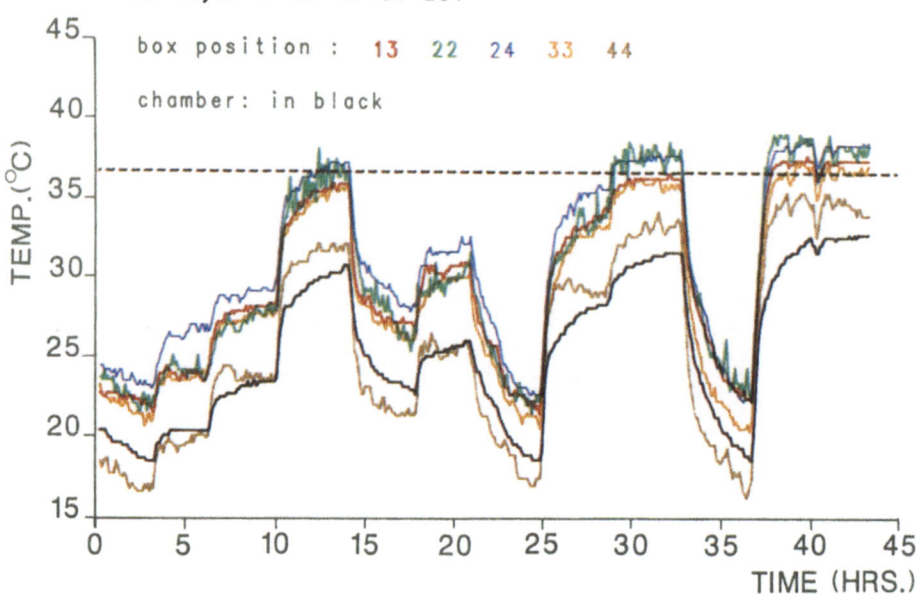

box position : 13 22 24 33 44

chamber: in black

Figure 13. Temperatures (dry bulb) as measured in type D boxes at layer 4 in trial 28.

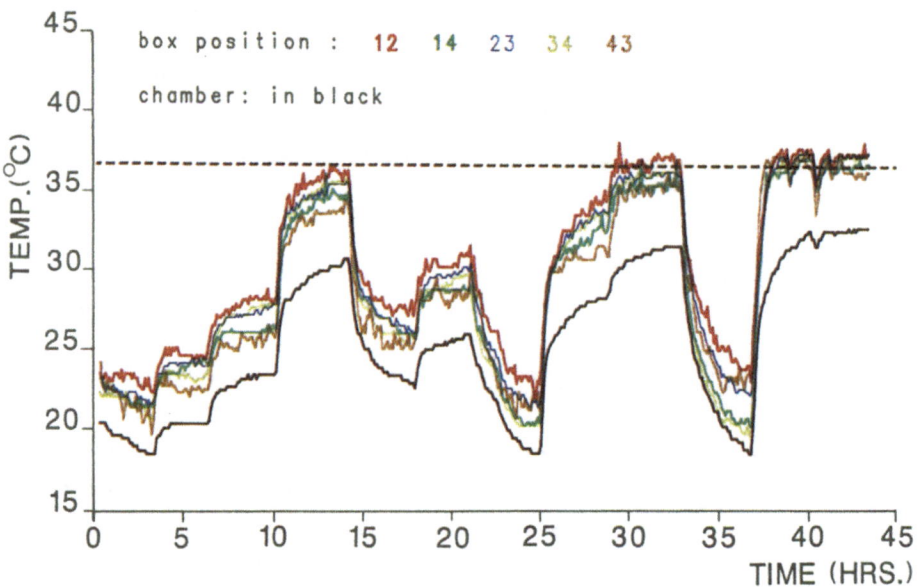

box position : 12 14 23 34 43

chamber: in black

283

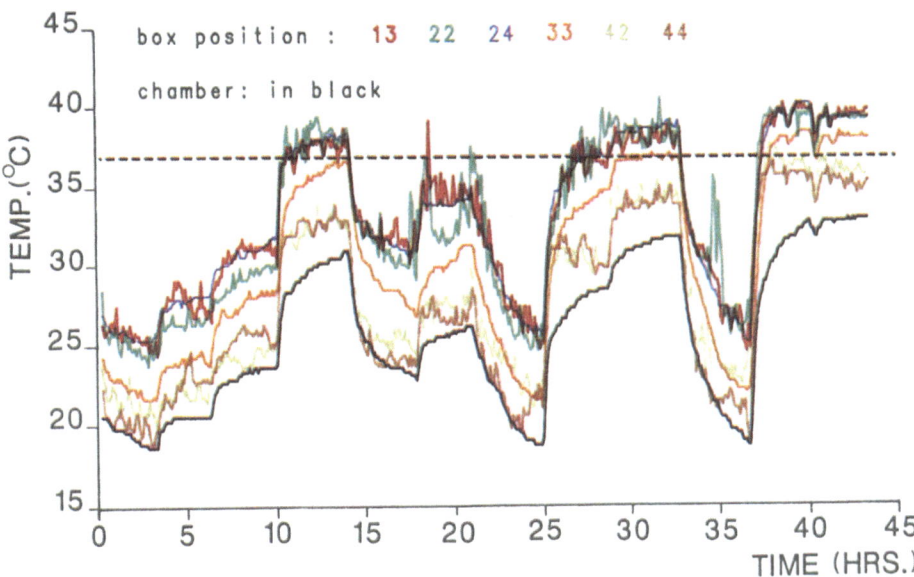

Figure 14. Temperatures (dry bulb) as measured in type B boxes just below the top in trial 28.

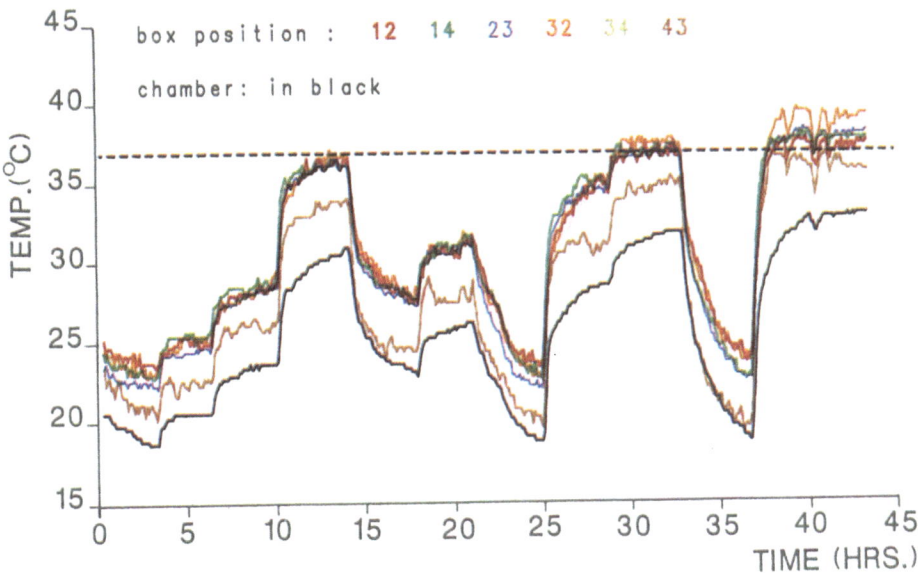

Figure 15. Temperatures (dry bulb) as measured in type D boxes just below the top in trial 28.

number of animals per unit surface. They even became worse than the 4-compartment type D boxes.

On average the chickens lost 16.6% of their initial weight (16.85 and 16.35% for chickens in type B and D boxes respectively). During period 11 at high temperatures they lost 0.29 $g.hr^{-1}$ and during the first 10 periods 0.12 $g.hr^{-1}$. The total mortality was 2.1% (1.84 and 2.36% for chickens in type B and D boxes respectively).

Comparing these results with those of trial 25, it can be concluded that spacing all around each stack has large advantages (weight loss, 16.6% vs 18.2%; mortality rate, 2.1% vs 16.3%). The figures of the observed temperatures show that spacing all around each stack decreases the difference between chamber temperature and box temperature (from maximally 14 to 8°C). Also the variation in temperature between boxes on the pallet is reduced.

Based on the results of this research, the KLM Royal Dutch Airlines adopted spacing all around as a routine procedure. Their preliminary results confirmed our findings.

CONCLUSIONS

The results obtained in the 28 trials performed lead to the following conclusions:

1. Based on heat production the effective temperature (T_{eff}) for one-day old chickens can be calculated as $0.81 * T_d + 0.19 * T_w$;

2. Heat production of the first day increases sigmoidally from 1.15 to 1.36 $kJ.an^{-1}.hr^{-1}$ when T_{eff} is raised from 27.8 to 38.1°C;

3. Heat production of the second day is on average 1.11 $kJ.an^{-1}.hr^{-1}$ and independent of T_d and T_w up to T_{eff_1} = about 37°C. Above T_{eff} = 37°C it falls to values below 1.0 $kJ.an^{-1}.hr^{-1}$, indicating that the chickens are very weak and that their lives are at risk;

4. Losses in body weight and water increase at higher T_d. They decrease when T_w is increased. The chickens cope with heat by evaporating water. A limit seems to exist in the total amount of water available for this purpose. They apparently succumb when about 17% of their initial body weight has been evaporated (and about 3% of dry matter oxidated). This result in mortality above about T_{eff} = 37°C;

5. Yolk loss decreases at higher T_{eff}, indicating that chickens use water originating from the remainder of their body to cope with heat;

6. Chickens kept at a T_d above 37°C (or T_{eff} above about 35°C) during the first two days) of their live do not fully recover within 14 days afterwards at normal conditions;

7. In view of performances, especially of broilers, after transport weight and/or water losses seem to be more appropriate to qualify the transport conditions than only mortality.

8. The temperature inside the chick boxes loaded on a pallet according to routine procedures is 8 to 14°C higher than that of the environment wherein the pallet is placed. The temperature in the central boxes is higher than that in the peripheral boxes. The highest temperatures were measured at layers just below the top;

9. The temperature inside the chick boxes depends not only on the outside temperature but also on the type of box and method of spacing. Boxes with two compartments for about 20 chickens each and vertical spacing all around each stack will give the lowest weight loss and mortality. They will reduce the increase in box temperature above outside levels. Also the variation in temperature between boxes on a pallet will be smaller. For routine procedure vertical spacing around each stack seems to be realistic. There is however a large variation in box types used for transport. This large variation may be an indication of the restricted knowledge on box characteristics. Further research to optimize transport quality seems to be justified.

REFERENCES

Bransen, A.M. and Kneepkens, H.E.M., 1982. De geitenrespiratiecellen (Respiration chambers for goats). Koeltechniek 75: 256-258.

Dixon, W.J., Brown, M.B., Engelman, L., Frane, J.W., Hill, M.A., Jennrich, R.I. and Toporek, J.D., 1983. BMDP Statistical Software. University of California Press, Berkeley, CA, 733 pp.

Egbunike, G.N., 1979. The relative importance of dry- and wet-bulb temperatures in the thermorespiratory function in the chicken. Zentralblatt für Veterinärmedizin A26: 573-579.

Freeman, B.M., 1963. Gaseous metabolism of the domestic chicken IV. The effect of temperature on the resting metabolism of the fowl during the first month of live. Brit. Poultry Sci. 4: 275-278.

Freeman, B.M., 1976. Thermoregulation in the young fowl (Gallus domesticus). Comp. Biochem. Physiol. 54A: 141-144.

Hel, W. van der, Mulder, L., Verstegen, M.W.A. and Ketelaars, E.H., 1987. Effect of high ambient temperatures on some aspects of thermoregulation in neonatal chickens. Poultry Sci.: submitted.

Hoogerbrugge, A. and Ormel, H.J., 1982. Transport of one-day old chicks by air. In: Transport of animals intended for breeding, production and slaughter (editor, R. Moss). Current topics in Veterinary Medicine and Animal Science, Martinus Nijhoff Publishers, vol. 18: 139-144.

Leeuwen, A.F.P. van, 1986. Ontwikkelingen in de export van eendagskuikens (Developments in the export of one-day old chickens). In: Seminar Luchttransport Eendagskuikens, gehouden op 28 mei 1986: 10 pp.

Macleod, M.G., 1982. The effect of travel on day-old chicks. Summary of papers from the conference on airfreighting hatching eggs and day old poultry. The West of Scotland Agricultural College, Auchincruive AYR, Scotland: 3-5.

Misson, B.H., 1976. The effects of temperature and relative humidity on the thermoregulatory responses of grouped and isolated neonate chicks. J. Agric. Sci., Camb. 86: 35-43.

Müller, W., 1985. Ventilation requirements during Air-Transport of Farm-Animals. A paper of the 11th International Conference Animal Air Transportation Assn., Inc. Tampa, Florida, March 18-21, 1985: 10 pp.

Osbaldiston, G.W., 1968. The effect of climate on the growth performance of populations of broiler chickens. Brit. vet. J. 124: 56-68.

Romijn, C. and Lokhorst, W., 1961. Some aspects of energy metabolism in birds. Proc. 2nd Symp. Energy Metab. Farm Anim., Wageningen: 49-58.

Verstegen, M.W.A., Hel, W. van der, Brandsma, H.A., Henken, A.M. and Bransen, A.M., 1987. The Wageningen respiration unit for animal production research: a description of the equipment and its possibilities. In: Energy Metabolism of Farm Animals, M.W.A. Verstegen and A.M. Henken (eds.). Martinus Nijhoff Publishers, Dordrecht.

Wekstein, D.R., and Zolman, J.F., 1970. Homeothermic development of young scaleless chicks. Brit. Poultry Sci. 11: 399-402.

CHAPTER IV. HEALTH AND ASPECTS OF ENERGY METABOLISM

ENERGY METABOLISM AND IMMUNE FUNCTION

J.M.F. VERHAGEN

ABSTRACT

Intensive husbandry systems have made it necessary to assess the environmental requirements of animals in order to minimize stress. Stressful stimuli from the environment are mediated through endocrinological alterations. These alterations may have effect on the immune system and energy metabolism, being important for animal health and productivity. It is described which endocrinological effects may occur and how they affect components of the immune system and energy metabolism. Data from literature that combine measurements of energy metabolism during the induction of an immune response are given.

INTRODUCTION

During the last decades production systems in animal husbandry were intensified. Animals in such intensive systems are provided with food and water and exposed to an artificial environment (social and climatic). Only limited possibilities are available to animals to alter their own environment. Moreover variation within the artificial environment and housing systems is less as compared with the natural environment. This means that it has become essential that the requirements of the animals towards their environment are met adequately. Through assessment of these requirements it is possible to assure optimal productivity, health and well-being. Immune defence and immune response are important mechanisms

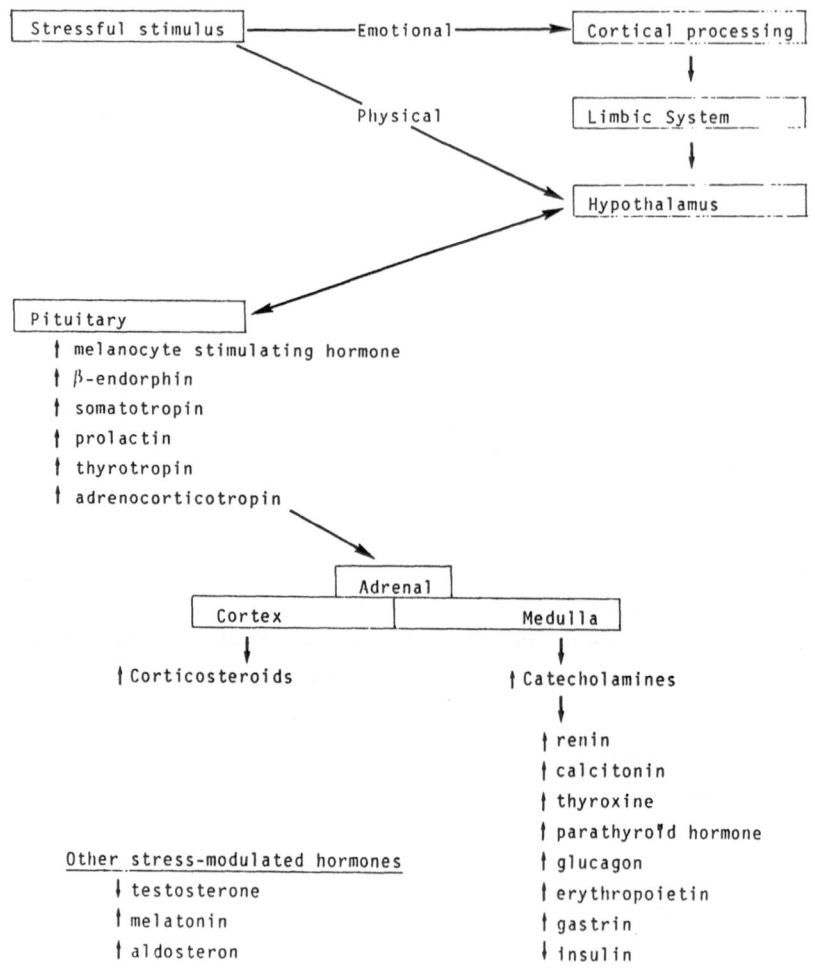

Figure 1. Neuroendocrine sequelae of stress (after: Borysenko and Borysenko, 1982).

for animals to maintain their health. Energy metabolism is closely connected with productivity traits. From research in different disciplines it has become evident that animals show an integrated response towards their 'external' environment. The underlying mechanisms that control the immune defence or immune response may also affect the mechanisms that regulate energy metabolism. Both the immune system and energy metabolism are usely studied in different areas of research. However, new disciplines in research are developing. Neurophysiology and psychoneuroimmunology are examples of such new areas of interest. The nomenclature of these areas of research does not only reflect the interdisciplinarity but also a 'causal' connection between the 'different' responses. In a simple form one might ask about the costs of an immune response as a result from infection or vaccination in terms of deposition of energy (fat or protein) and production. Moreover animals are subjected to a variety of stimuli from their environment. Besides antigenic challenges to which animals respond by their defence mechanisms, stimuli that in a broad sense are called stressors may be present. They elicit a wide range of events that occur through the sympatic nervous system and that are translated into endocrinological effects. Through the endocrinological pathways immunological and physiological mechanisms are affected and both are not mutually exclusive. Therefore the endocrinological effects of stress will be used here to describe the link that exists between immune response and energy metabolism. Some results of experiments performed to study the relation between energy metabolism and immune function will be presented.

GENERAL

The endocrinological effect of a stimulus as schematized by Borysenko and Borysenko (1982) is presented in Figure 1. It is noticeable that a variety of events in the body may occur as a result of the stimulus. The physical and the emotional component of the stimulus both affect the hypothalamus, although through a different pathway (Borysenko and Borysenko, 1982). The hypothalamus directly controls the secretion of pituitary tropic hormones, including those that act directly through the target glands (adrenal cortex, thyroid gland and gonads) or directly on target tissues (somatotrophin, prolactin) (Mclean and Reichlin, 1981). The adrenal cortex is stimulated and increases the output of corticosteroids whereas through the medulla the output of catecholamines is increased (Figure 1).

In general the catecholamines (adrenaline and noradrenaline) are responsible for an increased basal metabolic rate through the control of mitochondrial respiration, increasing lipolysis by activation of adenyl-cyclase and the stimulation of glycolysis due to increased cAMP activity (Johnson and Blanchard, 1974). Corticosteroids extend the metabolic effects of catecholamines (Dantzer and Mormède, 1983). The aim of these hormonal responses is to increase energy output (calorigenesis) needed by the animal for its reaction to various stressors. Therethrough a change in energy metabolism, dependent on these hormonal changes elicited by stressors, will result. The pattern of hormonal release and subsequently the effect on energy metabolism is a function of the type of stress, chronicity of stress and the ability of the organism to control the stress.

HORMONE LEVELS AND DURATION OF EXPOSURE TO STRESSORS

Dantzer and Mormède (1983) provided data on plasma cortisol levels in cattle that illustrate the dependency of the plasma levels on type of stressor (Figure 2) . Figure 2 shows that the response is different for gradual or acute exposure to either high or low environmental temperatures. In animals exposed to chronic heat or cold the emotional response is no longer present and the neuroendocrine changes depend on the physical quality of the stimulus (Dantzer and Mormède, 1983). Plasma cortisol level is different if cattle are exposed to either cold or heat. With beef cows Young (1975) found that chronic exposure to cold increased the resting metabolic rate resulting in a higher maintenance requirement. After acclimation to a high environmental temperature the plasma cortisol level is at a lower level whereas with cold it stays at a higher level (see Figure 2). Youssef and Johnson (1985) stated that these levels are beneficial for production of heat after acclimation.

Hormonal changes that result from exposure to stressors were also found to be present as a result from an immune response. Besedovsky et al. (1975) described experiments in which rats were immunized with SRBC (sheep red blood cells) and compared with non-immunized rats or rats injected with homologous rat erythrocytes. It was measured whether the immune responses had an effect on hormonal levels. Figure 3 shows

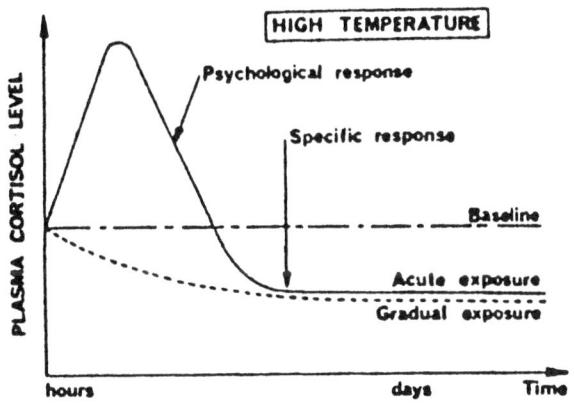

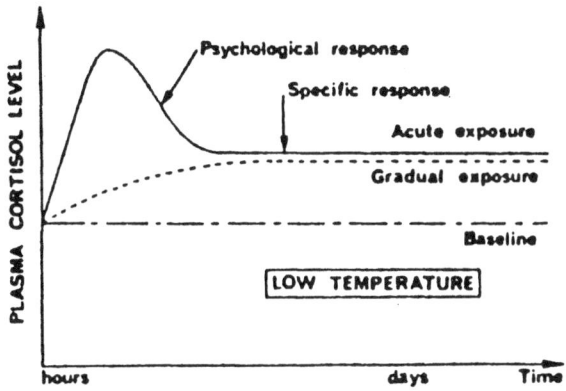

Figure 2.   Pituitary-adrenal responses of cattle to thermal stress.

the results (Besedovsky et al., 1975). It appears that at day 5-7 after
immunization with SRBC an increase in serum corticosterone is present.
Moreover the level of thyroxine in the serum was altered also. From day
3 to day 5-8 the decrease of the corticosterone level was approximately
30% (Besedovsky et al., 1975). This study shows that the immune response also brings about neuroendocrine changes. This includes effects
on noradrenaline concentration in lymphoid tissue (Besedovsky et al.,
1979). As shown in Figure 3 the change in the level of corticosterone
coincided with a change in thyroxin level. This indicates that in the
course of an immune response major changes occurred in the blood levels of these two hormones. According to Besedovsky and Sorkin (1981)
this implies that hormonal changes could regulate, at least in part by a

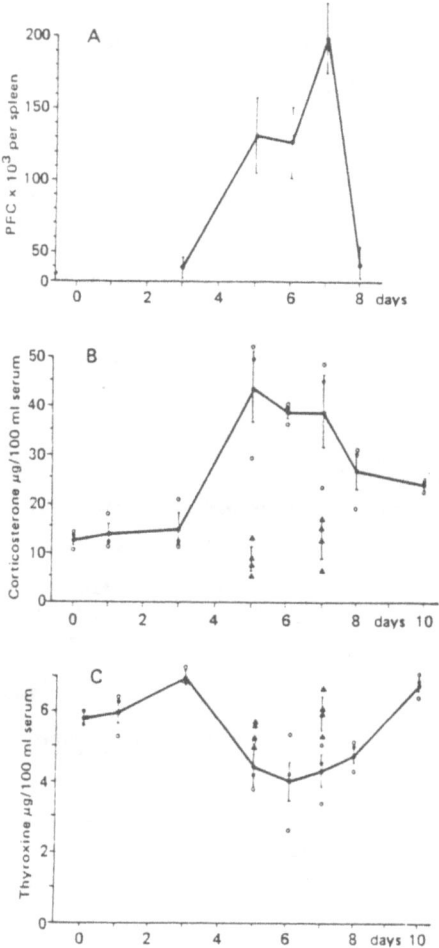

Figure 3. Changes in serum corticosterone and thyroxine levels during the immune response to sheep red cells (SRBC) in rats. (A) Plaque forming cells (PFC) x 10^3 per spleen. (B) Corticosterone levels in serum: O, animals immunized with SRBC: ▲, controls injected with rat red blood cells (RRBC). (C) Thyroxine levels in serum: O, animals immunized with SRBC; ▲ controls immunized with RRBC. From "Changes in blood hormone levels during the immune response" by Besedovsky, H. et al., Proceedings of the Society for Experimental Biology and Medicine, 1975, 150, 466-470.

feedback mechanism, the duration and, possibly, the magnitude of the immune response.

EFFECTS OF HORMONES ON IMMUNE PARAMETERS

Corticosteroids affect lymphoid tissue. They lyse lymphocytes in the thymic cortex of certains species (e.g. mouse, rat, rabbit) whereas in man and monkey they inhibit lymphocyte metabolism and interfere with proliferation. With chronic stress cell division is retarded and atrophy results. A decreased total number of circulating lymphocytes is then found (Borysenko and Borysenko, 1982). Lymphocytes also bear a number of membrane surface receptors for several hormones as catecholamines, growth hormone, insuline and prostaglandines. All these hormones stimulate cell membrane adenylcyclase and generate cAMP. Elevations in cAMP have differential effects on lymphocytes. The stage of maturation of the lymphocyte is important in this respect. In immature cells metabolism will be increased and maturation stimulated. Besides the stimulation of differentiation and proliferation these hormones also have an inhibitory effect on mature immunocompetent cells. Moreover, cAMP inhibits antibody formation to SRBC in mice. In addition to the effects on lymphocytes, epinephrine and corticosteroids also inhibit functions of macrophages, basophils, mastcells, neutrophils and eosinophils which all interact in immune function. Cortisol influences the function of macrophages by inhibiting the production of interleukine (Kelley, 1985). The neuroendocrine system and the immune system may thus not function independently but more likely interactively. Some of the examples described showed this. This means that the ultimate effects of an immune response in terms of costs of energy and protein can not be described in simple formulae.

It is however important to know whether increased demand for heat production alters the immune response, or whether the immune system itself requires extra energy. Siegel et al. (1982) determined total and activity-free heat production of pullets during the induction of an immune response to SRBC (Figure 4). At day 4 after immunization heat production was significantly increased, coinciding with a rise in haemagglutinin antibody titer to SRBC. In the immunized pullets the increase in heat production remained throughout the experimental period when

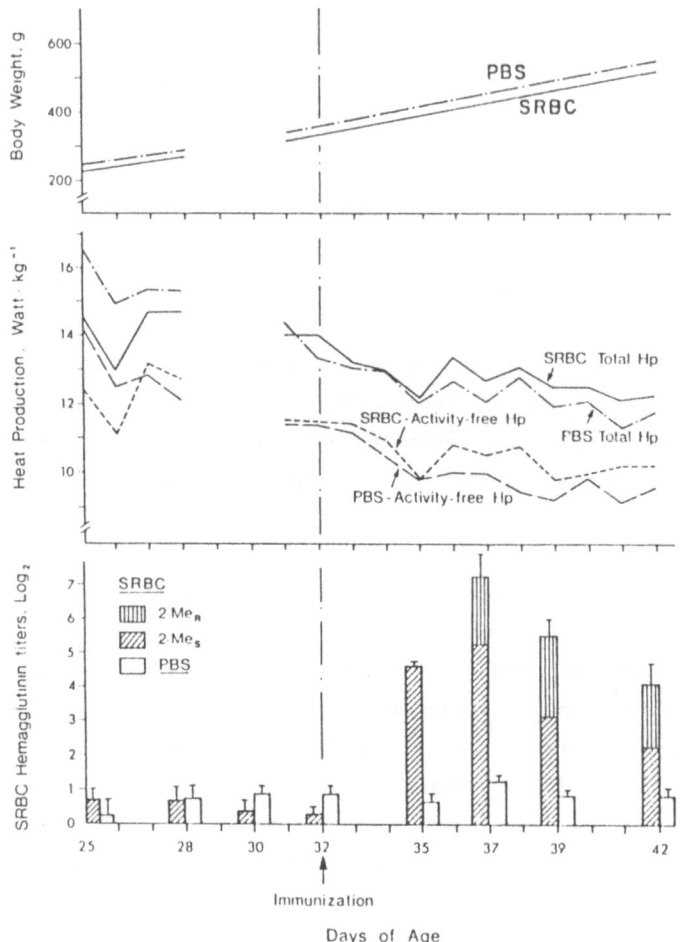

Figure 4. Haemagglutinin responses, heat production, and body weight of Warren pullets before and after immunization with sheep red blood cells. (from: Siegel et al., 1982).

compared with PBS-injected pullets (Figure 4). Henken (1982) in his work found evidence that suggested that immune responsiveness in pullets to SRBC may be reduced if animals cannot produce enough extra heat to cope with a cold environment. Data of Verhagen (1987) with pigs inoculated with <u>Haemophilus</u> <u>pleuropneumoniae</u> support this.

The endocrinological control of energy metabolism will be affected by endocrinological consequences of an immune response. Especially the work of Besedovsky et al. (1975) shows that the immune system is not purely functioning autonomously. In addition to this it must be remarked that most of the experiments that are conducted are using mainly corticosteroids as the markers in the blood. The regulatory or specific effects of other substances in the blood is not clearly understood yet. Schole et al. (1978) discussed the necessity to include the effects of ACTH and growth hormone apart from those of corticosteroids. The immune system or the resistance of cells of the organism may not only be regulated by the absolute levels of individual hormones but also by the relationship between them. In other words ratio's of hormones at a given moment may be crucial. Especially growth hormone and cortisol may act in a system of two components, being essential for the lymphatic tissue (Schole et al., 1978).

Hormones as thyroxine, growth hormone, triiodothyronine and glucocorticosteroids are known as calorigenic hormones. The sites of action of these hormones are, except for glucocorticosteroids, plasma membrane receptors of target cells. Transfer of information from receptor to appropriate intracellular sites is through the cyclic nucleotides (adenosine and guanosine 3',-5'-cyclic monophosphates; cAMP and cGMP respectively). They are important in cellular metabolism and mediate cell function under the influence of various humors in the extracellular environment (Monjan, 1981). As mentioned lymphocyte function is also modulated by this mechanism. Through the action of plasma membrane enzymes (adenylate cyclase and guanyl cyclase) the monophosphates (cAMP and cGMP) are derived from adenosine- or guanosine-triphosphate. The hormonal stimulus is thus amplified by activation of cell receptor and membrane enzymes through cAMP and cGMP that ultimately results in secretory availability of energy. Activation of adenylate cyclase by hormones and catecholamines increases cellular levels of cAMP and inhibit lymphocyte effector function. It is likely that especially the cAMP/cGMP ratio is important in the proliferative response (Monjan, 1981). The expression of receptor sites on the cells appear to change with maturation and antigenic history, thus altering the responses towards the internal milieu (Monjan, 1981).

METABOLIC CHANGES AND IMMUNE FUNCTION

To study metabolic changes due to stimulation of the immune system Klasing and Austic (1984a) exposed chickens to E. coli and SRBC. Both SRBC and E. coli resulted in pertubation of nitrogen metabolism as measured from ammonia, urea and uric acid output. Quantitatively changes due to E. coli were greatest, probably related to effects of endotoxin affecting phagocytosis and related processes (Klasing and Austic, 1984a). E. coli injection resulted in increased free amino acid concentrations in muscle (increased by 175%) and in reduced levels in liver, spleen and bursa (reduced by 25%). SRBC-injection induced similar effects in spleen and bursa but not in liver and muscle. Klasing and Austic (1984b) reported that after either an infection or a non-infectious bacterial challenge as well as after SRBC-injection the synthesis of protein in muscle was reduced.

Henken (1982) studied the effect of an immune response to sheep red blood cells (SRBC) on energy metabolism in pullets. Injection of SRBC was compared with PBS-injection (sham-immunization) in 32-day old pullets. During the first 5 days SRBC-immunized pullets produced less heat whereas energy retention and fat deposition were increased. As a result the maintenance requirements of the pullets were lowered (Figure 5). Total antibody titer against SRBC was increased 7-fold at day 5 after injection in SRBC-immunized pullets. During the second 5-day period (day 6-10) differences between SRBC-immunized and PBS-injected pullets were no longer significant. Antibody titer was lowered at day 10 (Figure 6). Verhagen (1987) conducted experiments in which young growing pigs were inoculated with Haemophilus pleuropneumoniae. Inoculation resulted in an increased energy retention and a lowered maintenance requirement during the first 7 days post infectionem (p.i.) when compared with periods before inoculation or compared with the period from day 7 to 14 p.i.. Antibody titer against Haemophilus pleuropneumoniae was increased 2-fold at day 14 p.i.

It may be obvious that the complexity of the pathways and the interaction between different pathways by which immunological and metabolical effects are determined are still far from being understood. New disciplines in animal and basic sciences that interdisciplinarily study the phenomena of immune function and metabolism will gain importance.

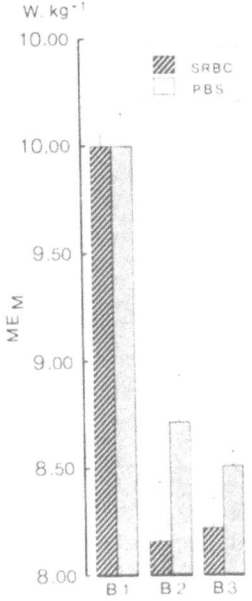

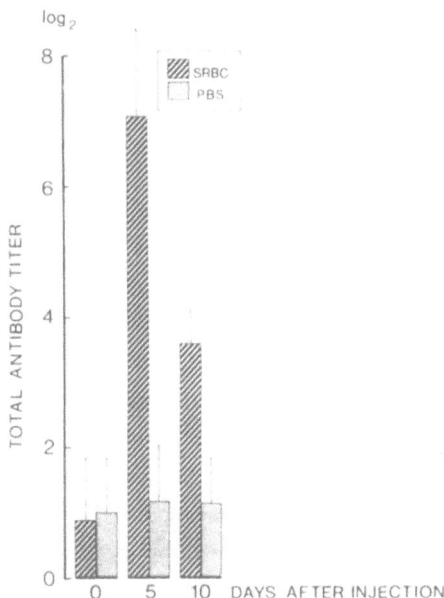

Figure 5. Maintenance requirement (ME_m, $W.kg^{-1}$) for SRBC- and PBS-injected pullets before (B1) and after injection (B2 and B3).

B1 = day -7 to -1
B2 = day 0 to 5
B3 = day 5 to 10

(from: Henken, 1982)

Figure 6. Anti-SRBC antibody titers (log_2) at day 0, 5 and 10 after injection.

(from: Henken, 1982)

several aspects should be considered. Infection of animals can change metabolism in several ways ranging from intake to partitioning over protein and fat gain, depending on the pathogenecity of the infectious organisms and/or its possible toxins. In some experiments it has been shown that infection resulted in depressed appetite and consequently depressed feed intake (Klasing and Austic, 1984a). The extent of the depression is largely influenced by the type of stimulus. However, even with non-infectious E. coli and non-microbial agents as SRBC feed intake was depressed (Klasing and Austic, 1984a). Sham-immunization with PBS showed that in chickens the depressed feed intake due to immunization with SRBC can be ascribed to the effect on the immune system (Henken,

1982).

REFERENCES

Besedovsky, H., Sorkin, E., Keller, M. and Müller, J., 1975. Changes in blood hormone levels during the immune response. Proc. Soc. Exp. Biol. Med. 150: 466-470.

Besedovsky, H., del Rey, A., Sorkin, E., Da Prada, M. and Keller, H.H., 1979. Immunoregulation mediated by the sympatic nervous system. Cell. Immunol. 48: 346-355.

Besedovsky, H. and Sorkin, E., 1981. Immunologic-neuroendocrine circuits: physiological approaches. In: R. Ader (Edt.), Psychoneuroimmunology. Academic Press Inc., New York: 545-571.

Borysenko, M. and Borysenko, J., 1982. Stress, behavior and immunity: animal models and mediating mechanisms. Gen. Hosp. Psych. 4: 59-67.

Dantzer, R. and Mormède, P., 1983. Stress in farm animals: a need for reevaluation. J. Anim. Sci. 57: 6-18.

Henken, A.M., 1982. The effect of environmental temperature on immune response and metabolism of the young chicken. Thesis, Wageningen, The Netherlands: 108 pp.

Johnson, D.E. and Blanchard, J.M., 1974. A literature review of some animal stress responses. Food & Drug Adm., Bureau Vet. Med., Colorado State University: 39 pp.

Kelley, K.W., 1985. Immunological consequences of changing environmental stimuli. In: Animal Stress. American Physiological Society: 193-223.

Klasing, K.C. and Austic, R.E., 1984a. Changes in plasma, tissue and urinary nitrogen metabolites due to an inflammatory challenge. Proc. Soc. Exp. Biol. Med. 176: 276-284.

Klasing, K.C. and Austic, R.E., 1984b. Changes in protein synthesis due to an inflammatory challenge. Proc. Soc. Exp. Biol. Med. 176: 285-291.

Maclean, D. and Reichlin, S., 1981. Neuroendocrinology and the immune process. In: R. Ader (Edt.), Psychoneuroimmunology. Academic Press Inc., New York: 475-508.

Monjan, A.A., 1981. Stress and immunologic competence: studies in animals. In: R. Ader (Edt.), Psychoneuroimmunologic. Academic Press Inc., New York: 185-217.

Schole, J., Harisch, G. and Sallmann, H.P., 1978. Belastung, Ernährung und Resistenz. Suppl. J. Anim. Phys. Anim. Nutr. 9: 1-84.

Siegel, H.S., Henken, A.M., Verstegen, M.W.A. and Hel, W. van der, 1982. Heat production during the induction of an immune response to sheep red blood cells in growing pullets. Poultry Sci. 61: 2296-2300.

Verhagen, J.M.F., 1987. Acclimation of growing pigs to climatic environment. Thesis, Wageningen, The Netherlands: 128 pp.

Young, B.A., 1975. Effects of winter acclimatization on resting metabolism of beef cows. Can. J. Anim. Sci. 55: 619-625.

Yousef, M.K. and Johnson, H.D., 1985. Endocrine system and thermal environment. In: M.K. Yousef (Edt.), Stress physiology in Livestock Vol I Basic Principles, CRC Press, Boca Raton, Florida: 133-141.

PARASITE WORRY AND RESTLESSNESS CAUSED BY SARCOPTIC MANGE IN SWINE

M.W.A. VERSTEGEN, J. GUERRERO, A.M. HENKEN, W. VAN DER HEL
AND J.H. BOON

ABSTRACT

Thirty-two crossbred pigs, weighing 13 to 28 kg each at the start,
were used to study the effects of sarcoptic mange infestations and subsequent treatment with ivermectin on metabolic rate, protein gain and
energy requirements of pigs. Pigs were paired by body weight and allocated to two identical environmentally controlled respiration chambers,
where they were penned in groups of eight. Amounts of feed provided
were measured and adjusted periodically according to weights of the
pigs. Treatment with ivermectin (300 mcg/kg) was administered to all
pigs in one chamber (selected as the one with pigs exhibiting more
severe clinical signs of mange) on day 21. Pigs in the second chamber
served as infested-nontreated controls. Weight gains, activity levels,
energy metabolism, parasitological data and leukocyte differentials were
monitored throughout the trial.

Mange mites were not identified in skin scrapings during the trial,
but clinical evidence of sarcoptic mange hypersensitivity was evident
(thickened and encrusted skin and restlessness) for these pigs. Skin
biopsies also revealed cell proliferation and hyperaemia, typically seen
with mange hypersensitivity. Mean weight gains were slightly greater for
controls at each sampling time, except at the final posttreatment evaluation. Similar trends were observed for feed conversion. In general,
pigs with heavy mange infestations did not eat well and had increased
activity levels, resulting in greater maintenance requirements and reduced efficiency in energy utilization. Substantial improvements were noted

following treatment with ivermectin. The trial provided useful data for examination of trends of effects of sarcoptic mange infestations in pigs.

INTRODUCTION

Performance of pigs is strongly determined by environmental and nutritional factors. Energy metabolism, and thus energy gain, is depended on the supply of nutrients in the ration. Under optimal conditions, energy and protein required for maintenance is minimal. Evidence that adverse conditions increase metabolic rate and reduce performance of production animals is available in the literature (Close, 1982; NRC, 1981). Factors associated with pathogenic organisms may increase nutritional requirements in growing animals also. This has been demonstrated in calves with gastrointestinal and pulmonary parasites (Kloosterman, 1971; Kroonen et al., 1986). Similar rates of gain were achieved in calves infected with Dictyocaulus viviparus, as compared with non-parasitized animals, only by the provision of additional nutrients (Verstegen et al., 1987). Results of feeding trials with pigs also suggest that nutritional requirements are increased when an endoparasite pathogen is present (Hale et al., 1981 and 1985; Stewart and Guerrero, 1986). Also ectoparasite presence may reduce performance (Cargill and Dobson, 1979b; Sheahan, 1979). Therefore, it is reasonable to expect that removal of the pathogen would eliminate the need for the increased nutrient levels to maintain a high level of performance in these animals.

The purpose of this trial was to measure the impact of sarcoptic mange and the effect of treatment with ivermectin on metabolic rate, protein gain and energy requirements of pigs.

MATERIALS AND METHODS

Animals

Thirty-two crossbred (Belgian Landrace and Landrace x Large White) piglets, weighing from 13 to 28 kg and showing clinical signs of mange, were purchased from two Belgian farms and shipped to the trial site five days before the start of the trial. Upon arrival, the pigs were weighed and separated into two weight classes due to a relatively large variation in body weights. Each pig was treated with thiabendazole before the start of the trial and with cambendazol 7 days later for removal of existing gastrointestinal parasitic infections.

Within the two weight classes, pigs were evenly allocated to housing within one of two identical respiration chambers. Animals were paired according to liveweight at the start of the trial (day 0). Inside each calorimeter, two pens were constructed for separation of pigs by weight class. In all, each pen contained eight pigs and each chamber housed a total of 16 pigs. Each pen had a non-toxic asphalt floor and measured approximately 9 m². Temperature and relative humidity within the pens were maintained at approximately 22°C and 70%. Light was provided each day from 8 A.M. to 8 P.M.

Water was provided ad libitum from water nipples from 4 P.M. to 9 A.M. each day. Animals were fed twice daily throughout the trial. Feed contained approximately 12 kJ (2.86 kilocal) metabolizable energy (ME) per gram and 16% crude protein. Feeding level was about 2.2 times maintenance requirement (70 to 80 g. $W^{-0.75}$ d^{-1}). Feeding was performed by confining each pig to the individual feeding box for approximately 15 minutes, after which the animal was permitted to join the group. Pigs in each pen were allowed to eat any uneaten feed remaining. This was done to have similar feed intakes in both chambers (same amount given and no feed left over). Amounts of feed given to individual pigs per pen were occasionally adjusted according to weight per week and consumption levels of previous days within weeks.

Allocation and treatments

Treatment against mange was assigned in such a manner that pigs in the chamber exhibiting the most severe clinical signs of sarcoptic mange would be assigned as treated group. Therefore, pigs in chamber 1 were not treated with any compounds during the trial but only with the carrier of ivermectin. Pigs in chamber 2 were treated with ivermectin* at a dosage rate of 300 mcg/kg body weight subcutaneously. Injections were administered three weeks after the start of the trial (day 21). On the

---

* Ivomec Injectable containing 1% Ivermectine w/v.

day of initiation (day 0) there were a total of 16 pigs per chamber. However, three pigs (one from chamber 1 and two from chamber 2) had to
be removed during week 1 because they refused to eat and two other
pigs (one from each chamber) were removed during week 2 and week 3
prior to treatment, due to acquired illnesses, leaving a total of 14 controls and 13 pigs in the treated group on the day of treatment (day 21).

Criteria for evaluation

Weight gains, energy metabolism, parasitological data and leukocyte
differentials were monitored for the pigs throughout the trial.

Body weights: each pig was individually weighed once each week, beginning on day 0 (3 weeks before treatment) until day 42 (3 weeks after
the treatment), for calculations of pen and group means.

Metabolizable energy (ME) was calculated weekly by monitoring energy
provided in the feed and energy voided in the urine and faeces. Energy
balances (RE) were calculated as the difference between ME intake and
heat production (H). Heat production was calculated by continuous measurement of CO_2 production and O_2 consumption over two 48-hour periods each week.

Protein gain was determined for each chamber by monitoring nitrogen
(N) in the feed and N voided in the urine and faeces. This was corrected for N escaping as NH_3 into the air.

Fat gain was calculated as RE minus energy deposited as protein,
using the equation fat gain (g) = (RE - protein × 23.6)/39.6, where
23.6 is the calorific value of 1 g of protein and 39.6 is the calorific
value of 1 g of fat.

Metabolizable energy is used for maintenance (MEm) and for production of protein and fat (MEp). Metabolizable energy used for production
of fat and protein was determined according to the ARC (1981), where
each kJ of protein gain required 1/0.54 kJ and each kJ of fat gain required 1/0.74 of ME (MEp = protein gain/0.54 + fat gain/0.74). The remaining portion of ME would be equal to the amount required for maintenance (MEm).

Activity level within each chamber was measured by a method similar
to that of Van der Hel et al. (1984), using a movement detector (Messl
Spaceguard type 15X) that responds to interference of ultrasonic waves.

Whole blood samples were collected from each pig (n=32) on the day of arrival at the test facility (day -5) and of the remaining pigs (n=23) on day 42. In addition whole blood was sampled from eight animals (four in each group) at day 7, 14, 21, 28 and 35. In the whole blood samples total serum protein, albumin and α-, β- and γ-globulins were measured.

Moreover heparinized blood was used to make smears. Stained cells (Giemsa) were differentiated according to Juncqueira and Carneiro (1980).

Skin biopsies were collected twice from each pig present at day 21 and day 42 for determinations of alterations resulting from mite infections. It was tested whether hyperkeratose, hyperaemia, dermatitis and acanthosis was present and each was graded from 0 to 3 or 4 (from no to serious signs).

RESULTS

Circumstances affecting conduct

In addition to the five animals previously mentioned that were removed from the trial before treatment, three other pigs were removed from the chambers (two controls during week 4 and one control and one treated during week 5). Since the pigs had been paired at the start of the trial, data were collected from 10 pairs of animals throughout the trial for treatment comparisons.

Body weights

Pen and group mean body weights are shown in Table 1. Total mean gains were generally similar, though slightly greater in the control group than in the treated group (17.3 vs 15.2 kg, respectively). Rates of gain within and between groups were similar in the weeks before treatment as during the weeks following treatment.

Feed conversion

Feed/gain ratios for the groups are presented in Table 2. During most of the weeks in the trial, feed/gain ratios were slightly to moderately higher in the treated group than in the control group. However, at the final evaluation (day 42), three weeks after treatment, the mean feed conversion ratio was lower in the treated group as compared to controls.

Table 1. Body weights and weekly gains of pigs remaining in pairs.

| | | | | Body weights (kg) | | | | | |
| | | | | Pretreatment | | | Posttreatment | | |
Group	Pen	n	Ini-tial	+ 1 week	+2 week	+3 week	+4 week	+5 week	+6 week
Con-	1	7	26.7	30.2	32.5	35.6	39.2	42.6	44.8
trol	2	3	22.1	24.9	27.0	29.6	31.7	34.6	37.2
Group Mean		10	25.3	28.6	30.9	33.8	36.9	40.2	42.6
Treat-	1	7	28.3	31.3	33.5	36.4	38.9	41.6	44.5
ed	2	3	21.3	24.0	26.0	26.9	29.9	31.8	34.3
Group Mean		10	26.2	29.1	31.2	33.5	36.2	38.6	41.4

| | Weight gains (kg) of pigs remaining in pairs | | | | | |
| | Pretreatment | | | Posttreatment | | |
Group	1 week	2 week	3 week	4 week	5 week	6 week
Control	3.3	2.3	2.9	3.1	3.3	2.4
Treated	2.9	2.1	2.3	2.7	2.4	2.8

Energy metabolism

Results of energy metabolism calculations are presented in Table 3. Metabolizable energy intakes are generally similar for the groups throughout the trial, with no apparent differences in the weeks before and after treatment. Heat production in the treated group tended to run slightly higher than for controls during the pretreatment weeks, after

Table 2. Weekly feed conversion ratio (feed/gain).

Group	Pen	Pretreatment			Posttreatment		
		1 week	2 week	3 week	4 week	5 week	6 week
Control	1	1.62	2.89	2.81	2.66	3.48	3.90
	2	2.85	2.57	2.85	3.27	2.57	3.10
Group mean		2.00	2.76	2.83	2.85	3.09	3.56
Treated	1	1.91	3.30	2.66	3.17	3.17	3.03
	2	2.28	3.62	5.30	3.55	3.71	3.18
Group mean		2.06	3.42	3.24	3.31	3.34	3.08

which the means for groups became very similar. As a result of this trend to lower heat production by the treated pigs after treatment, net energy retention increased proportionately. During the pretreatment weeks, control pigs had somewhat higher mean retained energy values, but after treatment, there were no differences in mean values for the groups.

Composition of gain

Rates of gain in body weight, protein and fat for all animals are shown in Table 4. Before treatment, body weight gains tended to be greater in the control group than in the treated group. This difference was diminished in the weeks following treatment and the treated pigs had greater daily weight gains than controls by the final week of the trial. Protein gain was generally greater for control pigs than for the treated group before and after treatment. By the final evaluation, protein gain for the treated pigs began to approach that of the controls. Fat gain was greater for controls before treatment and was similar for the groups

Table 3. Metabolizable energy intake, heat production and retained energy (all: kcal.kg$^{-0.75}$d^{-1}).

Time	Group	ME	H	RE
Week 2	Control	237	150	87
	Treated	239	153	86
Week 3	Control	243	152	92
	Treated	243	158	85
Pretreatment	Control	240	151	89
	Treated	241	156	86
Week 4	Control	236	153	84
	Treated	238	154	83
Week 5	Control	236	154	82
	Treated	239	158	81
Week 6	Control	239	157	82
	Treated	239	157	83
Posttreatment	Control	237	155	82
	Treated	239	156	82

in the weeks after treatment.

Metabolic rates for maintenance and activity

Calculations of energy maintenance requirements and activity expenditures are summarized in Table 5. Before treatment, pigs in the treated group had greater maintenance requirements, heat production, and heat expenditures from higher levels of activity. In the weeks after treatment, there is a reversal of these trends, and all requirements and ex-

Table 4. Composition of gain in pre- and post-treatment $(g.d^{-1}.an^{-1})$ in control (C) and treatment (T) group.

	Rate of gain			Fat gain			Protein gain		
	C	T	Dif.	C	T	Dif.	C	T	Dif.
Pretreatment									
Week 2	333	251	+82	73.5	70.0	+3.5	65.6	66.2	-0.5
Week 3	363	307	+56	85.8	77.1	+8.7	66.3	62.1	+4.2
Mean	348	279	+69	79.6	73.6	+6.0	66.0	64.1	+1.9
Posttreatment									
Week 4	367	282	+85	80.8	79.6	+1.2	72.0	62.7	+9.3
Week 5	356	311	+45	81.4	82.3	-0.9	77.4	70.0	+7.4
Week 6	348	361	-13	92.6	94.4	-1.8	73.6	71.9	+1.7
Mean	357	318	+39	84.9	85.4	-0.5	74.3	68.2	+6.1

Table 5. Maintenance requirements (MEm) and heat production associated with activity (HAC), both in $kcal.kg^{-0.75}.d^{-1}$.

	ME_m			Hac		
	C	T	Dif.	C	T	Dif.
Pretreatment						
Week 2	104	108	-4	17.0	25.3	-8.3
Week 3	105	114	-9	15.1	24.6	-9.5
Mean	104	111	-7	16.0	24.8	-8.8
Posttreatment						
Week 4	109	112	-3	14.3	19.6	-5.3
Week 5	111	116	-5	20.6	21.5	-0.9
Week 6	115	114	+1	19.4	19.6	-0.2
Mean	111	114	-3	18.2	20.3	-2.1

penditures are generally similar for the two groups. Clinical observation of the pigs supported the reduced level of activity in the treated group after treatment.

Differential counts

Results of leukocyte differential counts are presented in Figures 1, 2, 3 and 4. Percentages of adult neutrophils, lymphocytes, monocytes and eosinophils tended to decrease in the treated group. beginning one week after treatment. Percentages of juvenile neutrophils remained relatively constant throughout for both groups and eosinophils tended to increase

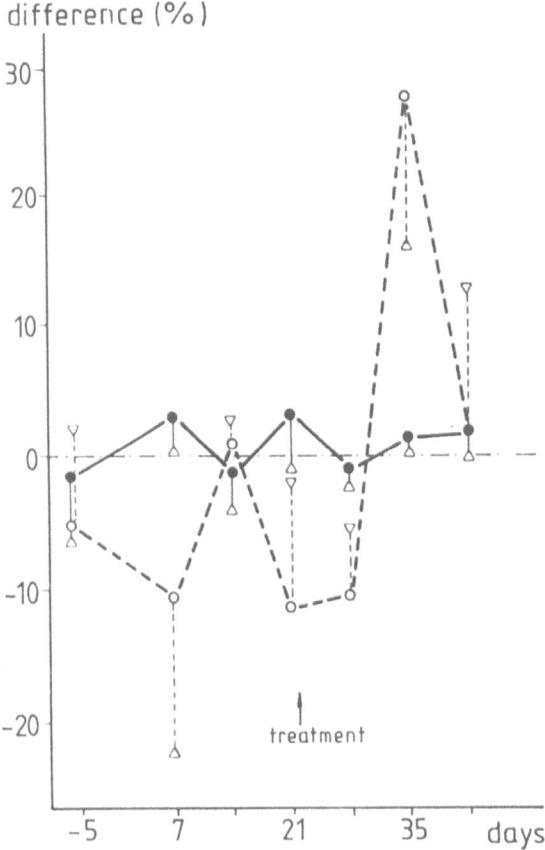

Figure 1. Differences in % of juvenile (———) and adult (- - -) neutrophils in the blood of 4 pairs of animals of the T minus C animals at various times (vertical bars = SEm).

314

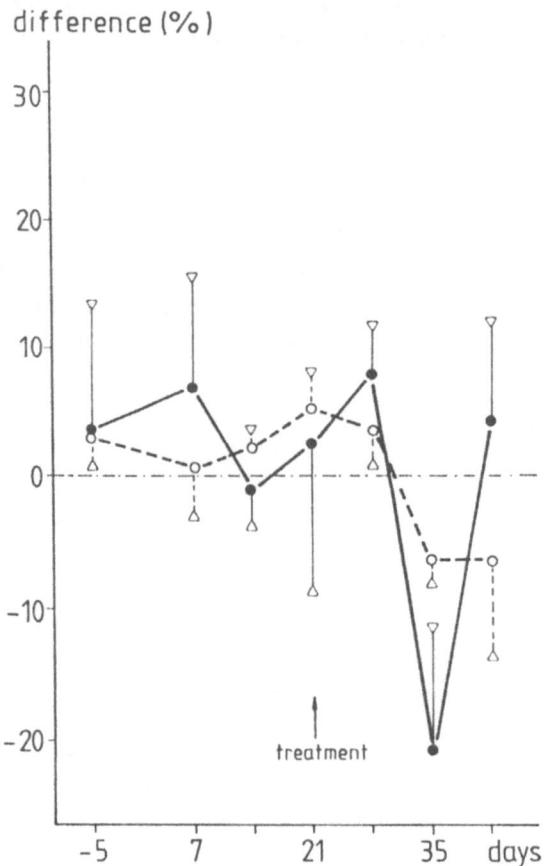

Figure 2. Differences in % of monocytes (- - -) and lymphocytes (————) in the blood of T minus C animals at various times (vertical bars = SEm).

during the posttreatment weeks for the control group. Total γ-globulin values increased and were similar for both groups (Fig. 5).

Skin biopsies

In Table 6 results of skin biopsies are presented. Mean data show that total level of skin keratosis, hyperaemia, acanthosis and dermatitis increased in control animals and decreased in treatment animals. No mites or mite eggs were observed in any of the samples examined at the end of the trial. Capillary hyperaemia was the most frequent observation in pigs

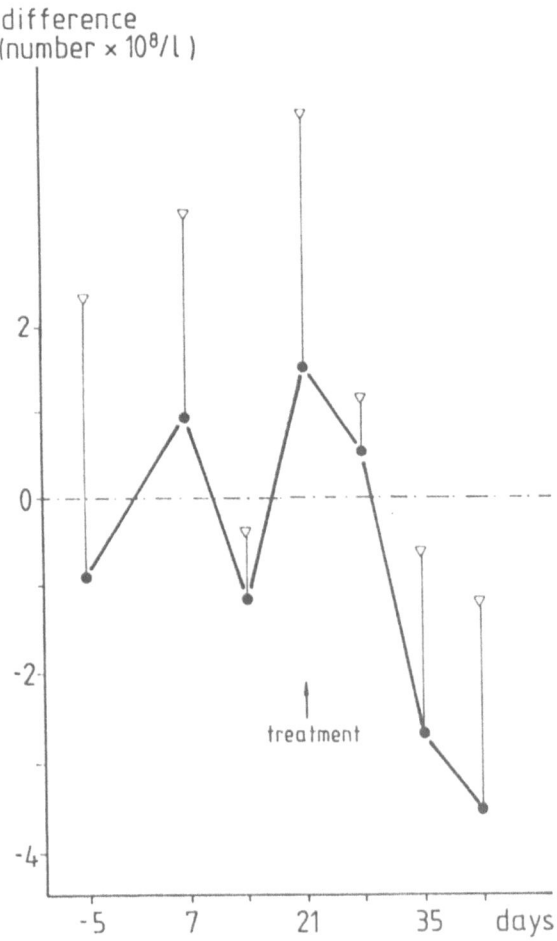

difference
(number × 10^8/l)

Figure 3.   Differences in numbers of eosinophils in blood of T minus C
animals at various times (vertical bars = SEm).

of both groups. Eosinophilic infiltration was noted only in one sample
taken from one pig of the treatment group before treatment.

DISCUSSION AND CONCLUSIONS

Results showed that sarcoptic Scabei was not identified in our pigs.
However clinical examination indicated similar signs as with sarcoptes;

316

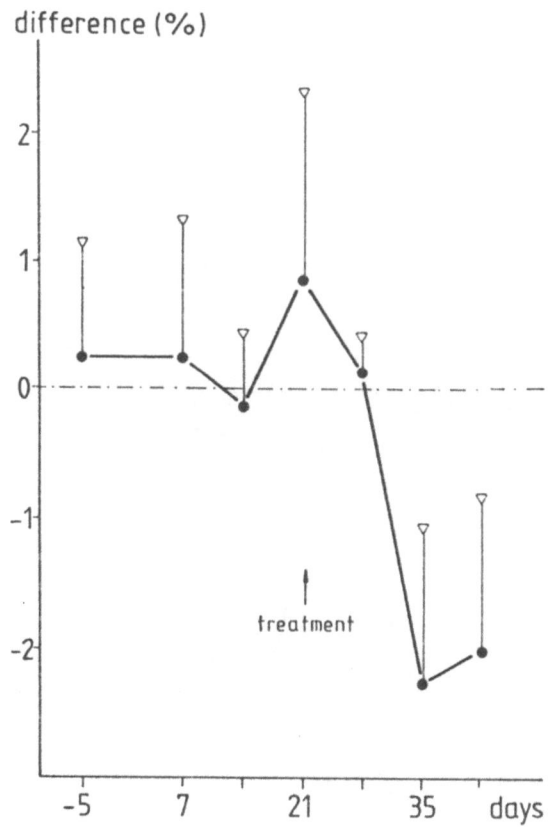

Figure 4. Differences in % of eosinophils in blood of T minus C animals at various times (vertical bars = SEM).

Table 6. Hyperkeratosis, parakeratosis, hyperaemia, acanthosis and dermatitis (mean judgement).

Start		End	
Control	Treatment	Control	Treatment
2.66	2.90	4.10	2.14

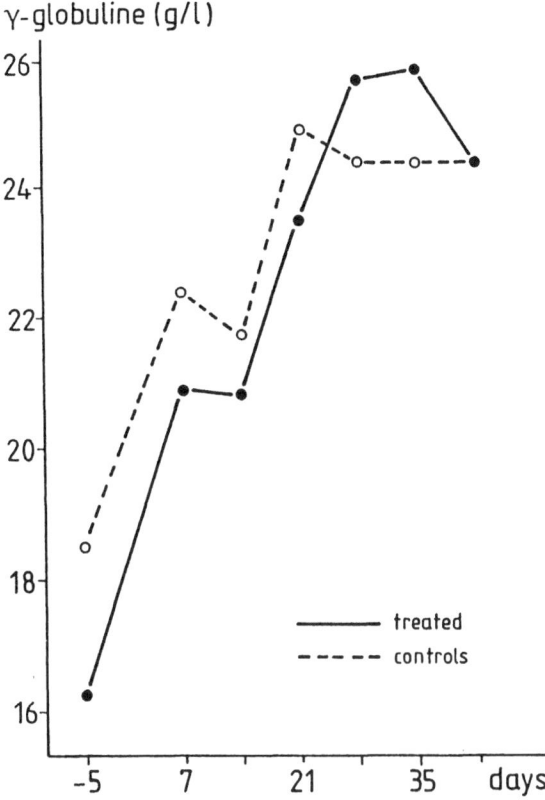

Figure 5.   Course of the γ-globulines in 4 pairs of pigs (controls and treated).

thickening of the skin, encrusted skin and restlessness (Soulsby, 1968). Cargill and Dobson (1979a) found that piglets were hypersensitized to sarcoptes and piglets showed mange. They also could not identify the mite. In our experiment pigs had mange (encrusted skins) similar to the pathological signs for sarcoptic Scabei (Nieberle and Cohrs, 1938). More- over blood traits after treatment (decreased levels of eosinophils) sup- ported these findings. Also skin biopsies (Table 6) showed that the level of proliferation of cells and hyperaemia are similar to hypersensitivity (Ryan and Wills, 1986).

After treatment with ivermectin there was an improvement of skin lesions in the treatment group. This improvement is also seen in Figure

4 in which eosinophils in the blood are reduced after treatment. Data of the skin biopsies of the control animals were judged and indicated that the conditions of these animals probably diminished.

Data available from this trial indicate that energy requirements and utilization are adversely affected by the presence of "mange like symptoms" and that administration of ivermectin to rid the animals of the infestation will initiate a reversal of these effects. Increased maintenance requirements and poor production performance in this trial could be related to increased levels of activity during the period of active mange infestation. In general, pigs with heavy mange infestations did not eat well, gained less weight and showed reduced feed efficiency as compared with non- or marginally infested animals. Substantial improvement, resulting from reductions in activity levels, was noted following treatment.

In an attempt to extrapolate impact on economics of production from these findings, it was determined that if the normal average maintenance requirement for growing pigs during the whole fattening period is approximately 800 grams of feed per pig per day, the presence of sarcoptic mange appears to increase the requirement by approximately 3.5%. This translates to approximately 30 grams of feed per pig per day or a total of 3,5 kg feed per pig for the entire fattening period.

Data collected from the trial, while useful for examination of trends, are unintentionally biased against the treated group, who exhibited more severe signs associated with mange infestations. This greater severity of signs, recognized at the start of the trial, was selected for treatment because it was thought that more distinct signs would be required to demonstrate differences between and within groups before and after treatment. In retrospect, it was more difficult to demonstrate these differences. Probably artificial infestations of normally growing pigs need to be preferred for a correct assessment of the costs.

REFERENCES

ARC, 1981. The Nutrient Requirements of Pigs. Commonwealth Agricultural Bureaux.

Cargill, C.F. and Dobson, K.J., 1979a. Experimental Sarcoptes scabei infestation in pigs: (1) Pathogenesis. Vet. Rec. 104: 11-14.

Cargill, C.F. and Dobson, K.J., 1979b. Experimental Sarcoptes scabei infestation in pigs: (2) Effects on production. Vet. Rec. 104: 33-36.

Close, W., 1982. The climatic requirement of the pig. In: Environmental aspects of housing for animal production (ed. A. Clark). Butterworths, Londen: 149-166.

Hale, O.M., Stewart, T.B., Marti, O.G., Wheat, B.E. and McCormick, W.C., 1981. Influence of an experimental infection of nodular worms (Oesophagostomum spp.) on performance of pigs. J. Anim. Sci. 52: 316-322.

Hale, O.M., Stewart, T.B. and Marti, O.G., 1985. Influence of an experimental infection of Ascaris suum on performance of pigs. J. Anim. Sci. 60: 220-225.

Hel, W. van der, Verstegen, M.W.A., Baltussen, W. and Brandsma, H., 1984. The effect of ambient temperature on diurnal rhythm in heat production and activity in pigs kept in groups. Int. Journal of Biometeorology 28: 303-316.

Juncqueira, L.C. and Carneiro, J., 1980. Lange Medical Publications, Los Altos, California.

Kloosterman, A., 1971. Observations on the epidemiology of trichostrongylosis of calves. Thesis Agr. Univ. Wageningen.

Kroonen, J.E.G.M., Verstegen, M.W.A., Boon, J.H. and Hel, W. van der, 1986. Effect of infection with lungworms (Dictyocaulus viviparus) on energy and nitrogen metabolism in growing calves. British Journal of Nutrition 55: 351-360.

Nieberle and Cohrs, 1938. Lehrbuch der Spezieller Pathologische Anatomie der Haustieren, Judar Tischer Jena: 751.

NRC, 1981. Effect of environment on nutrient requirements of domestic animals. Nat. Academic Press, Washington D.C.

Ryan, W.G. and Wills, P.D., 1986. Pre-farrowing use of ivermectin for the control of mange in piglets. Proceedings of the International Pig Veterinary Society: 9th congress, July 15-18, 1986, Barcelona, Spain: p. 365.

Sheahan, B.J., 1979. Experimental Sarcoptes scabei infections in pigs: Clinical signs and significance of infection. Vet. Rec. 94: 202-209.

Soulsby, E.J.L., 1968. Helminths, Arthropods and Protozoa of domestical animals, Baillière, Tindall and Cassell, London, p. 506.

Stewart, T.B. and Guerrero, J., 1986. The economic significance of pig

endoparasites. Proceedings of the International Pig Veterinary Society: 9th congress, July 15-18, 1976, Barcelona, Spain: 371.

Verstegen, M.W.A., Hel, W. van der., Boon, J.H., Kessels, M., Meu-
lenbroeks, J. and Mellink, H.M., 1987. Dictyocaulus viviparus infection and energy metabolism of calves. Submitted.

RESPIRATORY DISEASES IN PIGS: INCIDENCE, ECONOMIC LOSSES AND PREVENTION IN THE NETHERLANDS

M.J.M. TIELEN

ABSTRACT

Respiratory diseases in pigs are very common in The Netherlands, because the disease germs are continuously present in the population. These diseases result in enormous economic losses for the pig farmer. The occurrence and severity of the diseases depend strongly on farm conditions. Especially with regard to respiratory diseases climatic conditions play an important role. This is shown by results of field studies performed at regular farms. To assess the effect of each individual climatic factor separately experimental research in climate controlled pighouses has to be done. The development of pre-clinical parameters to determine the health status of the pig will contribute very much to the conclusions of such research.

INTRODUCTION

Respiratory diseases in pigs are very common in The Netherlands. Most of the affections of the respiratory tract result in clinical and subclinical signs situated in the lung (pneumonia and pleuritis) and the nose (atrophic rhinitis). Due to the high incidence of respiratory diseases, the economic loss caused by these diseases is substantially. Many factors are involved in the aetiology of these diseases, especially where pig production is intensive as in The Netherlands.

That means, that prevention of outbreaks and reduction of economic loss can only be achieved by improving conditions in the pig industry.

RESPIRATORY DISEASES

Pneumonia

Pneumonia can be caused by many bacterial and viral agents. Some are more important than others. The following agents are of primary importance in the Dutch pig industry: Mycoplasma hyopneumonia; Aujeszky disease virus; Haemophilus pleuropneumoniae serotype 2 and 9; and Porcine influenza virus N_1H_1 and N_3H_2. In the appearance of secundary infection a large scale of very common pathogens can be detected like Pasteurellae, Streptocococcae and Staphylococcae. The incidence of pneumonia in slaughterpigs in The Netherlands was studied by different researchers. Koopman (1962) reported an incidence of pneumonia of 74% in 1120 pigs. Sybesma and Zuidam (1966) found in a sample taken ad random that pneumonia was present in 56% of the animals. In an extensive field study (25418 slaughterpigs) Truijen (1967) noted an increase from 29.4 to 41.5% in the incidence of pneumonia within 3 years (1962 to 1964). Tielen (1974) found in 50.8% of slaughterpigs pneumonia. He also noted significant differences between fattening farms and between stables within farms. Since April 1975 a great part of the slaughterpigs in The Netherlands is examined on pneumonia. A lung is considered to be affected with pneumonia, when more than 5 cm^2 of the surface is affected. The results are given in Figure 1. From 1981 onwards there was a decrease in pneumonia. However, in 1985 a big difference was recorded in the frequency of pneumonia between individual farms. About 45% of the farms delivered less than 10% pigs with pneumonia, while 25% of the farms had more than 20% pigs with pneumonia.

Pleuritis

In many cases of pneumonia the lung inflammation is so severe, that the inflammation spreads to the pleura. A pleuritis may be the result. A great part of the lung and the pleura may get sticked together in severe cases. In that case examination of the total surface of the lung is no longer possible. Such cases are noted as pleuritis in the standard examination at the slaughter house. It has been found, that cases of pleuritis

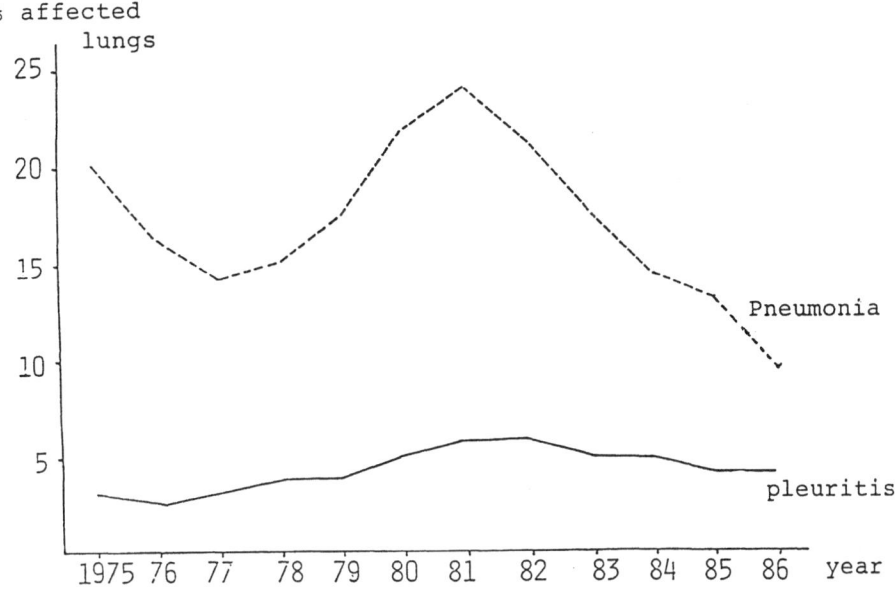

Figure 1. Incidence of pneumonia and pleuritis in slaughterpigs as found at the slaughterhouse, Encebe, Boxtel, the Netherlands (number of pigs per year: about 1.000.000).

often develop in the late weaning period or at the beginning of the fattening period (Tielen, 1978).

Especially in case of a clinical outbreak of Haemophilus pleuropneumoniae there is a clear increase in pleuritis cases (Hunneman, 1983). Since 1973 there is a firm increase in Haemophilus pleuropneumoniae cases in The Netherlands. The post mortem findings at the Animal Health Service-laboratory during 1973 to 1986 increased from 17 to 963 cases. Figure 1 shows that simultaneously there was also a slight increase in pleuritis cases.

Atrophic Rhinitis (AR)

The clinical signs of atrophic rhinitis culminate in deformation of the nose. In the aetiology of AR two bacteria species play an important role: Bordetella bronchiseptica and Pasteurella multocida. By taking nose-swabs it has been shown, that one or both of these bacteria are present

on most of the breeding and multiplier farms (about 85%) in The Netherlands (Van Nistelrooy, 1979). Recently more attention is given to especially the toxine producing <u>Pasteurellae</u> in view of the development of
clinical nose deformation (De Jong, 1985). Screening the breeding farms
on the presence of these <u>Pasteurellae</u> by taking nose-swabs from the
piglets seems to result in a lower percentage of positive farms (De Vries,
1986).

In a subclinical stage atrophic rhinitis can find expression in an atrophy of the conchae (CA). In a field investigation of 5444 slaughterpigs
from 18 fattening farms in 1979 (Tielen and Van Marle, 1980) 55% of the
pigs had a slight and 21% a severe atrophy of the conchae. Pigs with
conchae atrophy showed a greater incidence of pneumonia. The frequency of CA on the individual farms ranged from 8.0 to 49.2%, between individual pig houses within a farm from 1.1 to 49.2%.

Disease prevention

Surely outbreaks of respiratory diseases can only appear when the
microbial agents are present. In the high density pig population in several parts of The Netherlands most of these microbial agents are continuously present. The size of the pig-farming operations and the short
distances between farms/stables make it practically no longer possible to
eliminate these agents. In spite of the presence of the microbial agents,
outbreaks and economic losses seem to depend on the conditions in the
pig environment and on management. Especially conditions leading to an
increase in the supply of pathogen agents or reducing the resistance of
the pigs can disturb the balance between pathogen supply and resistance
of the animal. Prevention against these, so called multifactorial, diseases
by animal care can only be achieved by optimizing the conditions of
operation.

ECONOMIC LOSSES

Economic losses due to respiratory diseases depend on the age of the
pig, the severity of the outbreak and the course of the disease after the
first outbreak. Therefore it is very difficult to estimate the losses in all
individual cases. Results presented here are averages over different

Table 1. Daily gain and carcass quality in slaughterpigs in relation to lung affection (Tielen et al., 1978).

lung affection	number of pigs	Weight (kg) at the end of the breeding period	fattening period	Daily gain (g) breeding period	fattening period	Carcass quality % EAA+IA
None	93,437	24.6	85.4	335	642	67.0
Pneumonia	19,049	24.7	83.7	334	614	63.8
Pleuritis	2,711	24.3	85.4	311	614	54.4
Total/ mean	115,197	24.6	85.1	334	636	64.2

animals and farms.

Pneumonia and pleuritis

A great part of the pigs examined for lung affections at the slaughter house is also used to obtain data on daily gain and carcass quality (Brus et al., 1972). So it is possible to compare the data between groups with different affections within farms. In Table 1 the results on 115,197 pigs are presented. There is a clear decrease of 28 g per day in daily gain during the fattening period of pigs affected with pneumonia. In pigs with pleuritis there is no additional reduction in daily gain during the fattening period. The clear reduction in daily gain in the pleuritis group during the breeding period is remarkable. It seems, that pleuritis cases recorded at the slaughter house very often find their background in respiratory diseases in the period before fattening. These results are similar to those of Hunneman (1983). In spite of an increase in pleuritis he found no difference in daily gain during the fattening period between farms with and without an Haemophilus pleuropneumoniae outbreak. Most of the economic losses after an H. pleuropneumoniae infection are due to a higher mortality and increased costs for medical treatments.

Table 1 shows that there are additional economic losses due to a decrease in carcass quality of pigs having pneumonia and pleuritis.

It is possible that part of the pigs having unaffected lungs at the slaughter house had pneumonia earlier in live. The signs of such an early affection may have disappeared by regeneration. Therefore the economic losses caused by pneumonia were also calculated by ranking average daily gain during the fattening period at a farm in relation to the percentage of pigs with pneumonia within that farm. The results are presented in Table 2. The differences in daily gain between the categories of pigs affected in Table 2 are greater than what can be calculated from the differences between individual pigs in Table 1. There is a continuous decrease in daily gain when the percentage of pigs affected on a farm increases. Part of this relation may be caused by an interaction between incidence of pneumonia and farm conditions. Sub-optimal farm conditions will result in a greater risk to get affected pigs, but also lead to lower gains in the absence of affection.

Table 2. Average daily gain per fattening farm in relation to the percentage of pigs with pneumonia (Tielen and Brus, 1985).

% pigs with pneumonia	number of fattening farms	average daily gain (g) during the fattening period
< 10%	83	682
10-20%	172	664
20-30%	92	653
> 30%	28	628

Atrophic Rhinitis

The economic losses due to atrophic rhinitis are caused by less favourable production characteristics of pigs with nose deformations. The differences in production between affected and unaffected pigs depend on the feeding system, but daily gain was always more than 100 g lower

Table 3. Differences in production results and post mortem findings between pigs with and without AR-deformation of the nose within one fattening farm (Paridaans et al., 1981).

nose deform.	feeding system	number of animals	daily gain (g)	feed conver- sion	mortal- ity (%)	pneu- monia (%)	pleu- ritis	CA[1] (%)	SN[2] (%)
+	ad lib	37	512	3.51	7.5	24.2	12.1	96.7	40.0
	restricted	76	492	3.42	5.0	21.7	4.3	96.0	74.6
-	ad lib	39	680	3.66	2.5	13.5	8.1	45.5	9.1
	restricted	80	621	3.26	0	12.7	3.8	29.5	6.8

[1] CA = pigs with a severe atrophy of the conchae
[2] SN = pigs with a wry septum in the nose

(Table 3). The affected group showed a higher mortality rate. In pigs with clinical signs of AR a high incidence of atrophy of the conchae (CA) and wry septum in the nose (SN) was found. The groups with deformed noses also proved to contain more pigs with pneumonia. Because most of the clinical deformation of the nose appear at the end of the breeding period (8-10 weeks), the reduction in daily gain on the breeding farm itself is very small. However, breeding farms with clinical signs of atrophic rhinitis in part of the animals can no longer sell breeding animals. They even have great difficulties to sell the unaffected piglets to fattening farms. Thus, indirectly there is a great loss. To combat the disease all sows and piglets on the breeding farm have to be treated with drugs during a long time. The total economic losses therefore due to atrophic rhinitis are very large in The Netherlands.

ANIMAL HUSBANDRY

Prevention against respiratory diseases is possible by breeding for resistance. In practice however measures taken against infection intend to optimize housing and management conditions. At optimal conditions animals may live in coexistence with the disease germs without allowing them to cause disease and damage.

Farm conditions in general

The data obtained in the lung examination routinely performed at the slaughter house allow estimation of the effects of conditions at fattening farms on the incidence of pulmonary aberrations. In a comparison between 251 farms with a high percentage of pigs with pneumonia (over 25%) and 251 farms with a low incidence (below 10%) a clear influence of some farm conditions on the incidence of pneumonia was found. The incidence of pneumonia is lower at fattening farms:

1. receiving piglets from only one breeder;
2. practicing an all in all out system;
3. having small compartments (less than 100 PU);
4. keeping the pigs in the same sty during the whole fattening period; and
5. keeping the pigs on a closed or half slatted floor (Table 4).

These results confirm those of Hunneman (1983). He concluded, that incidence and severity of H. pleuropneumioniae depend on some farm conditions. Fattening farms with a H. pleuropneumoniae infection; received the piglets from more different multiplier-farms; practiced less the all in all out system; and moved the pigs often one or more times during the fattening period. In an investigation on 127 breeding farms with a clinical atrophic rhinitis outbreak, Deenen (1982) found also a clear relationship between percentage of piglets with nose deformations and farm conditions. The percentage was clearly lower at farms using all in all out per farrowing house consisting of small compartments with 8-12 farrowing pens, wherein the piglets stayed until leaving at about 24 kg. The all in all out system was combined with a thorough cleaning and desinfection. The incidence of nose deformation was higher at farms with straw bedded concrete floors than at farms having half slatted floors in the farrowing house.

Climatic conditions

In the literature about respiratory diseases in pigs climatic condition

Table 4. Influence of conditions at fattening farms on the frequency
of pneumonia in slaughterpigs (Tielen et al., 1978).

Farm condition	Number of farms	Pneumonia ($\%$)	x^2-test
Origin of piglets			
- own farm	119	15.5	
- from one breeder	184	16.3	$p < 0.005$
- from two breedes	50	24.6	
- from more than 2 breeders	61	23.0	
Fattening system			
- all in all out	83	14.2	$p < 0.005$
- continuously	401	19.3	
Replacements			
- none	162	15.9	
- one	279	19.2	$p < 0.005$
- more than one	61	21.9	
Size of the compartments			
- < 100 PU[1]	112	15.4	
- 100-200 PU	108	18.2	$p < 0.005$
- 200-300 PU	30	20.8	
- > 300 PU	20	28.9	
Floor			
- closed	35	12.5	
- half slatted	186	18.2	$p < 0.05$
- total slatted	47	21.8	

[1] 1 PU = pig unit = the place for 1 fattening pig till 100 kg

330

Table 5.　Influence of climatic conditions on the severity of H. pleuropneumoniae infection on fattening farms (Hunneman, 1983).

Climatic conditions	Number of farms	Number of com-partments	drug treatment by injec-tion (%)	by feed (days/pigs)	Mortal-ity (%)
Temperature					
≥ 17°C	32	80	32.3	1.4	3.1
< 17°C	37	100	35.4	2.3	3.2
CO_2-content					
≤ 0.15 vol %	55	158	31.2	1.8	3.1
> 0.15 vol %	20	33	37.4	1.6	3.5
NH_3-content					
≤ 10 ppm	55	166	34.4	1.9	3.1
> 10 ppm	17	25	19.7	1.1	3.5
Air-velocity					
≤ 0.15m/sec	49	125	32.9	1.7	3.2
> 0.15m/sec	33	66	31.1	1.9	3.2
Draught[1]					
TW ≤ 100	38	87	25.6	1.5	3.0
TW > 100	37	37	41.9	2.2	3.3

[1]　TW = draught index (Tielen, 1974).

in the pig house is frequently pointed at as the most important item
within the multifactorial aspects (Straw, 1986). In our studies mentioned
above, a clear relationship was found between the incidence of respiratory diseases at farms and the outdoor climate. Pneumonia, H. pleuropneumoniae and atrophic rhinitis show a clear higher incidence in winter
than in summer. The influence of the outdoor climate can only be due to

the influence of the outdoor on the indoor climate. The influence depend on the ventilation system in the pighouse. In pighouses with an indirect way of air inlet, the incidence of pneumonia in the slaughterpigs was lower than in pighouses with a direct way of air inlet (difference was 5.5%). Indirect ventilation causes the outdoor air to pass pre-space. Here the air is heated. The effect of wind is eliminated.

However, no information about specific climatic factors was obtained. Therefore, the relationship between specific factors and disease incidence was studied separately in some cohort studies on breeding and fattening farms. The climatic conditions within a pighouse were measured frequently. The results were related to the frequency of some clinical or subclinical signs of disease. A clear relationship was found between temperature, temperature fluctuation and air velocity in the pighouse on one hand and frequency and severity of pneumonia in the pigs at the slaughter house on the other hand (Tielen, 1974). Hunneman (1983) studied the severity of a H. pleuropneumoniae outbreak in relation to climatic conditions in the pighouse. In each pighouse at least two measurements of the indoor climate were performed. His results are given in Table 5. The data are divided into two groups for each individual climatic factor. In general the severity of the diseases as judged on basis of drug treatment and mortality was less at more favourable conditions. The most clear influence is seen when some individual factors are combined into a draught index (TW), the difference between out- and indoor temperature multiplied by the air velocity inside. The relationship between some climatic conditions and the severity of atrophic rhinitis infections on breeding farms is shown in Table 6. Especially the climate in the farrowing pen seems to be important. There is a tendency, that the severity of the disease is much less at good climatic conditions. Only on a part of the farms the piglets are replaced after weaning to a separate weaning house. There the relationship is not so clear and even in some cases in contrast with the relations found in the farrowing house. This may be due to interactions between climatic conditions and diseases in farrowing and weaning house. The influence of the indoor climate on the occurrence of respiratory diseases can be proved in field studies. However, it is not possible to assess in such studies the effect of each individual climatic factor separately.

Nevertheless insight in the role of each individual climatic factor is

Table 6.   Influence of climatic conditions on the severity of an Atrophic rhinitis infection on breeding farms (Deenen, 1982).

Climatic condition	Farrowing houses		Houses for weaned piglets	
	number of farms	% N.D.[1]	number of farms	% N.D.[1]
Temperature				
≥ 18°C	41	16.2	22	11.9
10-17°C	10	30.1	15	20.5
Relative humidity				
40-60%	23	16.5	6	19.5
≥ 60%	15	23.5	20	17.1
CO_2-content				
≤ 0.15 vol %	9	17.7	32	17.8
> 0.15 vol %	48	24.4	13	10.2

[1] N.D. = piglets with nose deformation

necessary for a good advice to farmers. Therefore research must be done in climate chambers or climate stables to assess the effect of each individual climatic factor. In such a recent study Verhagen (1987) demonstrated, that temperature, temperature fluctuation and air velocity can influence the physiological activity and the course of disease outbreaks in pigs after artificial infection with H. pleuropneumoniae germs.

Since April 1985 the Animal Health Service in North Brabant (The Netherlands) also possesses a climate controlled pighouse to investigate effects of individual climatic factors on the occurrence of diseases. The pighouse, built in cooperation with some farmer-organisations consists of two compartments completely separated from one another (Tielen, 1986). Each compartment can accommodate 5 sows with piglets or 60 weaned piglets or 40 fattening pigs. An artificial draught system is developed which

Table 7. Provisional results on the influence of individual climatic factors on the health status of pigs (Tielen, 1986).

Animals	Climatic treatment	Number of animals	Clinical signs Cough	Diarrh.	Post mortem[1] findings CA	Pneum.	Prod. results Daily gain (g)	Feed conv.
I weaned piglets	temp. fluctuation 22-16°C/day	60	59	9	33	10	578	1.81
(7-12 weeks of age)	constant temp.22°C	60	54	17	37	10	558	1.89
	difference		+5	-8	-4	0	+20	-0.08
II weaned piglets	draught[2] $A^3=49\times10^{-6}$ $T=22°C$	60	9.9	1.3	21	32	359	1.66
(5-11 weeks of age)	no draught $A=23\times10^6$ $T = 22°C$	60	6.6	0.5	22	21	399	1.63
	Difference		+3.3*	+0.8**	-1	+11	-40**	+0.03
III fattening pigs	draught $A=66\times10^6$ $T=10°C$	40	39	0	9.2	31	828	3.30
(20-26 weeks of age)	no draught $A=38\times10^6$ $T=10°C$	40	32	0	17.1	18	872	3.14
	Difference		+7	0	-7.9	+13**	-44**	+0.16*

[1] % of the animals showing signs

[2] draught is applied 4 times one hour per day

[3] A is cooling index according to Kleiber (1961) in $kJ/cm^2/sec$

* $p<0.05$

** $p<0.005$

can provide the lying areas in one compartment with a separate air stream having a lower temperature and a higher air velocity than the ventilation flow in the remainder of the compartment. Experimental pigs are always obtained from commercial farms, which had no clinical outbreaks of atrophic rhinitis or Haemophilus pleuropneumoniae for the last two years. Parameters used for comparison between compartments are: clinical signs, pathologic anatomical findings and production results. Climatic conditions tested are similar to those encountered in practice. In the first investigations attention was focussed on temperature and air velocity.

The provisional results of the first three trials in this climate controlled pig house are presented in Table 7. Definite conclusions can not yet be drawn. The trials will be repeated at least once. The results obtained so far show that a daily temperature fluctuation of 6°C has no influence on health status and production results of weaned piglets, when compared to a constant temperature regime. This trial was already repeated with a greater temperature fluctuation of 8°C. Again there were no significant differences between the results of experimental and control group. Conform to the results of Verhagen (1987), draught seems to result in a remarkable worsening of health status and production results of the pigs.

This was due to a much more severe outbreak of Coli-enteritis (weaned piglets) and a higher incidence of pneumonia (fattening pigs). A more detailed description of this research is given by Tielen (1986).

Studies as described must continue and be done intensively, preferably at different research institutes. They must partly concentrate on finding pre-clinical parameters which give information on resistance and health status of pigs under different housing conditions. Then determinations can already be done in an earlier stage. It will not be longer a necessity to examine only animals with clinical or subclinical signs.

REFERENCES

Brus, D.H.J., Truijen, W.T. and Tielen, M.J.M., 1972. Cijfermatige beoordeling van de gezondheidstoestand op varkensbedrijven. I. Tijdschr. Diergeneesk. 97: 1483.

Deenen, F., 1982. Evaluatie van de begeleiding op AR-bedrijven, aangemeld in de periode 1 augustus 1980 tot 1 april 1981. Rapport Gezondheidsdienst voor Dieren in Noord-Brabant.

Hunneman, W.A., 1983. Voorkomen, economische betekenis en bestrijding van Haemophilus pleuropneumoniae-infecties bij varkens. Thesis. Rijksuniversiteit Utrecht, 129 pp.

Jong, M. de, 1985. Atrofische rhinitis bij het varken. Thesis Rijksuniversiteit Utrecht, 260 p.

Kleiber, M., 1961. The fire of life. An introduction into animal energetics. John Wiley and Sons Inc., New York, London, 454 pp.

Koopman, J.J., 1962. Invloed van het klimaat op viruspneumonie. Tijdschr. Diergeneesk. 57: 1291-1304

Nistelrooy, van A., 1979. Atrofische rhinitis in rassen en standen. Report of the University of Utrecht.

Paridaans, H., Voets, M.Th., Tielen, M.J.M, Hendriks, H. and Paridaans, L., 1981. De economische betekenis van atrophische rhinitis op mestvarkensbedrijven. Bedrijfsontwikkeling 12: 795-800.

Sybesma, W. and Zuidam, L., 1966. Longontstekingen bij varkens en de gevolgen daarvan voor de vleeskwaliteit. Tijdschr. Diergeneesk. 91: 1727-1735.

Straw, B., 1986. A look at the factors that contribute to the development of swine pneumoniae. Veterinary Medicine, August: 747-757.

Tielen, M.J.M., 1974. De frequentie en de zoötechnische preventie van long- en leveraandoeningen bij varkens. Meded.Landbouwhogeschool, Wageningen 74-7, 142 pp.

Tielen, M.J.M., 1978. Buildings, environmental conditions and diseases. Proceed. of the 29th congress of the Eur. Ass. Animal Prod.: 747.

Tielen, M.J.M, Truijen, W., Groes, C. v.d., Verstegen, M., Bruin, J. de and Corbey, R., 1978. De invloed van bedrijfsstructuur en stalbouw op varkensmestbedrijven op het voorkomen van long- en leveraandoeningen bij slachtvarkens. Tijdschr. Diergeneesk. 103: 1155-1165.

Tielen, M.J.M., and Marle, A. van, 1980. Umwelt bedingte Conchea Atrofic bei Schlachtschweinen. Proc. 3nd Int. Kongress für Tierhygiene, Wien: 1-5.

Tielen, M.J.M. and Brus, D.H.J., 1985. Praktische Erfahrungen mit Hygienemassnahmen in der Tierproduktion. DLG-Forschungsbericht über Tierernährung nr. 538027.

Tielen, M.J.M., 1986. The influence of temperature and air velocity on the occurrence of respiratory diseases in pigs. Proc. of the 37th Animal Meeting of the E.A.A.P., Budapest.

Truijen, W.T., 1967. Enige zoötechnische aspecten van enzoötische pneumonie bij varkens. PhD-thesis, Rijksuniversiteit Utrecht.

Verhagen, J.M.F., 1987. Acclimation of growing pigs to climatic environment. PhD. Thesis, Agricultural University Wageningen, The Netherlands, 128 pp.

Vries, Th. de, 1986. De toxinevormende Pasteurella multocida als indicatie voor bedrijfsmatige atrofische rhinitis. Report Agricultural University Wageningen, The Netherlands.

MASTITIS IN DAIRY COWS WITH SPECIAL REFERENCE TO DIRECT AND INDIRECT EFFECTS OF CLIMATOLOGICAL FACTORS

F.J. GROMMERS

ABSTRACT

In mastitis there may be climatological effects on the cow and on the micro-organisms involved.

With regard to the direct and indirect climatological effects on the cow the effects on the teatskin are most obvious. Teatskin and teatcanal orifice lesions may harbour pathogenic bacteria and promote the incidence of intramammary infections. The effects of climatological stressors on the natural defence mechanisms are still hypothetical.

Climatological effects on the micro-organisms are especially valid for pathogens that are able to survive and multiply in the environment of the cow. Source of the bacteria; extent of contamination; type, humidity and temperature of the substrate (bedding material) are interacting factors, resulting in varying bacterial counts.

With the exception of (severe) clinical mastitis, the relationship between nutrient requirements and mastitis is mainly a direct function of the amount of milk secreted.

INTRODUCTION

Mastitis is a disease complex rather than a simple inflammation (-itis) of the mammary gland (mastos).

To understand the possible environmental influences on incidence and prevalence of mastitis in cattle, some basic knowledge of epidemiology

and terminology is required. For a short introduction in the host-patho-
gen relationship use has been made of the reviews by Schalm et al.
(1971), Bramley et al., (1981), Poutrel (1982), Dodd (1983) and Grommers (1984).

The micro-organisms

Physical or chemical insults may cause an acute or chronic reaction of
the mammary tissue, but mastitis commonly is caused by pathogenic bacteria penetrating the teatduct. There are however differences in habitat
or source and in pathogenicity of these bacteria. This leads to differences in: infection risk, defence response of the mammary gland and efficiency of control measures.

Historically the most important mammary pathogen is Streptococcus
agalactiae. It is the only mastitis pathogen with milk or mammary gland
as sole habitat. Survival of this bacterium is therefore strongly influenced by milking hygiene, which was often poor in former days. S. agalactiae is susceptible to penicillin. These characteristics make it possible
to eradicate S. agalactiae from dairy herds.

Other streptococci commonly associated with mastitis are Streptococcus
uberis and Streptococcus dysgalactiae. These streptococci can also be
found on other sites of the animals body and are better able to survive
in the environment. Especially S. uberis is often regarded as an "environmental" pathogen, whereas S. dysgalactiae infections show a relative
high incidence in case of teat injury.

Mastitis caused by Staphylococcus aureus may be more severe than
streptococcal mastitis since it produces toxins. The sources of S. aureus
are the udder and many sites on the body of animals. Once penetrated
into the mammary gland it may be difficult to cure while it can form foci
deep in the tissue, moreover some strains are penicillin resistant.

Escherichia coli is a typical "environmental" bacterium and the major
source is manure. If penetrating bacteria are killed by the defence system of the gland, then a potent endotoxin is released. E. coli mastitis
often is an acute and severe clinical mastitis. It is possible that E. coli
infections are enhanced by mastitis control programmes with use of post-
milking teat disinfection and dry cow therapy with antibiotics. This is
due to the fact that E. coli infections occur more readily in udder quarters that are free from any infection. At best incidence of E. coli infec-

tions and mastitis are not influenced by mastitis control programmes.

Pseudomonas aeruginosa and Klebsiella pneumoniae are also environmental bacteria, with dirt, soil and sewage as possible source. Klebsiella may originate from soil and sawdust under improper storage. Infections with those organisms are however relatively rare. If present they are difficult to detect and cure.

Mastitis where Corynebacterium pyogenes is involved is most times a more chronic and severe mastitis with purulent exudate. It is almost always a mixed infection with other micro-organisms and especially with Micrococcus indolicus. Pyogenes mastitis is typical for traumatized glands and non-lactating glands. In the latter case it occurs in the dry period or in heifers and is referred to as "summer mastitis". This summer mastitis needs specific preventive measures at pasture on wet sandy soils in areas with bush or wood. The fly Hydrotaea irritans plays an important role as a vector. Preventive measures are directed against flies or try to protect the gland (teat sealers or antibiotics).

The bacteria mentioned sofar (except M. indolicus) are the most important bacteria associated with mastitis in cattle and usually designated as major or primary pathogens. It is possible to make a distinction between contagious pathogens (S. agalactiae and S. aureus) and environmental pathogens (E. coli, Klebsiella pneumoniae, S. uberis, S. dysgalactiae) based on main reservoir and way of transmission (Smith et al., 1985).

On bacteriological examination of aseptically taken quarter milk samples, several other bacteria can be isolated: other streptococci, coagulase negative staphylococci, micrococci, corynebacteria and others. They usually are relatively harmless commensals and called minor or secondary bacteria.

Natural defence of the mammary gland

Signs of inflammation caused by invading micro-organisms are in fact signs of defence of the udder or the entire animal and/or a product of the inflammatory process. The defence response depends partly on the properties of the micro-organisms and partly on the quality of the various components of the defence.

It is generally recognized that the teat duct is the animals' first line of defence. Funnel shaped orifices and wide (over 0.5 mm) teatducts promote penetration of bacteria. The keratin lining of the duct is not only important for closure of the teatduct between milkings, but this keratin contains also bactericidal and bacteriostatic proteins and lipids. Leucocytes may contribute to the defence against bacterial penetration via the tissues of Fürstenberg's rosette.

Once bacteria have passed the teatduct a number of cellular and humoral resistance factors determine the success or failure of the defence. It is the accepted view that polymorphonuclear leucocytes (PMN) are the second major defence factor. In uninfected quarters the numbers of PMN vary from 50.000-500.000 per ml. milk depending on age and stage of lactation. Following intramammary infection numbers may increase tenfold or more. Cellcount of milk is therefore often used as a criterion for udder health or infection, on quarter, cow or herd (bulkmilk) level.

Other components contribute to phagocytosis and/or killing of bacteria such as: macrophages, lymphocytes, lactoferrin, lactoperoxydase system, complement and immunoglobulins.

In the course of the inflammatory process there are many changes in the composition of the cows' milk: increase of Na, Cl, protein and decrease of K and lactose. Clots in milk may be coagulations of secretory products or debris from tissue repair. Depending on the severity of the inflammation milk secretion is decreased.

Field investigations and experimental infections have shown that the efficiency of the defence mechanisms depends on lactation stage and age. The rate of new infections is highest at parturition, in early lactation and at drying off. Infection rate increases with age.

Duration of intramammary infections varies from a few days to several months. Spontaneous elimination takes place in 40-60% of the cases. Cases of mastitis requiring therapy are in 30-40% resulting from pre-existing intramammary infections.

Definitions and prevaling bacteriological causes

The classical signs of inflammation (-itis) are: redness (rubor), heat (calor), pain (dolor), swelling (tumor) and disturbed function (functiolaesa). Only in case of severe (acute) mastitis all these symptoms are present with additional systemic signs of disease like fever and loss of appetite. In many cases the inflammatory response is less severe or pronounced and some symptoms may be absent. Nevertheless, if some of these symptoms are observed this type of mastitis is called: clinical mastitis. In the absence of the symptoms mentioned, laboratory tests may reveal presence of pathogenic micro-organisms and changes in milk composition, including increased cell count. In this case the diagnosis is: subclinical mastitis. If no subclinical abnormalities can be demonstrated notwithstanding the isolation of a pathogen, then it is called: a latent infection. In case of a raised cellcount, in the absence of clinical or subclinical signs and without isolation of a pathogen, the diagnosis may be: secretion disturbance. This term however is confusing, since there are many fysiological and environmental influences on the day to day variation in cellcount of milk.

The three streptococci en S. aureus are isolated in about 90% of the cases of subclinical mastitis and latent infections. These bacteria only account for 40-50% of the clinical mastitis cases. E. coli and Cbt. pyogenes are isolated in 20-30% of the clinical cases depending on circumstances. In 20-30% of the clinical cases no pathogenic bacteria are isolated. In the latter case it is supposed that the pathogens have been present in an earlier stage of the inflammation.

PREVALENCE OF SUBCLINICAL MASTITIS AND LATENT INFECTIONS AND INCIDENCE OF CLINICAL MASTITIS

The prevalence of mammary infections, which do not lead to clinical symptoms, is determined by the new infection rate and the duration of these infections. Especially the new infection rate shows variation from herd to herd depending on the quality of management and the adoption of mastitis control measures.

In The Netherlands the National Mastitis Survey in 1980 (Vecht et al., 1982) has shown at that that time 9.6% of 33, 240 quartermilk samples were at that time bacteriologically positive. Secretion disturbance (more than 500.000 cells/ml) was present in 14.9% of the quarters and 75.4% were regarded "normal".

In a similar survey in the United Kingdom it was found that 14.1% of the quarters were infected (Wilson and Richards, 1980).

The incidence rate of clinical mastitis has a wide range. Remmen (1986) reported 13.8% Meuse-Rhine-IJssel cows and 10.6%. Dutch Friesian cows with mastitis in one year. Variation between farms was 0.0-27.3 and 1.4-26.7 respectively.

Wilesmith et al. (1986) found an average of 41.2 cases per 100 cows in 1982 in Britain. Much higher incidence rates may occur. Smith et al. (1985) for instance analyzed a herd in the USA with 194.4 cases per 100 cows per year. Robinson et al. (1983) recorded on six British farms a variation from 40 to 165 cases per 100 cows per year in different mastitis control treatment groups.

ENVIRONMENTAL INFLUENCES ON MASTITIS

Bramley and Dodd (1984) state: "Cattle have always been exposed to micro-organisms that cause bovine mastitis and their resistence mechanisms against udder infection have evolved over countless generations by natural selection". "It is to be expected that commercial dairy cattle management will increase the chances of cows contracting in intramammary infections, mainly through increasing the exposure of teats to pathogens but also by increasing the probability of penetration of the teat duct". "Cows are kept in herds and are communally milked, thus transmitting pathogens between cows even with the best hygiene." "Winterhousing methods concentrate the cows and increase exposure to faecal pathogens and those that can multiply in bedding materials". "Since mastitis control systems based upon teat disinfection and dry cow antibiotic therapy have been widely tested and applied for more than 10 years it is timely to consider the success of these measures under commercial conditions and determine the need for any change in direction". "It is clear, however, that S. uberis and coliform mastitis constitute major unsolved problems for the dairy industry". "In the longer term the major research effort going into the investigation of the immune system of the mammary gland and the genetics of susceptibility to mastitis may pay dividends."

These (selected) quotations make clear that:
- milking procedures, including hygiene and technology, are very important in the epidemiology of mastitis,
- control systems based on use of disinfectants and antibiotics are beneficial, but do not solve all problems,
- especially pathogens surviving in the environment may cause pro-

blems,
- the natural defence systems of the animals need attention.
Milking procedures and mastitis control programmes are hardly influenced by climatological factors. We therefore concentrate on the possible environmental or climatological influences on host and mastitis pathogens.

Pasture environment

As indicated in the name of "summer mastitis", it occurs particularly in the summer months. There is a remarkable similarity in habitat and activity of the fly Hydrotaea irritans at pasture and the incidence of summer mastitis (Sol, 1983). Where the fly is not present there is no risk for typical summer mastitis, which also means that it does not occur inside buildings.

Sol (1983) reported a variation in affected herds over a period of six years from about 10 to 50%. The number of affected animals in the same years was about 1 to 6% of the heifers.

O'Rourke et al. (1984) found in the U.K. over six years 40 to 59% of the herds affected. The percentage of dry cows affected was on average 0.7 (0.4-1.2) and the percentage of heifers affected was on average 1.5 (1.1-2.0).

Variation in incidence between animals, months and places are ascribed to variation in the presence of the vector fly.

The effects of summer mastitis are severe. Most times the affected quarters are lost and often the animals do not fully recover or die.

Control measures may be: teat sealers, fly repellants, insecticidal sprays or eartags and for dry cows intramammary application of long-acting antibiotics.

Intramammary infections and clinical mastitis caused by pathogens, other than Cbt. pyogenes, show monthly or seasonal variation. This may mainly be due to the contamination of the teat ends from the lying surface. This is especially important at parturition and in the first months of lactation because of the high infection rate at that time.

Verhoeff et al. (1981) observed the incidence of clinical mastitis in cows without teat lesions. Of 604 cows housed during the first month of lactation clinical mastitis was reported in 11.4% of the animals, whereas of 185 cows at pasture in the first month of lactation 6.5% of the animals had clinical mastitis (P = 0.03).

Jackson and Bramley (1983) state that sporadic cases and, very occasionally, outbreaks of E. coli mastitis occur among pastured cattle but about 70% of the cases occur in housed cows. It should be kept in mind that this distribution will be influenced by the monthly calving distribution and management practices before and at calving.

An epidemiological study in France (Barnouin et al., 1986) showed that monthly rainfall of more than 120 mm increased the risk of clinical mastitis in the pasture period. Incidence rate was 37.2 of cows, whereas this was 16.0 and 13.0 for rainfall of less than 60 mm. or 60-120 mm respectively. Milk cellcount however was lowest in the high rainfall categorie.

Housing environment

General effects on the animals: As compared with the pasture environment, the (winter) housing environment as a rule may pose more stressfull factors on the animals. These include: quality of the floor to lie on or walk, available area per cow, group size, intergroup transfer of cows, manger space per animal, feeding methods and changes in composition of ration, air temperature, relative humidity of air, drafts, increased infection risk, technical malfunction of equipment and handling of stock.

Commonly most cows in a herd give birth to a calf and have their peak milk production during the winter (housing) season. Both these factors can be regarded as stressfull for the animal. Giesecke (1985) reviewed the effects of lactation stress and other stress factors on udder health of dairy cows. He concluded that the metabolic and immunological homeostasis of the dairy cow and its mammary gland depends on: intramammary epithelial integrity, somatic cellular defence and bacterial challenge. It is well known that most intramammary infections and clinical mastitis cases occur at parturition and in early lactation. The homeostatic factors are impaired at that time. There are however considerable differences between cows in susceptibility to mastitis. Van de Geer et al. (1979) compared two groups of cows with low or high rates of intramammary infection. Differences were greater for infections (78 vs. 202) than for clinical mastitis (26 vs. 34). Although the knowledge on resistance factors is increasing (Poutrel, 1982), it is not clear how they are modified by environmental circumstances and what their relative contribution is to the total defence.

Infection rate is also high at drying off and in the dry period and is also increased in older animals. Infections and clinical mastitis usually are higher during the housing period but monthly variations differ by pathogen (Funk et al., 1982; Francis et al., 1981). The varying susceptibility over the dry period may be determined by: bacterial loads on the teat skin, characteristics of the teat skin and internal protective mechanisms (Eberhart, 1986).

General effects on the micro-organisms: Under housing conditions contamination of the bedding material may be the most important factor in the risk of infection. This is particularly valid for bacteria that are able to survive or multiply in the environment: the environmental pathogens. Zehner et al. (1986) inoculated sterilized bedding materials with E. coli, K. pneumoniae and S. uberis. Samples were incubated at 37°C. Rapid growth was observed in chopped straw and recycled manure, there was some growth in hardwood chips, but bacterial counts declined rapidly in chopped newspaper and soft wood sawdust.

Fairchild et al. (1982) determined bacterial counts in untreated samples of bedding material and on teat ends. There was a high correlation in bacterial counts between bedding samples and teat end swabs. Coliform and Klebsiella counts were highest in softwood sawdust and paper and very low in sand and lime. There were significant differences among weeks. In sawdust Klebsiella counts were much higher in the first week (281.8×10^4) than in week 2 to 9 (31.6×10^4 to 0.6×10^4). Environmental temperature, pH of the material, availability of nutrients, contamination and other factors may strongly influence bacterial counts in practical conditions.

Robinson et al. (1985) found an association between poor cleanliness of cows and/or the cows environment and higher rates of coliform infections. This was not so for new S. uberis infections. Another observation was that S. uberis infections were more frequent in the dry and periparturient periods while coliform infections were more common during lactation. The authors stress the difference in epidemiology and pathogenesis between these two environmental pathogens.

Bramley and Neave (1975) observed an increase in coliform infections when coliforms per gram wet weight of bedding rose from 10^4 or 10^5 to

10^7.

In reviewing the possibilities to control coliform mastitis Eberhart et al. (1979) concluded that:

1. bedding materials, especially sawdust may be a source of coliform bacteria;
2. increasing area per cow, keeping stall clean and reduction of housing time may decrease exposure to coliforms;
3. keeping udders dry at milking and proper milking may reduce coliform mastitis.

They also point to the possible effects of weather.

Specific environmental effects: If a cow has to lie on a poorly insulated floor it may result in an elevated cell count in the quarters in contact with the floor (Ewbank, 1966).

Francis et al. (1981) investigated E. coli populations in bedding material in two cubicle buildings. During lying periods of on average two hours the bedding temperature rose to 25°C for low yielding animals and to 30°C for high yielding animals. When environmental (air) temperatures were somewhat lower the average duration of lying periods was extended to three hours. Under those conditions the temperature of the bedding material rose to 26°C and 35°C for low and high yielding animals respectively. Scores for cleanliness or appearance of cubicle bedding condition was somewhat better for the high yielding cows but the E. coli bacterial counts were higher. Bacterial counts showed a wide variation (10^2-10^8) and there was no relationship between visual scores and bacterial counts.

These findings may be explained by the fact that coliform bacteria are intolerant of desiccation but on the other hand they do not thrive in cold, anaerobic conditions which may be the case in grossly soiled bedding or slurry (Jackson and Bramley, 1983).

Climate inside stalls may have an effect on the animals as well as on the (microbes in the) bedding. Spikker (1977) found a relation between poor ventilation and bulk milk somatic cell count and mastitis on commercial farms. Poor ventilation resulted in an elevated air temperature, higher relative humidity and higher CO_2 concentration.

A dairy herd in total confinement (housed all year) was studied by Smith et al. (1985). Incidence rate of clinical mastitis was high (194.4)

and incidence of clinical cases was highest in the first 76 days of lactation and in summer. Cows were housed in cubicles (75%) or tie stalls
(25%) and for 7 days around parturition in individual box stalls. Bedding
material consisted of recycled manure, except the maternity boxes wherein pelleted corn cobs were used. In all stalls and bedding material coliform counts were highest in summer and fall.

If cows are at pasture teat chaps and so called "summer sores" may
occur from adverse weather conditions and/or flies (Francis, 1981). Teat
lesions in general promote intramammary infections and mastitis.

MASTITIS AND MILK PRODUCTION

Nutritional requirements may be changed as a consequence of mastitis.
The most direct relationship exists between level of production and required feed intake. This is especially the case if some secretory tissue is
lost and milk secretion is not compensated for by parallel quarters.

It is unlikely that metabolic efficiency is affected in case of subclinical (local) infections, since there are no systemic signs of disease and
the mammary gland exists of external secretory cells.

In case of acute or chronic clinical mastitis the situation may be extremely different. The animal may suffer from fever and pain, feed intake and digestion (rumen motility) may be disturbed and body energy
reserves may be used. The animal may eventually die or recover after
drying off.

Most cases of clinical mastitis however, if treated properly, are cured
within three to five days. Milk production following an episode of clinical
mastitis with succesful treatment is not necessarily lower than before.

Although clinical mastitis may be costly to the farmer due to discard-
ed milk and treatment costs, the long term relationship between subclinical or clinical mastitis and milk production is much more important, from
both practical and scientific point of view.

Whether high milk yield predisposes to the development of mastitis is
still subject to debate. Comparisons between farms or even between cows
are often biased by genetic-, phenotypic-, nutrition-, feeding- and
management conditions. Jones and Jones (1986) compared cows with
E. coli mastitis with unaffected animals in the same herd which calved

near the same date. Animals developing <u>E. coli</u> mastitis gave significantly higher yields than control animals (mean ± se 24.3 ± 1.01 kg vs 21.53 ± 0.72 kg) in the three weeks before mastitis occurred. An age effect could not be excluded.

Hirsch (1985) reported on the effect of clinical mastitis in the dry period on production 5-7 days post partum. If mastitis was cured and the affected quarter bacteriologically negative, then there was no decrease in production. Infected and not treated quarters showed a production decrease of 15.6%. Measurements were based on comparison with healthy not infected parallel quarters.

In a comparison of successive lactations with and without mastitis in a regression analysis, Lucey and Rowlands (1984) found that clinical mastitis before peak production influenced production significantly negative. Mastitis later than 10 weeks after peak yield however did not affect production.

If a quarter is affected (sub)clinically, production may be compensated by the parallel quarter. Woolford et al. (1984) performed an identical twin study with experimental infections in early lactation. From this experiment it was concluded that intramammary infections in first calf heifers result in a milk yield (kg's and fat) loss of about 8%, both in first and following lactation, even if the infection had been eliminated. For infected mature cows milk loss in the affected quarter may be substantial, but this can almost completely be compensated by parallel quarters.

It has been known for many years (Blackburn, 1966) that milk cell count usually increases with stage of lactation and in successive lactations. The latter is mainly due to an increase in polymorphonuclear leucocytes, which most probably is the result of a history of infections and mastitis.

The productivity loss commonly is for 70-80% attributed to milk yield loss due to subclinical mastitis (Dijkhuizen and Stelwagen, 1981; Jones et al., 1984). This figure however may be much lower in case of low prevalence of subclinical intramammary infections. Yield loss increases with increasing cell count, but is lowest in first lactation and less when average yield per herd is lower (Jones et al., 1984).

The correlation between bulk milk cell count and prevalence of infection is about 0.5 (Grommers, 1984). In quarter milk samples, however, the cell count shows variation by bacteria species (Vecht et al., 1982).

From the literature cited it may be clear that the relationship between infection, mastitis, cell count and milk production is rather complex.

REFERENCES

Barnouin, J., Fayet, J.C., Jay, M., Brochart, M. and Faye, B., 1986. Enquête éco-pathologique continue: facteurs de risque des mammites de la vache lattiere. II. Analyses complementaires sur données individuelles et d'élevage. Can. Vet. J. 27: 173-184.

Blackburn, P.S., 1966. The variation in the cell count of cow's milk throughout lactation and from one lactation to the next. J. Dairy Res. 33: 193-198.

Bramley, A.J. and Neave, F.K., 1975. Studies on the control of coliform mastitis in dairy cows. Br. Vet. J. 131: 160-169.

Bramley, A.J., Dodd, F.H. and Griffin, T.K., ed., 1981. Mastitis control and herd management, Technical Bulletin 4, Nat. Inst. Res. Dairying, Hannah Res. Inst., UK.

Bramley, A.J. and Dodd, F.H., 1984. Reviews of the progress of dairy science: Mastitis control-progress and prospects. J. Dairy Sci. 51: 481-512.

Dodd, F.H., 1983. Mastitis-Progress in control, in: Symposium Advances in understanding mastitis. J. Dairy Sci. 66: 1773-1780.

Dijkhuizen, A.A. and Stelwagen, J., 1981. De economische betekenis van mastitis bij huidig en gewijzigd landbouwbeleid. Tijdschr. Diergeneesk. 106: 492-496.

Eberhart, R.J., 1986. Management of dry cows to reduce mastitis. J. Dairy Sci. 69: 1721-1732.

Eberhart, R.J., Natzke, R.P., Newbould, F.H.S., Nonnecke, B. and Thompson, P., 1979. Coliform mastitis - a review. J. Dairy Sci. 62: 1-22.

Ewbank, R., 1966. A possible correlation, in one herd, between certain aspects of the lying behaviour of tied-up dairy cows and the distribution of subclinical mastitis among the quarters of their udders. Vet. Rec. 78: 299-303.

Fairchild, T.P., McArthur, B.J., Moore, J.H. and Hylton, W.E., 1982. Coliform counts in various bedding materials. J. Dairy Sci. 65: 1029-

1035.

Francis, P.G., 1981. Teat leasions and machine milking. In: Mastitis control and herd management, Bramley et al. (ed.). Technical Bull. 4, Reading-Ayr, UK.

Francis, P.G., Summer, J. and Joyce, D.A., 1981. The influence of the winter environment of the dairy cow on mastitis. Bovine Practitioner 16: 24-27.

Funk, D.A., Freeman, A.E. and Berger, P.J., 1982. Environmental and physiological factors affecting mastitis at drying off and postcalving. J. Dairy Sci. 65: 1258-1268.

Geer, D., van de, Grommers, F.J. and Houten, M. van, 1979. Comparison of dairy cows with low or high rate of udder infection. Vet. Quarterly 1: 204-211.

Giesecke, W.H., 1985. The effect of stress on udder health of dairy cows. Onderstepoort J. Vet. Res., 52: 175-193.

Grommers, F.J., 1984. Het begrip mastitis (The mastitis concept). Syllabus Themadag Mastitis, NRLO, Wageningen: 5-16.

Hirsch, H.P., 1985. Zur postpartalen Milchleistung von Kühen nach Eutererkrankungen in der Trockenperiode. Mh. Vet.-Med. 40: 469-471.

Jackson, E. and Bramley, J., 1983. Coliform mastitis. In practice, July 1983: 135-146.

Jones, T.O. and Jones, P.C., 1986. Cow milk yield and composition before development of Escherichia coli mastitis. Vet. Rec. 119: 319-321.

Jones, G.M., Pearson, R.E., Clabaugh, G.H. and Heald, C.W., 1984. Relationship between somatic cell counts and milk production. J. Dairy Sci. 67: 1823-1831.

O'Rourke, D.J., Chamings, R.J. and Booth, J.M., 1984. Summer mastitis surveys in England and Wales: 1978-1983. Vet. Rec. 115: 62-63.

Poutrel, B., 1982. Susceptibility to mastitis: a review of factors related to the cow. Ann. Rech. Vet. 13: 85-99.

Lucey, S. and Rowlands, G.J., 1984. The association between clinical mastitis and milk yield in dairy cows. Anim. Prod. 39: 165-175.

Remmen, J., 1986. Verslag van de begeleiding van 42 melkveebedrijven door de Stichting Gezondheidsdienst voor Dieren te Boxtel over de periode 1 mei 1984 t/m 30 april 1985. Boxtel.

Robinson, T.C., Jackson, E.R. and Marr, A., 1983. Within herd comparison of teat dopping and dry cow therapy with only selective dry cow therapy in six herds. Vet. Rec. 112: 315-319.

Robinson, T.C., Jackson, E.R. and Marr., A. 1985. Factors involved in the epidemiology and control of Streptococcus uberis and coliform mastitis. Br. Vet. J. 141: 635-642.

Schalm, O.W., Carroll, E.J. and Jain, N.C., 1971. Bovine mastitis, Lea and Febiger, Philadelphia.

Smith, K.L., Todhunter, D.A. and Schoenberger, P.S., 1985. Environmental mastitis: cause, prevalence, prevention; In: Symposium Environmental effect on cow health and performance. J. Dairy Sci. 68: 1531-1553.

Sol, J., 1983. Zomerwrang: de pathogenese, de schade, het voorkomen en de preventie. Tijdschr. Diergeneesk. 108: 443-452.

Spikker, J.W.M., 1977. Een orienterend onderzoek naar de relatie stalklimaat en diergezondheid op melkveebedrijven. Gezondheidsdienst voor Dieren in Noord-Brabant, Boxtel.

Vecht, U., Dam., H. van and Berg, J. van de, 1982. Verslag Landelijke Steekproef Mastitis 1980, Lelystad.

Verhoeff, J., Geer, D. van de and Hagens, F.M., 1981. Effects of a mastitis control programme on the incidence of clinical mastitis. Vet. Quarterly 3: 158-164.

Wilesmith, J.W., Francis, P.G. and Wilson, C.D., 1986. Incidence of clinical mastitis in a cohort of British dairy herds. Vet. Rec. 118: 199-204.

Wilson, C.D. and Richards, M.S., 1982. A survey of mastitis in the British dairy herd. Vet. Rec. 106: 431-435.

Woolford, M.W., Williamson, J.H., Copeman, P.J.A., Nupper, A.R., Phillips, D.S.M. and Uljee, E.J., 1984. The effect of mastitis on production in the following lactation. Proc. Ruakura Farmers' Conf., 36th Conf., Hamilton, New Zealand: 29-33.

Zehner, M.M., Farmsworth, R.J., Appleman, R.D., Larntz, L. and Springer, J.A., 1986. Growth of environmental mastitis pathogens in various bedding materials. J. Dairy Sci. 69: 1932-1941.

THE EFFECT OF GASTROINTESTINAL NEMATODES ON METABOLISM IN CALVES

A. KLOOSTERMAN AND A.M. HENKEN

ABSTRACT

The results of metabolic studies on calves infected with gastrointestinal nematodes are critically reviewed with some reference to work carried out in sheep. Special attention is given to feed intake, energy and nitrogen utilization. Impairment of all these processes may be contributing factors in the poor performance of parasitized animals. In practice the reduced feed intake is undoubtedly the most important factor.

INTRODUCTION

Gastrointestinal nematodes may negatively affect production of meat, wool and perhaps milk in cattle and sheep all over the world. In the field these production losses have mostly been estimated by comparison of anthelmintic treated animals with naturally infected controls. In the laboratory more detailed metabolism studies have been done mainly in sheep and less frequently in calves.

Among these trials there were relatively few in which complete energy balances were assessed by use of respiration chambers. Recent reviews on this area of research are given by Coop (1982), Symons (1985) and Holmes (1986). The present paper will focus on the work that has been done in calves. The topics feed intake, digestion and efficiency of the utilization of metabolizable energy and digested protein will successively be dealt with.

FEED INTAKE

In sheep, Sykes and Coop (1977) found a 20% reduction of voluntary feed intake by lambs daily infected with 4,000 larvae of Ostertagia cir- cumcincta. The same authors found a 10% reduction in lambs infected daily with 2,500 Trichostrongylus colubriformis (Sykes and Coop, 1976). Steel et al. (1980) found a 20% reduction in O. circumcincta infections at an infection rate of 120,000 larvae/week. Steel et al. (1982) reported a 55% reduction in T. colubriformis infections at an infection rate of 30,000/week.

In calves reduction in feed intake was reported by Armour et al. (1973) in Ostertagia ostertagi infections. Kloosterman (1971) found a 50% lower hay intake in heavily infected calves compared to lightly infected ones, after they had been housed and were offered 2 kg of concen- trates/day and ad libitum hay. Twelve days after anthelmintic treatment both groups were on the same (higher) level of hay consumption. These were natural infections, predominantly Cooperia oncophora, O. ostertagi and Nematodirus helvetianus in that order. Albers (1981) found after a single infection with 100,000 C. oncophora larvae 7% reduction of hay in- take during a very limited period around 4 weeks p.i. After secundary infections (both continuous infections and single doses) he found a more extended period of reduced intake, the reductions in hay intake varying from 7 to 20%. Here again the hay was given ad libitum on top of 2 kg of concentrates per day, so the reductions in total feed intake were much smaller. Jordan et al. (1977) working with feedlot steers that har- boured low level infections of Ostertagia spp. and Cooperia spp. found no reduction of feed intake. However they gave restricted rations. Be- sides a maintenance fed group there was a full fed group and these groups received approximately 740 kJ GE/kg$^{0.75}$/day and 1050 kJ GE/kg$^{0.75}$/day respectively. These intakes were much lower than in com- parable experiments with ad libitum feeding (Randall & Gibbs, 1981; Verstegen et al., in press). Randall & Gibbs (1981) studied the metabo- lism of calves comparing a clinically infected group (600,000 larvae of a mixture of O. ostertagi and C. oncophora) and a subclinically infected group (60,000 larvae of the same mixture) with an uninfected group. Taking the reduction of GE intake as a measure of inappetance, it can be calculated from their data that reduction of feed intake was for the clinically infected group 48% at 3 weeks post infection (3 wpi) when compared to preinfection level and 58% if compared to uninfected controls. At 5 wpi the feed intake of the clinically infected group had returned to pre-infection level, but was still 14% lower than that of the control group. The subclinically infected group had a reduced feed intake only at 3 wpi, the reduction being approximately 24%. Verstegen et al. (in press) used calves secundarily infected with 350,000 L3 of C. oncophora after the calves had been primarily infected with 20,000 larvae of the same species 7 weeks earlier (Exp. 1). They found a 6% reduction in GE intake compared to uninfected controls. This reduction was most prominent (10%) at 31-35 days after the secundary infection.

A reduced voluntary intake is by most authors held responsible for the poor growth results of infected animals when compared to ad lib fed controls. The cause of anorexia is still unknown. As possible factors are mentioned: abdominal pain, impaired digesta flow, impaired gastrointestinal motility and changes in concentration of gastrointestinal hormones like cholecystokinin (Symons, 1985; Holmes, 1986). Abbott et al. (1986), studying acute haemonchosis in sheep found a more severe anorexia in lambs given a low protein diet than in those given a high protein diet. The authors did not expect any difference in palatability of the two diets on basis of the ingredients. This leads us to a point that has been largely overlooked by most authors, the feed and the feeding method in relation to intake. At least some influence of palatability on voluntary intake may be expected. Differences in palatability might for instance explain differences in intake between experiments. Even within experiments where the same ration is offered to infected and control groups this factor may play a role. When concentrates plus roughage is offered it is always the roughage that is refused first and to a larger extent than the concentrates (Kloosterman, 1971; Albers, 1981). Also, when only roughage is given the animals select a part of the ration that differs from the refusals not only in composition, but more importantly, also in digestibility, the eaten part having a higher digestibility than the refused part (Zemmelink, 1980). This author also points to another factor affecting voluntary intake: the amount of excess feed offered to the animals. From this it may be concluded that three factors: palatability, selection and amount of excess feed need to be considered separately. They all may influence the respons of infected animals. In many studies they have not received the attention they deserve.

DIGESTIBILITY

It seems logical that at least at the site of infection the digestion and absorption of nutrients is impaired. Nevertheless it is accepted by most authors that the depression of these processes is not a very important factor in the poor utilization of nutrients by infected animals (Coop, 1982; Symons, 1985; Holmes, 1986). Probably a dysfunction of the anterior gastrointestinal tract, where most parasites live, can be compensated by the functional reserves of the parasite-free ileum (Symons, 1985). This may not be possible in the case of Oesophagostomum spp. infections, which are located in the colon and coecum (Bremner, 1969).

Conventional balance studies cannot provide answers on true digestibility because parasitized animals show:
1. an increase in losses of endogenous protein and other materials due to leakage of plasma proteins into the gut;
2. loss of erythrocytes by gastrointestinal haemorrhage and blood sucking parasites;
3. exfoliation of epithelial cells; and
4. increased production of mucus.

The first two processes have been the object of quantitative studies, for the latter two the losses cannot yet be quantified. It seems that a large part of the endogenous protein can be reabsorbed (Poppi et al., 1981). For further details the reader is referred to the reviews given by Holmes (1986) and Symons (1985).

From the foregoing it is clear that the results of balance studies must be interpreted in view of the methods used. This should be kept in mind if below some results of experiments are compared.

Digestibility of nitrogen

Data from various experiments are summarized in Table 1. The apparent digestibility of nitrogen in sheep infected daily with 4,000 larvae of O. circumcincta was reduced 25% when compared to pair fed uninfected controls (Sykes and Coop, 1977). No significant reduction in apparent digestibility of N was found in lambs infected daily with 2,500 larvae of

Table 1. Apparent digestibility of nitrogen (%) in various experiments.

Authors	Host	Parasite(s)	Time	Groups		
Sykes & Coop (1976)	sheep	T.colubr.		PF-cntr.	infect.	
			6- 7 wpi	57.0	55.9	
			12-13 wpi	64.1	60.9	
Sykes & Coop (1977)	sheep	O.circ.	2- 3 wpi	59.5	43.9	
			7- 8 wpi	63.8	54.7	
			12-13 wpi	66.8	61.4	
Randall & Gibbs (1981)	cattle	O.ost + C.onc.		uninf.ctr.	clin.inf.	subcl.inf.
			preinf.	66.7	71.0	71.7
			2 wpi	67.9	64.5	65.7
			3 wpi	68.1	54.2	65.5
			4 wpi	66.4	61.4	66.3
			5 wpi	68.6	67.6	69.5
Entrocasso et al. (1986)	cattle	O.ost.+ C.onc.		clean	MSRB	untreated control
			pre-type 2	53.9	52.5	49.1
			type 2	50.8	54.0	49.2
Verstegen et al. (in press) exp. 2	cattle	O.ost.+ C.onc.		PF-cntr.	infect.	
			2 wpi	68.1	68.0	
			3 wpi	69.7	68.1	
			4 wpi	68.5	67.3	
			5 wpi	66.8	66.4	
			6 wpi	66.5	63.7	
			7 wpi	68.0	64.9	
			8 wpi	69.7	63.8	

T. colubriformis (Sykes and Coop, 1976). In cattle Randall and Gibbs (1981) found a significant depression of apparent N digestibility. Three weeks post infection they found in the clinically infected group a reduction of 24% compared to preinfection level and of 20% compared to uninfected controls. The mean depression over the entire period of 5 weeks was 13% and 9%, compared to preinfection level and to uninfected controls respectively. These were mixed infections of O. ostertagi and C. oncophora. In the subclinically infected group they found a non-significant reduction over the 5-week period of 7% (compared to preinfection level) and 1.5% (compared to the uninfected control group). Entrocasso et al. (1986) found in naturally infected calves during the winter-housing period a depression of nitrogen digestibility of approximately 7% compared to calves that were regularly treated with fenbendazole and calves that had been treated with a Morantel Slow Release Bolus during the preceeding grazing season. These natural infections were predominantly O. ostertagi infections. In a respiration trial (Exp. 2), using pair fed controls and calves that were infected three times a week with doses corresponding to daily doses of 2,500 O. ostertagi and 25,000 C. oncophora, Verstegen et al. found during 8 weeks a mean depression of apparent N-digestibility of 3.2%. The depression became significantly stronger in the last 3 weeks; then it was 5.8%. It should be noted that the intake of roughage and concentrates was not the same among treatments within experiments. Therefore the effects on digestibility of nitrogen may be partly confounded with the composition of ingested feed ingredients.

Digestibility of energy

Table 2 summarizes the results of various experiments. Sykes and Coop (1977) found a two percentage units lower digestibility of energy in lambs infected daily with 4,000 larvae of O. circumcincta compared to pair fed uninfected controls. This corresponded to a 3.7% reduction. With infections of 2,500 larvae daily of T. colubriformis they found no reduced digestibility of energy (Sykes and Coop, 1976). MacRae et al. (1982) however, working with the same dosing regime of T. colubriformis found a significantly lower metabolizability 4 to 8 weeks after infection, while urinary and methane energy were little affected. From their figures the maximum reduction in digestibility can be estimated to be 13 and 9%

Table 2. Apparent digestibility of energy (%) in various experiments.

Author	Host	Parasite(s)	Time	Groups			
Sykes & Coop (1976)	sheep	T.colubr.		PF-cntr.	infect.		
			6- 7 wpi	56.4	56.2		
			12-13 wpi	57.9	55.9		
Sykes & Coop (1977)	sheep	O.circ.	2- 3 wpi	57.5	55.0		
			7- 8 wpi	54.9	53.1		
			12-13 wpi	52.2	50.4		
Jordan et al. (1977)	cattle	O. ost. + C. onc.		FFP	FFC	MFP	MFC
			pre-inf.	72.1	68.7	73.1	74.2
			post-inf.	72.3	71.3	74.7	77.2
Randall & Gibbs (1981)	cattle	O. Ost. + C. onc.		uninf. cntr.	clin. inf.	subcl. inf.	
			pre-inf.	66.1	66.8	74.6	
			3 wpi	68.6	39.4	61.0	
			5 wpi	67.8	69.2	67.2	
Entrocasso et al. (1986)	cattle	O. ost. + C. onc.		clean	MSRB	untreated control	
			pre-type 2	58.0	59.0	55.0	
			type 2	59.0	62.0	59.0	
Verstegen et al. (in press) Exp. 2	cattle	O. ost. + C. onc.		PF cntr.	infect.		
			2 wpi	70.9	68.9		
			3 wpi	70.2	68.6		
			4 wpi	70.2	67.1		
			5 wpi	67.7	67.9		
			6 wpi	68.3	64.8		
			7 wpi	70.6	67.0		
			8 wpi	71.0	66.7		

in the successive experiments. Jordan et al. (1977) working with feedlot steers with very low mixed infections of O. ostertagi and Cooperia spp. found no influence on energy digestibility. Their digestibility figures were on a very high general level, when compared to similar experiments by other authors, presumably as a result of the restricted feeding levels. Randall and Gibbs (1981) found in clinically infected calves a 41% reduction in energy digestibility, compared to preinfection levels and a 43% reduction when compared to uninfected controls, 3 weeks after infection. In subclinically infected calves these figures were 18% and 11% respectively. At 5 weeks p.i. both groups had returned to normal. This lowered digestion was even found while animals had significantly reduced their feed intake. Entrocasso et al. (1986), working with housed calves after natural infections on pasture, found a 5% reduction in digestibility during the following winter season. This was in untreated controls, compared with fenbendazole and MSRB treated animals on a restricted feeding regime. Verstegen et al. (in press) found an average reduction of 3.4% in energy digestibility, during 8 weeks of continuous dosing (Exp. 2). The depression was most pronounced during the last 3 weeks, when a 5.2% reduction was seen.

Summarizing the results on digestibility, it can be seen in Table 1 that from the 22 occasions that the N digestion was studied in infected animals, there is only one example (the subclinically infected group of Randall and Gibbs at 5 weeks p.i.) where the digestion was not impaired. The digestion of energy (Table 2) shows a similar trend, although less clear. Jordan et al. (1977) found in all their groups a higher digestion of energy after exposure than before. But as stated earlier, these experiments were done at a very low level of parasitism. In the other experiments there were 18 occasions where infected animals or their controls were subjected to balance studies. On only two occasions the digestibility of energy was not impaired in parasitized animals: Randall and Gibbs the clinically infected group at 5 weeks p.i. and Verstegen et al., exp. 2., the infected group at 5 weeks p.i.

It should once more be stressed that we are dealing with apparent digestibility figures. More data are needed from experiments with cannulated animals and labelled nutrients to estimate the influence on true digestion of nutrients in separate parts of the gastrointestinal tract.

UTILIZATION OF NUTRIENTS

The efficiency of utilization of metabolizable energy and digested nitrogen can be and has been estimated in two different ways. In a complete balance trial in respiration chambers (with measurement of O2, CO2, NH3 and CH4) the nitrogen and energy balance (NB and EB) can be assessed and utilization is expressed as NB/DN and EB/ME. This procedure was used by Jordan et al. (1977), Randall and Gibbs (1981) and Verstegen et al. (in press). This procedure can also be used for nitrogen in conventional balance trials, although here no measurement of NH3-loss is possible. The energy balance can also be estimated from whole carcass evaluation. This method was followed by Sykes and Coop (1976, 1977).

Utilization of nitrogen

The ratio NB/DN as a measure of N-utilization is presented for several experiments in Table 3. This ratio is directly related to the loss of N in urine, because NB = DN -urinary N, if we ignore the relatively small losses with NH_3 as is done by most authors. In experiment 2 of Verstegen et al. the loss of N by air was estimated to be 5.5% of that in urine, both for infected and for uninfected control animals.

Increased urinary N excretion has been observed by many authors a.o. Dargie (1973, Haemonchus contortus, sheep), Parkins et al. (1973, O. circumcincta, sheep) and Roseby (1977, T. colubriformis, sheep). In Table 3 the NB/DN ratio is decreased in many of the separate balance trials, but not in all. In the T. colubriformis experiment of Sykes and Coop (1976) the utilization of the pair fed control group is very low (and in fact inferior to that of the infected group) in the trial 12-13 weeks p.i. In the O. circumcincta experiment of the same authors (Sykes and Coop, 1977) the general level of utilization is low in both groups but in all trials the p.f. controls are superior to the infected group. In the experiment of Randall and Gibbs (1981) all NB/DN ratios were lower in infected than in uninfected animals, except that in the subclinically infected group at 5 weeks p.i. The utilization in the uninfected controls is remarkably high in this experiment. These animals were fed ad lib however, in contrast to all other uninfected control groups mentioned in the table. Presumably the feeding level per se has a great influence on the

Table 3. Utilization of nitrogen (NB/DN) * 100% in various experiments.

Authors	Host	Parasite(s)	Time	Groups		
Sykes	sheep	T. colubr.		PF-cntr.	infect.	
& Coop			6- 7 wpi	12.2	4.3	
(1976)			12-13 wpi	3.7	20.0	
Sykes	sheep	O. circ.	2- 3 wpi	4.7	-24.0	
& Coop			7- 8 wpi	13.6	12.4	
(1977)			12-13 wpi	12.6	10.8	
Randall	cattle	O. ost. +		uninf.ctr.	clin.inf.	subcl.inf.
& Gibbs		C. onc.	preinf.	54.2	45.4	44.2
(1981)			2 wpi	49.7	35.7	46.8
			3 wpi	54.3	6.4	40.8
			4 wpi	50.5	41.9	45.8
			5 wpi	40.0	49.8	39.7
Entrocasso	cattle	O. ost. +		clean	MSRB	untreated
et al.		C. onc.	pre-type 2	42.3	41.0	37.7
(1986)			type 2	30.9	31.0	15.6
Verstegen	cattle	O. ost. +		PF.ctr.	infected	
et al.		C. onc.	2 wpi	44.0	41.6	
(in press)			3 wpi	43.7	47.0	
Exp. 2			4 wpi	40.2	45.3	
			5 qpi	40.5	43.4	
			6 wpi	34.3	32.2	
			7 wpi	30.0	29.2	
			8 wpi	27.4	33.0	

utilization of nitrogen. The data of Verstegen et al., exp. 2, show an erratic pattern, the NB/DN ratio in infected animals being higher at four occasions and at three lower than that in controls. As the mean ratio is higher in infected animals, this is the only experiment from which it is

impossible to conclude that the utilization of nitrogen is impaired by infection.

A comparison between data from uninfected, pair fed controls in exp.
2 of Verstegen et al. with those from uninfected ad lib fed calves of
Randall and Gibbs is revealing. It appears that nitrogen utilization is not
only very much influenced by level of intake, but also by the protein/
energy ratio of the diet. The mean intake of DN in the controls of Randall and Gibbs was 1.514 g/kg 0.75/day, while it was 1.713 g/kg 0,75/
day in exp. 2 of Verstegen et al., i.e. 13% more. The intake of digestible energy (DE) was in the two groups 1205 and 1000 kJ/kg 0.75/ day
respectively, that is 17% less in the experiment of Verstegen et al. It
may be assumed that in the latter group protein has been used as a
source of energy by these animals, which is supported by the results of
urinary N output. This was in the Randall and Gibbs animals 0.735g/kg
0.75/day and in the Verstegen et al. animals 1.012 g/kg 0.75/day. So it
is clear that any results of N-utilization should be interpreted with cau-
tion, taking the feeding level and protein/energy ratio into account.

Utilization of energy

The utilization of energy, expressed as percentage of metabolizable
energy (ME) that is retained as Energy Balance (EB), is presented in
Table 4. ME is calculated as digestible energy (DE) minus energy lost in
methane and urine. Holmes (1986) referring to unpublished work with
subclinical haemonchosis in lambs, found an increased methane production. Randall and Gibbs (1981) and Verstegen et al. (exp. 1, in press)
found no significant differences but methane production in infected animals was at a lower level, presumably as a result of the lower feed intakes compared to the ad lib fed controls. In exp. 2 of Verstegen et al.
pair fed controls were used. There was a significant increase in methane
production during the period 2 to 8 weeks after commencement of infections in the pair fed control group, but not in the infected group,
where the course of methane production during the experiment was erratic.

EB was measured by Sykes and Coop (1976, 1977) and Coop et al.
(1982) from carcass analyses, in the other experiments it has been calculated as ME minus the Heat Production (HP) of the animals.

Table 4. Utilization of energy (EB/ME) * 100%, in various experiments.

Author	Host	Parasite(s)	Time	Groups				
Sykes & Coop (1976)	sheep	T. colubr.	Total 14 weeks	Adlib. contr. 26.2	PF-contr. 24.2	infect. 13.3		
Sykes & Coop (1977)	sheep	O. circ.	Total 14 weeks	19.4	20.4	14.0		
Coop et al. (1982)	sheep	O. circ.	Total 14 weeks	Adlib. contr. 22.6	1 (1000) 24.6	2 (3000) 18.2	3 (5000) 15.5	4 (5000) +anth. 17.7
Jordan et al. (1977)	cattle	O. ost. + C. onc.	pre-inf. post-inf.	FFP 40.0 32.2	FFC 37.7 39.5	MFP 21.9 25.8	MFC 22.2 22.0	
Randall & Gibbs (1981)	cattle	O. ost. + C. onc.	pre-inf. 3 wpi 5 wpi	Uninf. contr. 35.4 41.4 38.0	Clin. inf. 21.9 -179.3 22.5	Subcl. inf. 41.5 15.7 43.8		
Verstegen et al. (in press) Exp. 1	cattle	C. onc.	10-14 dpi 17-21 dpi 22-28 dpi 31-35 dpi	Adlib. contr. 29.4 31.0 26.4 23.7	Infect. 27.5 36.1 29.4 21.7			
Verstegen et al. (in press) Exp. 2	cattle	O. ost. + C. onc.	2 wpi 3 wpi 4 wpi 5 wpi 6 wpi 7 wpi 8 wpi	PF-contr. 18.1 27.2 21.2 19.5 6.6 13.1 17.6	Infect 17.3 22.0 14.3 17.7 7.2 3.0 11.3			

In all the experiments where respiration trials were done (Jordan et al., 1977; Randall and Gibbs, 1981; Verstegen et al., in press) a significant correlation was found between the level of intake (GE) and the Heat Production. Therefore differences can be expected in HP between infected animals and ad lib fed controls but not between infected animals and pair fed controls. However, in exp. 2 of Verstegen et al. a higher HP is seen in the infected animals despite the fact that they had a lower ME intake than the controls.

Furtheron, when comparing the EB/ME percentages in various experiments, it should be clear that a fixed part of the ME is necessary for maintenance (MEm), the amount being directly proportional to the metabolic weight of the animal. The remainder of the ME (MEp) is available for production, i.e. the deposition of fat and protein. Because MEm is fixed the ratio EB/ME is dependent on the level of ME intake. A reduction in feed intake will therefore result in a reduced utilization, and so will a reduced ME intake that is caused by impaired digestion.

If with this in mind, we compare the gross efficiencies (EB/ME) in Table 4, we see in the first place that the utilization depends heavily on the general intake level. In the experiment of Jordan et al. (1977) where parasite burdens were low and played no significant role in digestion, the efficiency was much higher in the full fed groups than in the maintenance fed groups where it should actually be zero, when defining maintenance as EB = 0. Ad lib fed controls (Randall and Gibbs, 1981; Verstegen et al., exp. 1) show higher efficiencies than pair fed controls (Verstegen et al., exp. 2). The efficiency in pair fed controls in the latter experiment shows a sharp decrease at 6 and 7 weeks p.i., when intake of infected counterparts was seriously affected. Secondly we see that in most of the infected groups the gross efficiency is impaired and that the decrease corresponds with the level of infection (Coop et al., 1982; Randall and Gibbs, 1981).

The results of Verstegen et al., exp. 1, are rather exceptional in that no decreased utilization of energy by infected animals was found. There are several possible reasons for this, for instance the species of parasite used, and the fact that it was a secondary infection: the calves had 7 weeks earlier experienced infections with 20,000 larvae of the same species. The most likely reason however is the feature that during the third and fourth week of infection the calves showed signs of decreased digesta flow. The faecal output was decreased while feed intake remained constant unit 4 weeks p.i. Clinical signs of infection (reduced weight gain, diarrhee) became apparent in the last week of the experiment.

In exp. 2 of Verstegen et al. we see in both groups a low gross efficiency if compared to the other calf experiments. The poor utilization in the uninfected controls is probably due to the pair feeding system: it falls to very low levels in weeks 6 and 7 p.i., when the intake of infected animals was seriously affected. From the high correlations between EB and ME (0.96 and 0.97 for uninfected and infected groups respectively) it can be concluded that the low utilization is predominantly caused by the low ME intake, but the somewhat higher HP in infected animals has also made a certain contribution (Verstegen et al., in press).

It is well known that infected animals are rather dull and inactive compared to parasite free animals. This was also seen in the experiment 2 of Verstegen et al. but in addition it was observed that the pair fed controls were restless and unquiet, especially in periods when the restriction of feed was severe. The HP, expressed as percentage of ME was on average 83% in the uninfected p.f. control group and 97% in the infected group. Both percentages were very high compared to those of other experiments: Jordan et al. found 64% in full fed animals and 67% in maintenance fed ones, Randall and Gibbs found 124% in infected animals and 65% in uninfected, and Verstegen (exp. 1) 71% and 72% in infected and uninfected animals respectively. The high percentage of 83% in p.f. controls in exp. 2 of Verstegen et al. might be explained by the increased activity of the animals. The 93% of the infected animals in the same experiment can only be explained by an increased use of energy for the increased protein synthesis that occurs in infected animals to compensate for plasmaprotein leakage and loss of other constituents of endogenous protein (Holmes, 1986). It must however be noted that HP as a percentage of ME will increase when the difference between ME intake and maintenance becomes smaller.

Summarizing it is clear from Table 4 that the gross efficiency of energy utilization is impaired in infected animals and that this is largely due to a reduced ME intake. When infected animals are compared to ad lib fed controls this reduced ME intake must be attributed to inappetance of infected animals. If comparisons are made with p.f. controls, the reduced ME intake stems from impaired digestion.

The net efficiency of utilization of energy is the conversion of MEp into deposited fat and protein. MEp is that part of ME that is left for production after MEm (the energy necessary for maintenance) has been

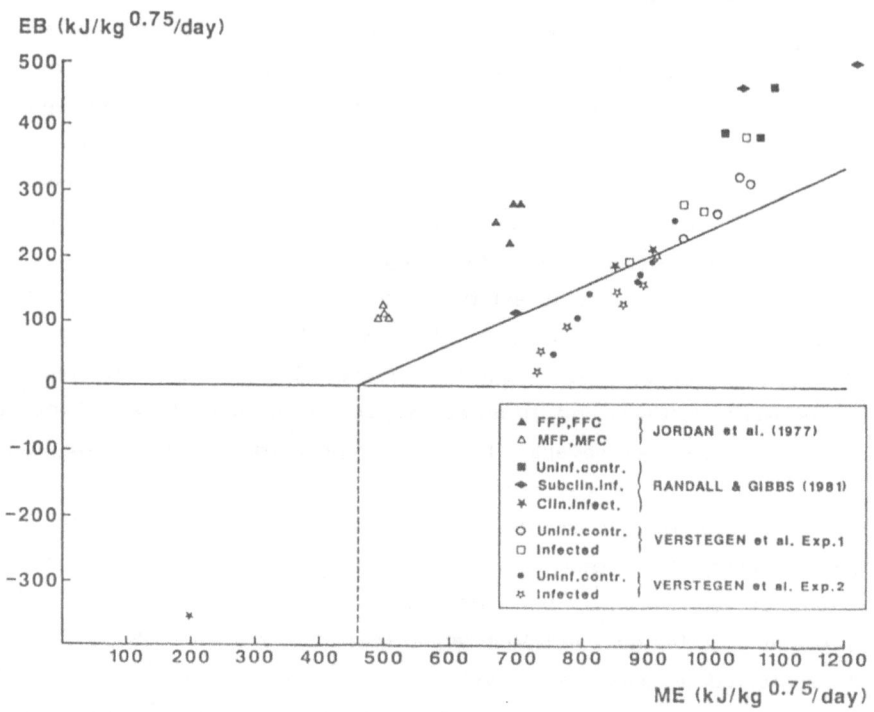

Figure 1. The relation between metabolizable energy (ME) intake and energy balance (EB).

subtracted. MEm is estimated at about 460 kJ/kg 0.75/day for warm blooded animals (Van Es, 1972). The solid line in Fig. 1 gives the relation between ME and EB; it is based on the equation Kg = 0.0078 * q + 0.006, wherein q is an average of all trials. In this equation Kg is the net efficiency coefficient for growth and q = ME/GE in % (Van Es, 1978). The MEm that can be estimated for the various experiments varies considerably (Table 5). A very low maintenance level can be calculated for the animals of Jordan et al. (1977). However in their study ME- and HP-measurements were not done at the same moment (ME before HP) and same place (stalls vs. chambers). This may have caused an underestimation of HP and, thus, an overestimation of EB. The HP of their full-fed animals (calculated as ME - EB) was 432 kJ/kg 0.75 and of the main-

Table 5. The net efficiency of energy utilization calculated directly as $Kg = 0.0078*q + 0.006$, or derived (under the assumption that ($Mem = 460$ kJ/kg$^{0.75}$), $Kg = EB/(ME-460)$, and the derived MEm (under the assumption that $Kg = 0.0078*q + 0.006$), $MEm = ME - EB/(0.0078*q + 0.006)$.

Authors	Exp. groups	EB (kJ/kg$^{0.75}$)	ME (kJ/kg$^{0.75}$)	Kg direct	Kg derived	MEm derived (kJ/kg$^{0.75}$)
Jordan	FFP and FFC	258	690	0.505	1.124	179
et al.	MFP and MFC	114	495	0.529	3.278	279
Randall Gibbs	Group 1 (clin. inf.)*	196	880	0.470	0.466	463
	Group 2 (subc. inf.)	358	988	0.469	0.636	225
	Group 3 (adlib contr.)	407	1063	0.471	0.676	199
Verstegen et al. Exp. 1	Adlib contr.	285	1035	0.450	0.494	414
	Infected	280	980	0.465	0.538	378
Verstegen et al. Exp. 2	Pair fed contr.	151	850	0.470	0.387	529
	Infected	109	818	0.452	0.304	576

*: leaving out the negative values at wpi 3

tenance-fed animals 381 kJ/kg 0.75. This is very low indeed compared to the mean HP found by Randall and Gibbs (642 kJ/kg 0.75) and by Verstegen et al.: 725 kJ/kg 0.75 (exp. 1) and 704 kJ/kg 0.75 (exp. 2). Also the calculated maintenance in the experiment of Randall and Gibbs (1981) seems to be too low, at least for the parasite free control group and the subclinically infected group. Reasons for this are also not clear, but it might be a result of their correction of HP data for differences in activity. As stated above, differences in activity between parasitized and parasite free animals may be large. To obtain accurate estimations of EB, activity related HP should not be corrected for, the measured HP should be used regardless the amount of activity.

Both from Fig. 1 and Tabel 5 it appears that in the first three experiments the EB, and thus the utilization is higher than can be expected on the basis of GE and ME intake. Only in the last experiment (Verstegen et al., exp. 2) the utilization is relatively low in both the infected and the pair fed control group, but lowest in the infected group. This is perhaps caused by the fact that in this experiment the calves were infected more or less continuously.

CONCLUSION

When the energy and nitrogen metabolism of calves is studied after infection with gastrointestinal nematodes, several factors appear to contribute to the poor growth performance. The most important factor in practical situations is undoubtedly the reduced feed intake of infected animals. In addition impaired digestion resulting in lower N and energy balances are seen in most cases.

Reduced utilization of nitrogen was shown in various experiments except in one where it could be assumed that nitrogen utilization in the pair fed control group was impaired because protein was used as energy source.

Utilization of energy expressed as EB/ME (gross efficiency) was reduced in most cases. It was however difficult to show a reduction in net efficiency (EB/MEp). This occurred only in the situation of continuously infected calves and during a limited period after a single, clinical disease producing infection.

REFERENCES

Albers, G.A.A., 1981. Genetic resistance to experimental Cooperia oncophora infections in calves. Meded. Landbouwhogeschool, Wageningen, 81-1.

Armour, J., Jennings, F.W., Murray, M and Selman, J., 1973. Bovine
ostertagiasis. Clinical aspects, pathogenesis, epidemiology and control. In: Helminth Diseases of Cattle, Sheep and Horses in Europe
(G.M. Urquhart and J. Armour, eds.), University Press, Glasgow,
pp. 11-20.

Bremner, K.C., 1969. Pathogenic factors in experimental bovine oeso-
phagostomosis. IV. Exudative enteropathy as a cause of hypoprotein-
aemie. Exp. Parasitol. 25: 382-394.

Coop, R.L., 1982. The impact of subclinical parasitism in ruminants.
In: Parasites - their world and ours. Proc. 5th Intern. Congress of
Parasitology, Toronto, Canada, 7-14 August 1982 (D.F. Mettrick and
S.S. Desser, eds.), Amsterdam, Elsevier Biomedical Press (1982),
pp. 439-450.

Coop, R.L., Sykes, A.R. and Angus, K.W., 1982. The effect of three
levels of intake of Ostertagia circumcincta larvae on growth rate, food
intake and body composition of growing lambs. J. Agric. Sci. (Cambridge) 90: 247-255.

Dargie, J.D., 1973. Ovine haemonchosis. Pathogenesis. In: Helminth
Diseases of Cattle, Sheep and Horses (G.M. Urquhart and J. Armour,
eds.), University Press, Glasgow, pp. 63-71.

Entrocasso, C.M., Parkins, J.J., Armour, J., Bairden, K and McWil-
liam, P.N., 1986. Metabolism and growth in housed calves given a
morantel sustained released bolus and exposed to natural trichostron-
gyle infection. Res. Vet. Sci. 40: 65-75.

Es, A.J.H. van, 1972. Maintenance. In: Handbuch der Tierernährung
(W. Lenkeit and K. Breirem, eds.), Paul Parey, Hamburg, Vol. II:
1-54.

Es, A.J.H. van, 1978. Feed evaluation for ruminants. I. The system in
use from May 1977 onwards in The Netherlands. Livest. Prod. Sci. 5:
331-345.

Holmes, P.H., 1986. Pathophysiology of nematode infections. In: Parasitology - Quo vadit? Proc. 6th Intern. Congr. Parasitology (M.J.

Howell, ed.), Austr. Acad. Sci., Canberra, 1986, p. 443-451.

Jordan, H.E., Cole, N.A., McCroskey, J.E. and Ewing, S.A., 1977. Influence of Ostertagia ostertagi and Cooperia infections on the energetic efficiency of steers fed a concentrate ration. Am. J. Vet. Res. 38: 1157-1160.

Kloosterman, A., 1971. Observations on the epidemiology of trichostrongylosis of calves. Meded. Landbouwhogeschool, Wageningen, 71-10.

MacRae, J.C., Smith, J.S., Sharman, G.A.M., Corrigall, W. and Coop, R.L., 1982. Energy metabolism of lambs infected with Trichostrongylus colubriformis. In: Energy metabolism of Farm animals (A. Ekern and F. Sundstol, eds.), E.A.A.P. Publ. no. 29, The agricultural University of Norway, 1982, pp. 112-115.

Parkins, J.J., Holmes, P.H. and Bremner, K.C., 1973. The pathophysiology of ovine ostertagiasis: some nitrogen balance and digestibility studies. Res. Vet. Sci. 14: 21-28.

Poppi, D.P., MacRae, J.C. and Corrigall, W., 1981. Nitrogen digestion in sheep infected with intestinal parasites. Proc. Nutr. Soc. 40: 116a.

Randall, R.W., and Gibbs, H.C., 1981. Effects of clinical and subclinical gastrointestinal helminthiasis on digestion and energy metabolism in calves. Am. J. Vet. Res. 42: 1730-1734.

Roseby, F., 1973. Effects of Trichostrongylus colubriformis (nematoda) on the nutrition and metabolism of sheep. I. Feed intake, digestion and utilization. Austr. J. Agric. Res. 24: 947-953.

Steel, J.W., Jones, W.O. and Symons, L.E.A., 1982. Effects of a concurrent infection of Trichostrongylus colubriformis on the productivity and physiological and metabolic responses of lambs infected with Ostertagia circumcincta. Austr. J. Agric. Res. 33: 131-140.

Steel, J.W., Symons, L.E.A. and Jones, W.O., 1980. Effects of level of larval intake on the productivitiy and physiological and metabolic responses of lambs infected with Trichostrongylus colubriformis. Austr. J. Agric. Res. 31: 821-838.

Sykes, A.R. and Coop, R.L., 1976. Intake and utilization of food by growing lambs with parasitic damage to the small intestine caused by daily dosing with Trichostrongylus colubriformis larvae. J. Agric. Sci. (Cambridge) 86: 507-515.

Sykes, A.R. and Coop, R.L., 1977. Intake and utilization of food by growing sheep with abomasal damage caused by daily dosing with

Ostertagia circumcincta larvae. J. Agric. Sci. (Cambridge) 88: 671-677.

Symons, L.E.A., 1985. Anorexia: occurrence, pathophysiology and possible causes in parasitic infections. Adv. Parasitol. 24: 103-133.

Verstegen, M.W.A., Hel, W. van der, Albers, G.A.A. and Kloosterman, A., in press. Effect of trichostrongylid infection on feed intake, metabolic rate and protein gain of calves.

Zemmelink, G., 1980. Effect of selective consumption on voluntary intake and digestibility of tropical forages. Agric. Res. Reports 896, PUDOC, Wageningen, 1980.

ENERGY AND NITROGEN METABOLISM OF GROWING CALVES CONTIN-
UOUSLY INFECTED WITH DICTYOCAULUS VIVIPARUS

J.H. BOON AND M.W.A. VERSTEGEN

ABSTRACT

The influence of a moderate level of continuous lungworm infections on energy and nitrogen metabolism of calves was investigated. A pair-feeding and a pair-growing experiment were performed. The metabolism parameters were related to clinical, parasitological, haematological and serological parameters. It was shown that metabolism changed during the clinical phase of the infection. The maintenance requirements of the infected calves increased with about 5% above that of non-infected controls.

INTRODUCTION

Infections of cattle with <u>Dictyocaulus</u> <u>viviparus</u> (D.v.) are very common in regions with a temperate climat (Jørgensen and Ogbourne, 1985).
The incidence in the Netherlands is estimated to be at least 80% (Boon et
al., 1984b). The infection causes often an obstructive bronchitis-bron-
chiolitis (Jarrett et al., 1960; Lekeux et al., 1985) resulting in a reduced elasticity and damaged epithelial tissue of the lungs and a high
activity of macrophages and eosinophilic granulocytes. The gross- and
histopathology caused by D.v. is described by Jarrett et al. (1960).
They divided the disease process in 4 phases: 1. the penetration phase
with penetration of the ingested infective larvae (L_3) into the body and
their migration to the lungs, 1-7 days post infectionem (p.i.); 2. the
prepatent phase with development of the worms in the lungs, 7-25 days

p.i.; 3. the patent phase with egg producing mature worms in the bronchi, 25-55 days p.i.; and 4. the post-patent phase with reconvalescence of diseased animals, 55-77 days p.i. Clinical symptoms most frequently observed are: 1. an increased respiration rate; 2. coughing; and 3. a reduced growth rate (Boon et al., 1984a). These symptoms are related to the pathological changes in the lungs which are most pronounced at the end of phase 2 and during phase 3. The reduced growth rate may be caused by a reduced appetite (Boon et al., 1984a) as is observed in other animals (Abbott et al., 1986). In addition infection may also influence the utilization of nutrients. Effect of appetite and utilization can be be estimated separately in experiments with pair gaining and pair feeding (Kroonen et al., 1986). In gastro-intestinal parasitic infections these techniques have been used by Steel et al. (1980) and Abbott et al. (1986).

In two recent experiments performed in Wageningen the effects of D.v. on energy and nitrogen metabolism of growing calves were measured. In experiment I D.v. infected animals received similar amounts of feed as non-infected controls, i.e. pair feeding based on the average of the infected group (Kroonen et al., 1986). In experiment II the feed allowance of the control animals was decreased in such a way that growth of these animals was similar to that of infected ones, i.e. pair gaining based on individuals (Verstegen et al., 1987).

MATERIALS AND METHODS

Animals, infections, feeding and experimental procedures

In each experiment ten Dutch Friesian male animals were used. They were reared worm-free and housed individually in wooden boxes from about 5-7 days of age onwards. During the rearing period they received liquid milk substitute and from 9 weeks of age onwards ad lib good quality hay and water. Concentrates were supplied in restricted quantities. During the whole rearing and experimental period calves were weighed individually each week. Schemes of the respective experiments are given in Figure 1.

Experiment I: At 13 weeks of age the animals (about 100 kg) were allocated to either a control (C) or an experimental group (I) of five ani-

374

Experiment I:

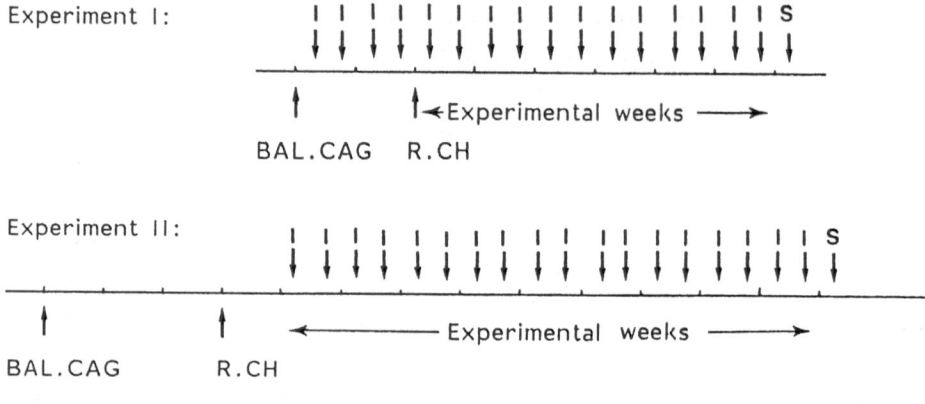

<Experimental weeks ──►

BAL.CAG   R.CH

Experiment II:

──────── Experimental weeks ──►

BAL.CAG   R.CH

Figure 1.   Scheme of Experiment I and II (I = artifical oral infection with 640 infective larvae of D.v.; S = slaughtering; BAL.CAG = placed on balance cages; R.CH = transferred to respiration chambers).

mals each and placed in individual metabolism cages. The groups were balanced for weight and rate of gain during the rearing period. From this week onwards until week 21 I calves were infected twice weekly with 640 infective larvae (L_3) of D.v. in a gelatine capsule. These L_3 were developed and harboured from a donor calf (Boon et al., 1984a). In week 2 post initial infection (P.I.I.) the cages with the I and C calves were transferred to two respiration chambers. Cages with C calves were placed in one chamber, cages with I calves in the other. During the first 3 weeks P.I.I. both I and C calves received hay ad lib and 1 kg concentrates per day. In the last 5 weeks control animals received 1.5 kg concentrates and 1.3 kg hay in two portions per day. These amounts were similar to the average ad lib intake of the infected animals.

Experiment II: The procedure of allocating the calves (about 110 kg) to an I or C group and method and level of infection were the same as in Experiment I. However, the calves were placed in the metabolism cages 4 weeks before the start of the experiment. The experimental period started at first infection (Fig. 1). Feed was provided to each calf twice daily during the experimental period. Infected animals received 2 kg concentrates per day. The feed allowance of the controls was diminished over time to equalize their rate of weight gain with that of their infected pair-mates. Hay was provided to each calf on the following rates: 9-13 weeks of age, 1.25 $kg.d^{-1}$; 14-17 weeks of age, 1.50 $kg.d^{-1}$; and 18-21 weeks of age, 1.75 $kg.d^{-1}$.

Table 1.   Composition of the feed used in the experiments.

	Experiment I	Experiment II
Hay		
dry matter (%)	80.0	88.8
crude protein (%)	11.0	8.5
Energy in dry matter (kJ/g)	20.7	17.7
Concentrates		
dry matter (%)	88.8	80.4
crude protein (%)	16.0	16.9
Energy in dry matter (kJ/g)	18.8	20.3

Compositions of the feed used in both experiments are given in Table 1. In both experiments the environmental temperature was maintained at 20°C and relative humidity at about 65%. Water was supplied ad lib. Light was on from 8 a.m. to 6 p.m. At the end of the experiments all calves were slaughtered.

The following will deal especially with the metabolism parameters. The non-metabolic parameters used are given in Table 2.

Digestibility of Dry Matter (DM), Nitrogen (N) and Gross Energy(GE)

The amounts of feed ingested and faeces excreted by each animal were recorded and collected daily in each balance period. Feed and faeces (mixed samples of 1 week) were analyzed for DM, N- (Kjeldahl) and GE content (Bombcalorimetry). Digestibility of dietary DM, N and GE were thus measured on a per animal per week basis.

Table 2. Non-metabolic parameters used in Experiment I and II and differences between infected (I) and control (C) calves during the first 8 weeks P.I.I.

Parameter	References	Exp.	Week 1 I	Week 1 II	Week 2 I	Week 2 II	Week 3 I	Week 3 II	Week 4 I	Week 4 II	Week 5 I	Week 5 II	Week 6 I	Week 6 II	Week 7 I	Week 7 II	Week 8 I	Week 8 II
Clinical																		
respiration rate	Boon et al. (1984a)		0	0	←	0	←	←	←	←	←	←	←	←	←	←	←	←
coughing	Boon et al. (1984a)		0	0	←	0	←	←	←	←	←	←	←	←	←	←	←	←
Parasitological																		
faecal larval count	Bearmann (1917)		0	0	0	0	0	0	0	←	←	←	←	←	←	←	←	←
eggs in sputum	Boon (1979)		0	X	0	X	0	X	0	X	←	X	←	X	←	X	←	X
lungworm burden	Oakly (1980)							0	0	←								
Haematological																		
peripheral eosinophils	Romeis (1967)		X	0	X	←	X	←	X	←	X	←	X	←	←	←	←	←
Serological																		
specific antibodies	Boon et al. (1984a)		0	0	0	0	0	0	←	←	←	←	←	←	←	←	←	←
angiotensin converting enzyme	Neels et al. (1983)		X	0	X	0	X	0	X	0	←	0	←	0	←	0	←	0
Cytological																		
number of cells in broncho alveolar lavage fluid (BAL)	Trigo et al. (1984)															X		←
eosinophils in BAL	Trigo et al. (1984)															X		←
macrophages in BAL	Trigo et al. (1984)															X		←
lymphocytes in BAL	Trigo et al. (1984)															X		←
neutrophils in BAL	Trigo et al. (1984)															X		←
Post mortal																		
weights of liver, lungs heart and pulmonary lymphnodules	Verstegen et al. (1987)															X		←

0: no difference between mean values of I and C calves
←: I-C = positive
←: I-C = negative
X: not done

Metabolizability

The amount of urine voided by each animal was collected with hydrochlorid acid and sampled daily. The samples for 1 week of each animal were mixed and analyzed for N and GE. Metabolizability of GE (ME%) was determined per animal subtracting energy in urine (UE) from digestible energy (DE). The CH_4 production was measured per chamber (for 5 animals together) and divided over animals by assuming it to be related to GE intake.

N balance

The N balance (Nbal) was determined from N ingested and from N excreted in faeces and urine. It was corrected for N escaping as NH_3 assuming the correction to be equivalent to the N excreted in the faeces and urine per animal. This assumption had to be made because NH_3 production was measured per chamber (over 5 animals together).

Energy balance

Energy gain was calculated from ME and heat production. The heat production was measured by determining gaseous exchange for 48 hours on days 2 + 3 and 5 + 6 in each week P.I.I. Weekly energy gain for each animal was calculated by subtracting heat production (H) from the individual ME values. H was divided over animals by assuming that each animal in the group had the same H, on a per unit metabolic weight and a per unit ME intake basis. Fat gain was calculated by subtracting energy deposited in protein estimated from the N balance from the total energy gain.

Statistics

In Experiment I differences within weeks between the two treatment groups were subjected to analysis of variance according to the model:

$$y_{ij} = \mu + A_1 + e_{ij}$$

and in Experiment II according to the model (also within weeks):

$$y_{ijk} = \mu + A_i + B_j + AB_{ij} + e_{ijk}$$

where $y_{ij(k)}$ = trait observed or calculated

A_i = treatment effect (i=1,2)

B_j = pair effect (j=1,5)

AB_{ij} = interaction effect

$e_{ij(k)}$ = remainder

In the model for Experiment I no pair effect was used because pairing was done on a group level.

The overall means per treatment were tested by a Student's t-test (Exp. I). In Experiment II data from week -4 to 0 (= Pre-infection period) were compared to those of week 1 to 6 (= infection period). All analyses were done using the SPSS Statistical Package (Nie et al., 1975).

RESULTS

Intake of DM

In Experiment I the control animals consumed on average 100 g DM. d^{-1} more than the infected animals (about 4.5% of the daily DM intake). The infected animals in Experiment II ingested more DM than the controls (on average 135 g concentrates. d^{-1}, 5% of the average daily intake).

Table 3. Group means of live weight (kg: $\mu \pm$ s.e.m.) on various days of the experiment.

Age of calves (days)	93		99		105		112		150	
	I	C	I	C	I	C	I	C	I	C
Exp. I	102.2	102.2			112.4	113.9	109.8	111.4	122.4	126.4
	(1.7)	(1.9)			(2.5)	(1.9)	(1.2)	(1.2)	(2.5)	(1.6)
Exp. II			110.1	108.4			119.0	118.6	136.6	138.5
			(0.8)	(1.8)			(1.3)	(1.7)	(1.5)	(2.1)

Performance

Mean live weight of the calves during the experimental periods is given in Table 3. The I calves in Experiment I showed a reduced rate of weight gain during the time in the respiration chambers compared to that of C calves ($\mu \pm$ S.D.: 359 ± 65 vs 429 ± 27 g.d^{-1}). In Experiment II the mean daily weight gain during this period was 523 ± 67 and 602 ± 59 g.d^{-1} for I and C calves respectively.

Digestibility and Metabolizability

No significant influences of the infection on digestibility of E and N was found. In Experiment II there was a little difference in mean metabolizability of GE over the whole infection period between I and C calves (49.5 vs 52.4, $p < 0.05$) while this difference was not significant in Experiment II (64.1 vs 65.3).

Table 4. Mean intake of Gross Energy (GE), metabolizable energy (ME), heat production (H), energy retention (RE) and maintenance (MEm) during the weeks after first infection (P.I.I.) of infected (I) animals and controls (C) in Experiment I and II. The data are given as group means (KJ/kg live weight$^{0.75}$.d^{-1}).

| | Experiment I | | Experiment II | |
	I	C	I	C
GE	1277	1308	1596	1519
ME	819	855	837	753
H	611	604	678	624*
Estimated RE	210	235*	159	129*
MEm[1]	491	460*	464	441*

1) MEm (ME at maintenance) = ME-ME$_p$, ME$_p$ = RE/k$_g$, k$_g$ is the partial efficiency of energy gain from ME above MEm

* Within experiments values of parameters superscripted with * differ significantly p < 0.05

Energy balance

In Table 4 the GE intake, ME intake, H and estimated RE are given. It appears that the differences in H and estimated RE per kg live weight$^{0.75}$ are significant in Experiment II (I < C, p < 0.05) while in Experiment I only the estimated RE per kg live weight$^{0.75}$ was different (C > I, p < 0.05).

The pattern of GE intake, ME intake and estimated RE of calves in Experiment II is shown in Figure 2. There was a sharp decrease of values of the parameters of C calves in week 7. The difference between I and C calves continues until week 12.

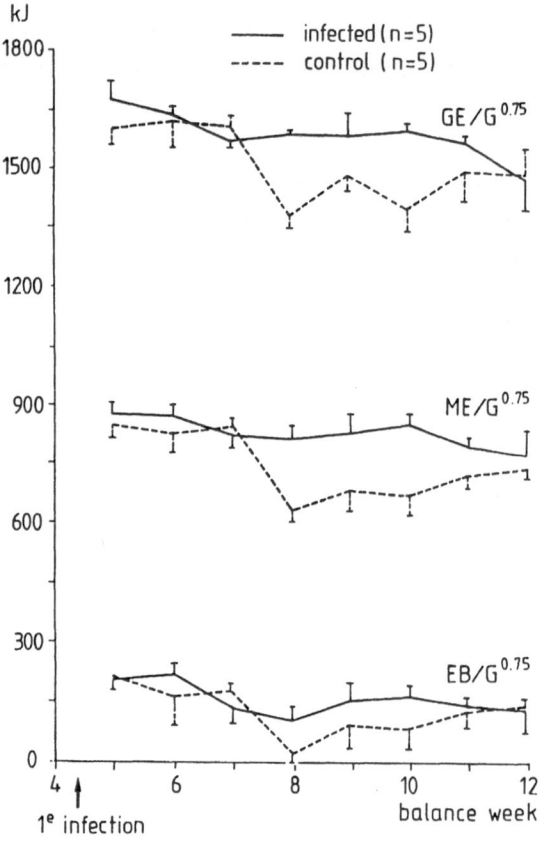

Figure 2. Intake of gross (GE) and metabolizable (ME) energy and the energy retention (EB) of the infected calves and the non-infected controls per balance week in Experiment II (kJ.kg$^{-0.75}$.d^{-1}).

N-balance

In Experiment I there was only a small difference in N ingested between C and I calves throughout the experimental period ($p < 0.10$). In
Experiment II this difference was much higher and significant ($p <
0.05$). It began during week 3 P.I.I. and reached a maximum during
week 4 P.I.I. (Figure 3).

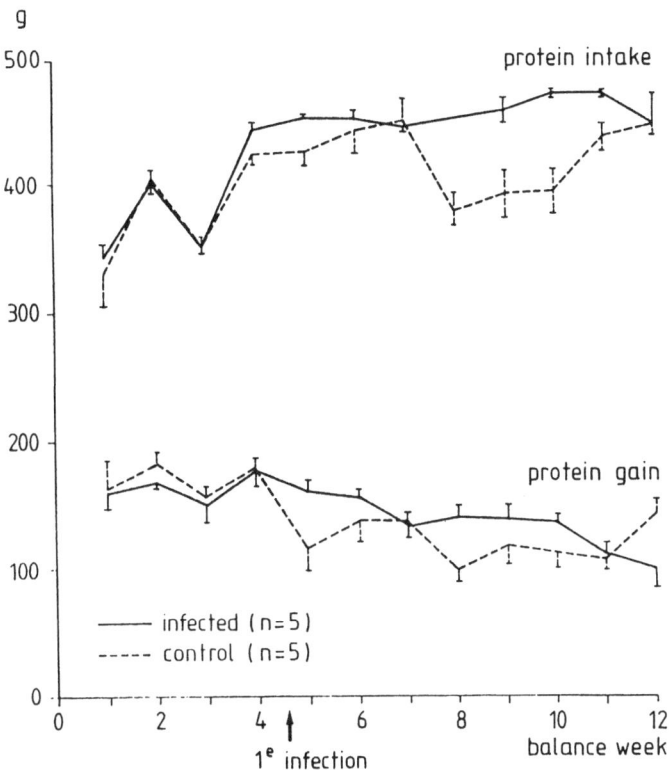

Figure 3. Protein intake and protein gain of the infected calves and the
non-infected controls per balance week in Experiment II ($g.d^{-1}$).

In Experiment II the I calves deposited 138.7 g protein per day vs 125.0
g for the C calves ($p < 0.05$). In Experiment I this was 78.5 and 91.3
respectively ($P < 0.10$). Expressed as % of N ingested the I and C
calves of Experiment I retained 19.7 and 23.2% respectively. These data
of the calves in Experiment II were 29.3 and 30.0% respectively (Table
5). The urinary N excretion in % of the apparently digested N for C and
I animals was 57 and 63% in Experiment I and 30.4 and 34.2% in Experi-

Table 5. Mean nitrogen intake, digestible N, digestibility of N, urinary excretion of N, N retention and N retention as a percentage of N ingested in control (C) and infected (I) animals of Experiment I and II during the weeks after first infection (P.I.I.). Data are given as group means.

	Experiment I		Experiment II	
	I	C	I	C
N intake $(g.d^1)$	60.8	63.0	74.0	68.3*
Digestible N $(g.d^1)$	36.3	38.1	41.5	37.2
Digestibility of N (%)	59.9	60.6	56.1	54.5
Urinary excretion of N $(g.d^1)$	23.8	23.0	18.0	16.3
N retention $(g.d^1)$	12.0	14.6	22.2	20.0
N retention as a percentage of N intake	19.7	23.2	30.0	29.3

* Within experiments values of parameters superscripted with * differ significantly (p < 0.05)

ment II. So, in both experiments I calves excreted relatively more N than C calves (Table 5). This started in week 4 P.I.I. and remained fairly constant until the end of the experiments.

Non-metabolic parameters

Differences between mean values of clinical, parasitological, haematological, serological and post-mortem parameters observed during the experimental period are given in Table 2. In both experiments the standard deviations of the mean values of the clinical parameters of the I calves were higher than those of the controls.

DISCUSSION

Parasitological parameters and symptoms showed that a clinical D.v. infection was evident. The ELISA titercounts were as high as those of a moderate experimental continuous infection (Boon et al., 1984a). The pattern of the angiotensin converting enzyme (Experiment II) is comparable to that of man with an obstructive bronchitis (Neels, 1983). Parasitic bronchitis is obstructive too (Jarrett et al., 1960). The increase of eosinophils in the peripheral blood and cells in the broncho alveolar lavage fluid of infected calves (Experiment II) is thought to be associated with the defence mechanism of the calves against D.v. This was found earlier (Boon et al., 1987). Jarrett et al. (1960) already indicated that the cellular immune response is very important in the defence of calves against D.v. Also the increase in weight of some parenchymetic organs as liver, heart and lungs and the very large increase in weight of the bronchial lymphnodes of D.v. infected calves are indications of a strong systemic reaction of the calf against D.v.

The goal of the experiments was to investigate the possible alteration in maintanance requirements of D.v. infected calves: on the one had with pair feeding and on the other with pair growing. With other words: what does a D.v. infection cost a calf or what has a calf to pay to cope with lungworms. Different parameters were used.

The performance of the calves differed between the experiments (578 g in Experiment II vs 359 g in Experiment I). The level of gain in Experiment I was low. This may be caused by the low feeding level (1.23 kg hay and 1.47 kg of concentrates vs 1.75 kg hay and 2.00 kg of concentrates in Experiment II). In Experiment I the control calves showed at nearly the same food intake an increase in performance compared to the D.v. infected calves, while in Experiment II it was succeeded with less food to get the same performance in control calves as in infected calves. These are indications that a D.v. infection costs feed. This is in agreement with Kloosterman (1971) who found that farmers may compensate the negative effects on rate of gain of calves with gastro-intestinal parasites by giving extra concentrates. Although performance as such can be measured fairly accurately, still correction for some factors is necessary (body weight, feed intake, gut content). It might therefore be better to use specific metabolism parameters.

In both experiments the difference between I and C calves in digestibility of GE was not significant. This is in contrast with results obtained by Symons (1982) in experiments with sheep parasitized with gastro-intestinal worms. However, a comparable effect of D.v. is hardly to expect since D.v. mainly affects the lungs. The damage to the epithelial cells of the gastro-intestinal tract caused by penetrating L_3 at the level used is considered to be very small (Jarrett et al., 1960). In Experiment I energy gain was decreased in the infected calves possibly influenced by the somewhat lower E-intake. However, despite the lower E-intake these calves showed an increased H production indicating a high level of metabolism. In Experiment II the I calves showed an increased RE (23,5%) and H production (8,5%). This increase may be related to the increased E-intake but despite this the performance of I calves did not differ significantly from that of the C calves. So this is also an indication of increased level of metabolism.

In Experiment I N retention tended to increase in control animals. In Experiment II the I animals had an increased N retention. Both differences may have been determined by the differences in N intake. As in energy digestion there was no influence on digestibility. In both experiments the infected animals excreted more N in the urine voided. However, in Experiment II it may be related to the increased amount of N ingested while in Experiment I the C calves had an increase of N intake. The increase of urine N may be the result of digestion of N produced in the lungs of calves as a result of the reaction of the immune system after D.v. infection in producing large quantities of immune cells and protein which will be swallowed and ingested and thereafter digested.

It can be expected that if infected animals use more ME per kg live weight$^{0.75}$ and produce more $H.kg^{-0.75}$ (as shown in both experiments) the maintenance requirement per live weight$^{0.75}$ (MEm) is increased. MEm is calculated as follows: The partial efficiency for converting ME above maintenance into RE was calculated by $k_g = 0.006 + 0.0078 * ME\%$ (Van Es, 1978). Then the feed needed for gain is RE/k_g (= MEp). Subsequently the maintenance ME required is calculated by: MEm = ME - MEp. Results show that MEm in control calves is similar to that derived by Van Es (1972): about 460 $kJ.kg^{-0.75}$. In both experiments MEm of infected calves was increased; in Experiment I with 31 $kJ.kg^{-0.75}$ (= 6.7%) and in Experiment II with 22.4 $kJ.kg^{-0.75}$ (= 5.1%). The difference between the experiments may be explained by the lower production level in Experiment I. An increased MEm may be related to a high E demand because of an increased respiration rate of D.v. infected calves. Howeever, the increase should not be more than 4%. H related to breathing is 2% of the total H production. At a serious decrease in elasticity, which may appear in chronical obstructive bronchitis, maximally a doubling of H related to breathing might be possible (Guyton, 1981; Lekeux et al., 1985). However, the increase in MEm is higher than 4%. There-fore, altered breathing may only explain a part of the increase.

It is obvious that significant changes of haematological and clinical parameters related to D.v. infection started in the third week P.I.I. At this stage of the infection L_4 and L_5 larvae of D.v. develop and migrate through the interstitial lung tissue to the alveoli and bronchioli. This is the period of a developing parasitic broncho-pneumonia initiating cellular and humoral immune responses (Jarrett et al., 1960). Ten weeks P.I.I. immunity is normally reached and parameters return to normal values. Changes in metabolism parameters occured during the same time wherein larvae were migrating through the lungs and immune responses develop-ed. It clearly appears in both experiments that these immune responses, resulting in a decrease in values of parasite related parameters and a decrease of clinical signs is accompanied by another level of E metabolism (e.g. H production). This might induce an extra feed requirement. This is in agreement with Van Miert (1986) who found a high N demand of animals injected with micro-organisms.

REFERENCES

Abbott, E.M., Parkins, J.J. and Holmes, P.M., 1986. The effect of di-etary protein on the pathophysiology of acute bovine Haemonchosis. Vet. Parasitology 20: 281-306.

Baermann, G., 1917. A simple method for the recovery of Ankylostomum (Nematode) larvae in soil samples. Eine einfache Methode zur Auffin-dung von Ankylostomum (Nematoden-larvae) in Erdproben. Geneeskun-dig Tijdschrift voor Nederlandsch-Indië 57: 131-137.

Boon, J.H., 1979. An investigation into possible causes of coughing in calves at pasture. Ph.D. thesis. State University, Veterinary Faculty Utrecht: 121 pp.

Boon, J.H., Grondel, J.L., Hemmer, J.G.A. and Booms, G.H.R., 1987. Relationship between cytologic changes in broncho alveolar lavage fluid and weight gain in calves with gastro-intestinal nematodes and lungworms. Vet. Parasitology: in press.

Boon, J.H., Kloosterman, A. and Breukink, M., 1984a. Parasitological, serological and clinical effect of continuous graded Dictyocaulus viviparus inoculations in calves. Vet. Parasitology 16: 261-272.

Boon, J.H., Kloosterman, A. and Lende, T. van der, 1984b. The incidence of Dictyocaulus viviparus infection in cattle in the Netherlands. II. Survey in the field. Vet Quarterly 6: 13-17.

Es, A.J.H. van, 1972. Maintenance. In: Handbuch der Tierernährung (W. Lenkeit and P. Breirem, eds.). Paul Parey, Hamburg, vol. II: 1-54.

Es, A.J.H. van, 1978. Feed evaluation for ruminants I. The system in use from May 1977 onwards in the Netherlands. Livestock Production Science 5: 331-343.

Guyton, A.L., 1981. Textbook of Medical Physiology. Saunders & Co Ltd, Philadelphia-London-Toronto: 478-480.

Jarrett, W.F.M., McIntyre, W.I.M., Jennings, L.W., Mulligan, W., Sharp, N.C. and Urquhart, G.M., 1960. Symposium on Husks. I. The disease process. Vet. Rec. 72: 1066-1067.

Jørgensen, R.J. and Ogbourne, G.P., 1985. Bovine Dictyocaulis: A review and annotated bibliography. Miscellaneus Publication No. 8 of the Commonwealth Institute of Parasitology (C.I.P): 104 pp.

Kloosterman, A., 1971. Observations on the epidemiology of trichostrongylosis of calves. Ph.D. thesis Agricultural University Wageningen: 114 pp.

Kroonen, J.E.G.M., Verstegen, M.W.A., Boon, J.H. and Hel, W. van der, 1986. Effect of infection with lungworm (Dictyocaulus viviparus) on energy and protein metabolism in growing calves. Brit. J. Nutrition 55: 351-360.

Lekeux, F., Hayer, R., Boon, J.H., Verstegen, M.W.A. and Breukink, H.J., 1985. Physiological effects of Experimental Verminuous bronchitis in Friesian calves. Can. J. Comp. Med. 49: 205-207.

Miert, A.S.J.P.A.M. van, 1986. Inflammation and febrile conditions. The role of interleukine I. Proceedings IVth Intern. Symposium of Vet. Lab. Diagnosticians Amsterdam: 122-130.

Neels, H.M., 1983. Biochemical and clinical aspects of ACE. Ned. Tijdschr. Geneesk. 127: 1871.

Neels, H.M., Sande, N.E. van and Schorpé, S.L., 1983. Sensitive colorimetric assay for ACE in serum. Clin. Chemistry 29: 1399.

Nie, N.H., Hull, C.H., Jenkins, J.G., Steinbrenner, K. and Bent, D.H., 1975. Statistical Package for the Social Sciences. McGraw-Hill,

2nd Edn.: 675 pp.

Oakly, G.A., 1980. The recovery of <u>Dictyocaulus viviparus</u> from bovine lungs by lungperfusion: a modification of Inderbitzinn's method. Res. Vet. Sci. 29: 395-396.

Steel, J.W., Symons, L.E.A. and Jones, W.O., 1980. Effects of level of larval intake on the productivity and fysiology and metabolic respons- es of lambs infected with <u>Trichostrongylus colubriformis</u>. Austr. J. Agric. Sci. 31: 821-828.

Symons, L.E.A., 1982. Gastro-intestinal pathology incidence by enteric parasites in: Parasites. Their world and ours. Elseviers Biomedical Press, Amsterdam: 233-241.

Romeis, B., 1967. Mikroskopische Technik. Oldenbourg Verlag Mün- chen-Wien: 757 pp.

Trigo, F.J., Liggilt, H.O., Breeze, R.D., Leid, R.W. and Silflow, R.M., 1984. Bovine pulmonary alveolar macrophages, Ante mortem recovery and in vitro evaluation of bacterial phagocytosis and killing. Am. J. Vet. Res. 45: 1842-1847.

Verstegen, M.W.A., Boon, J.H., Hel, W. van der, Kessel, M.H., Meu- lenbroeks, J. and Mellink, H.M., 1987. Estimation of the effect of <u>Dictyocaulus viviparus</u> infection on energy metabolism of calves. (submitted).

RESPIRATORY DISEASES IN CALVES

P. FRANKEN, C. HOLZHAUER AND L.A. VAN WUIJCKHUISE-SJOUKE

ABSTRACT

A literature review is given about various agents related to infectious diseases of the respiratory tract of calves. It appears that many agents are involved in affections of this tract. The consequences of various pathogens in terms of energy and protein metabolism however are not clearly described or unknown. The literature mostly describes only effects on morbidity and mortality. In the description presented here reduced feed intake and/or reduced performance are mentioned, but not quantified. In some recent studies these latter aspects can be derived from the data given by the authors. This is illustrated in a study with beef calves. Besides a mortality of 3.1% and culling of 4% of affected calves, the growth period of about one year of the remaining calves is on average lengthened by 14 and 7 days due to clinical and subclinical respiratory diseases respectively.

INTRODUCTION

Respiratory diseases are thought to cause major economic losses in calves (Willadsen et al., 1977; Andrew, 1978; Roy, 1980). However, their aetiology, incidence and costs are still poorly defined.

High incidences of infections of the respiratory tract of calves are reported by several authors (Holzhauer, 1978; Bryson et al., 1978 and 1979; Boon, 1979; Verhoeff, 1983; Postma, 1985; Espinasse, 1986;

Franken et al., 1986; Ploeger et al., 1986; Postema et al., 1987).

Infections with respiratory pathogens in youngstock may appear as a high incidence of clinical disease but also as general unthriftiness without specific signs. Both circumstances however may result in considerable economic losses because of serious growth retardation although mortality rate may not be increased (Andrews, 1978; Roy, 1980; Pirie, 1982).

Respiratory diseases and/or growth depression are related to unfavourable conditions by the epidemiological principal: host × agent × environment. This results in a considerable difference in morbidity and mortality between farms, herd groups and years (Espinasse, 1986).

In this paper attention is focussed on:
- aetiology and pathogenesis of infections of the respiratory tract;
- the major pathogenic agents involved in respiratory diseases in calves;
- the impact of respiratory diseases for production.

AETIOLOGY AND PATHOGENESIS OF INFECTIONS OF THE RESPIRATORY TRACT IN CALVES

An important number of the pathogenic organisms (virusses, bacteria, mycoplasms etc.) which cause respiratory disease in calves will reach the respiratory tissue (the upper respiratory tract and the lungs) by airborn transmission (droplet infection).
The haematogenic route is known in calves for infections with Salmonella dublin and some secondary invading bacteria. The oral infection route is known for lungworm infections. Here special attention is given to the airborn transmission.

After aerogenic contamination of the respiratory tissue colonisation by pathogenic micro-organisms depends on the present or induced defence mechanisms of the host, the number and virulence of the pathogen and secondary, environmental, factors. The infection may run different courses:
a) Each day many antigens and micro-organisms are eliminated by the defence mechanisms of the host without causing any further damage;

b) The subclinical course. If infection develops many animals may not show clinical symptoms. The subclinical infection can be demonstrated for instance by serological techniques. However a subclinical course can result in growth retardation by decreased food consumption, in body temperature rise, etc. Subclinical disease may result in a decrease in host defence mechanisms, making the animal more susceptible to secondary infections (Pirie, 1982; Yates, 1982 Houghton and Gourlay, 1983; Yates et al., 1983; Frank et al., 1986; Kimman et al., 1986). In a herd the subclinically diseased animal increases the burden of infection and influences in this way the balance host × agent × environment;

c) The clinical course. Specific signs of respiratory disease appear: e.g. increased respiration rate, coughing and nasal discharge. Systemic clinical signs may also appear: a.o. increased body temperature and increased pulse rate. Production loss orginates from mortality rate, growth retardation, increased energy need, body temperature rise, increased pulmonary resistance, decreased food conversion, etc. (Willadsen et al., 1977; Andrews, 1978; Pritchard et al., 1981; Pirie, 1982; Verhoef, 1983; Slocombe et al., 1984; Postuma, 1985; Lekeux et al., 1986; Zimmer and Wierenga, 1982.

Sometimes the host cannot overcome an infection and the animal will become a carrier. The balance between micro-organism and immunity will be very delicate and the animal may shed respiratory pathogens intermittently or continuously.

Host defence mechanisms can be defined in nonspecific and specific components. Both types of components can be strongly influenced by secundary circumstances.

The upper respiratory tract has an important nonspecific defence system. The conchae in the nose do enlarge the surface area. The inhaled air is moistered and raised in temperature. The surface of the upper respiratory tract is covered with mucous on a brush border. This mucociliair system cleans the inhaled air: droplets, dust and micro-organisms are excreted with the mucous. Deeper in the lungs the mucociliar barrier is absent and the nonspecific defence depends largely upon alveolar macrophages and neutrophils (Andrews, 1978).

As far as the specific immunity is concerned, in aerogenic infections the local IgA level is most important.

The humoral antibodies, in cattle most of the IgG class, protect the animal against a spread of the pathogens through the animal, i.e. prevention of systemic infection (Pirie, 1982; Kimman et al., 1986; Holzhauer, 1987). In parasitic bronchitis however the cellular immunity is predominantly important (Bos et al., 1986). In the first months of a calves' life maternal (colostral) antibodies provide a passive, humoral, immunity. Older calves start making their own antibodies after contact with micro-organisms. The maternal immunity of the calf depends on the mothers' contact with the pathogenic agent and on the colostral management. However there is only a slight negative correlation between incidence of respiratory diseases and passive, maternal, immunity as measured by IgG serum concentration one week after birth (Postema et al., 1987).

In The Netherlands a large number of calves have specific maternal antibodies against Infectious Bovine Rhinotracheitis (IBR), parainfluenza type III (PI 3), Bovine Virus Diarrhoea (BVD) and Bovine Respiratory Syncytial Virus (BRSV) as reported by Holzhauer (1978 and 1987) and Terpstra et al. (1982).

Both the specific and non-specific defence mechanisms of calves can be influenced strongly by secondary, environmental, factors. Because respiratory infections are mostly of aerogenic origin, climatic conditions (and thus ventilation) are very important (Roy, 1980; Pritchard et al., 1981; Webster, 1981; Verhoeff, 1983; Wathes et al., 1983; Postema, 1985; MacVeau et al., 1986). Though ventilation controls the relative humidity, the number of airborn pathogens and the concentration of harmfull gases (NH_3, CO_2, H_2S), it can initiate draught (air speed and temperature fluctuations) which may lead to stress and immunodepression (Andrews, 1978; Elazhary and Derbyshire, 1979a and b; MacVeau et al., 1986). Droplet/dust size is important with respect to depth of penetration in the respiratory tract (Andrews, 1978). Transport plays a crucial role in shipping fever. Mortality and Morbidity are related to time and distance of the transport. Transport induces stress and imposes unfavourable climate conditions. Usually the animals are not fed nor watered, inducing even more immunosuppression. (Andrews, 1978; Espinasse, 1986).

The presence of diseased animals in the herd increases the risk of further infection. The multiplication and shedding of micro-organisms in a diseased animal induces a strong horizontal infection to other animals in the group. This spreading is exponential. It is of special importance when groups of calves, especially of different ages, are mixed (Ploeger et al., 1986).

The high feeding level in veal calf production may stress the respiratory tract due to the required high respiration rate (Postema, 1985).

THE MAJOR PATHOGENIC AGENTS INVOLVED IN RESPIRATORY DISEASES IN CALVES

Respiratory diseases in calves are usually initiated by infections with pathogenic micro-organisms of different groups (virusses, bacteria, mycoplasms and parasites).

Virusses

Bovine Respiratory Syncytial Virus (BRSV) is wide spread and is a primary pathogen initiating viral pheumonia, lungoedema and emphysema. Youngstock is especially susceptible. The calves will be affected at first infection and a high mortality may result (Holzhauer, 1978; Bryson et al., 1979; Van Nieuwstadt et al., 1982; Van Nieuwstadt and Verhoeff, 1983; Verhoef, 1983; Anonymous, 1984; Ploeger et al., 1986).

Parainfluenza III virus (PI 3) is also a primary pathogenic micro-organism infecting the lower respiratory tract. Infections are also seen after, or in combination with, other respiratory infections. PI 3 virus is wide spread and many types of different virulence are known. The first infection can induce a high morbidity in young calves (Holzhauer, 1978; Bryson et al., 1979; Elazhary and Derbyshire, 1979b; Van Nieuwstadt et al., 1982; Anonymous, 1984).

Bovine Herpes Virus I (BHVI) is known as a causative agent of Infectious Bovine Rhinotracheitis (IBR) and is a primary pathogenic virus of the upper respiratory tract. The virus is wide spread and after a first infection the animals become virus carriers. Secondary factors influence strongly the course of infection (Stott et al., 1982; Van Nieuwstadt et al. 1982; Terpstra et al., 1982; Yates, 1982; Yates et al., 1983; Franken et al., 1986).

Bovine Virus Diarrhoea virus (BVD) is pathogenic for the respiratory tract in young calves. The infection becomes systemic and is immunosuppressive resulting in serious secondary bacterial infections. Persistent

virus infections do occur. The virus is wide spread (Reggiardo, 1975;
Terpstra et al., 1982; Potgieter et al., 1984; Bolin et al., 1985; Duffel
and Harkness, 1985; Brownlie, 1986; Franken et al., 1986).

Bovine adeno-, rhino- and reovirusses are known as respiratory
pathogens in calves causing a "cold". These virusses probably induce
colonisation of secondary pathogens as pasteurellae (Reed et., 1978;
Anonymous, 1984).

Bacteria

Pasteurellae (P. haemolytica and P. multocida) are important primary
and secondary pathogens of the respiratory tissue in calves. They are
frequently cultured from (pleuro-)pneumonic lungs. Many reports deal
with a synergism with other respiratory agents. P. haemolytica is known
to be more and more resistant to antimicrobial drugs (Gibbs et al., 1982;
Wellemans, 1982; Yates, 1982; Houghton and Gourlay, 1983; Anonymous,
1984; Filion et al., 1984; Slocombe et al., 1984; Espinasse, 1986; Frank
et al., 1986).

Salmonella dublin is a primary pathogen for the respiratory tract.
Other salmonellae are encountered as secondary invaders. Haemophilus
somnus is in the Anglo Saxon countries well known in infectious trombo-
embolic meningitis but it also induces respiratory disease. Probably the
pathogen is conditional (Corboz, 1982; Humphrey and Stephans, 1983;
Anonymous, 1984).

Corynebacterium pyogenes, Escheria coli and Fusobacterium necro-
forum are secondary pathogens only infecting damaged lungtissue.
Pseudomonas, Proteus and Streptococcus spp. are found sometimes at
post mortem in pneumonic lungs, but their role in respiratory diseases is
unclear.
Actinobacillus spp. and Mycobacterium tuberculosis in the Netherlands
are no longer of importance.

Mycoplasms

Mycoplasma mycoides, the causative agent of Infectious Bovine Pleur-
opneumonia is an important primary pathogen of the bovine respiratory
tract. However the disease is not present in the Netherlands. Other mycoplasms as M. bovirhinis, M. dispar and Ureoplasms are probably wide
spread and synergism with other respiratory pathogens is reported

(Andrews, 1978; Houghton and Gourlay, 1983; Anonymous, 1984; Stalheim, 1983).

Parasites

Dictyocaulus viviparus can induce severe parasitic bronchopneumonia
in outdoor čalves. The disease largely depends upon the number of infectious larvae present on the pasture which relate to infected herd-
mates, pasture management, vaccination and anthelmintic treatments. The
infection is wide spread in the Netherlands and of economic importance
(Boon, 1979; Bos et al., 1986).

THE IMPACT OF RESPIRATORY DISEASES FOR PRODUCTION

The impact of respiratory diseases on production in calves can be
divided into:
- decreased daily live weight gain;
- reduced feed conversion efficiency;
- increased mortality rate and premature slaughter rate;
- increased costs for veterinary treatment.

To prevent respiratory diseases costs are made for vaccination, medica-
tion, adaptation of ventilation and/or housing. Verhoeff (personal communication, 1986) estimated the total costs of respiratory diseases at
about DFL 130 per calf. This estimation is based upon research on
19 dairy farms with 28 groups of youngstock (in total 442 calves). Not
included are the costs of preventive measures and of chronic diseased
animals ("poor doers").

A quantitative approach for beefcalves based on data of Zimmer and
Wieringa (1987) on 8 groups of 80 animals is presented in Table 1.

From the data given in Table 1 and those on rate of gain and feed
intake an estimation can be made of the economic losses.

It was assumed that 50% of the calves were clinically diseased, while
the other calves experienced a subclinical infection. From the data of
Zimmer and Wierenga (1987) and those used by Verhoeff we concluded
that calves with a clinical respiratory disease had a reduced rate of gain.
During a growth period of about one year 14 more days were needed on

Table 1. Clinical data on respiratory disease in beef cattle.

Parameter	mean	range
Calves with respiratory disease (%)	49.3	26-71
Treatment per animal (n)	2.1	0.9-3.6
Mortality (%)	3.1	0-11
Calves with chronic pneumonia (%)	4.0	

Table 2. The effect of respiratory diseases on rate of gain and feed intake (maize silage and concentrates) as compared to normal animals during a growing period of about 12 months (all in $g. an^{-1}.d^{-1}$).

Class of infection	Effect of respiratory disease		
	Gain	Feed intake	
		Maize	Concentrates
Clinical	-36	-40	-80
Subclinical	-18	-20	-20

average for reaching the same body weight as unaffected calves. The subclinically diseased animals needed about 7 days extra. In Table 2 data are given on the reduction in rate of gain and on the concommittent effect on feed intake. It was calculated that the growth rate of clinically diseased animals was about 36 $g.d^{-1}$ less than that of unaffected animals. This reduction was related to a somewhat reduced feed intake. This loss in production comprises about 35-45% of the total costs due to respiratory diseases. In addition also mortality and extra culling will contribute to the loss in production.

REFERENCES

Andrews, A.M., 1978. Some factors influencing respiratory diseases in growing bulls and the effect of treatment on live weight. Seminar Edinburgh. Respiratory diseases in Cattle, ed. Martin, The Hague. Martinus Nijhoff, ISBN 902421342.

Anonymous, 1984. Central Veterinary Institute, Lelystad, The Netherlands. Year report: 13-18.

Bolin, S.R., McClurkin, A.W. and Coria, M.F., 1985. Frequency of persistent bovine viral diarrhea virus infection in selected cattle herds. Am. J. Vet. Res. 461: 2385-2387.

Boon, J.H. 1979. An investigation into possible causes of coughing in calves at pasture. Thesis Utrecht: 142 pp.

Bos, H.J., Beekman Boneschanscher, J. and Boon, J.H., 1986. Use of Elisa to assess lungworm infection in calves. Vet. Rec. 118: 153-156.

Brownlie, J., 1986. Clinical aspects of the bovine virus diarrhoea/ mucosal disease complex in cattle. In practice Vet. Rec. 7: 185.

Bryson, D.G., McFerran, J.B., Ball, H.J. and Neill, S.D., 1978. Observations on outbreaks of respiratory disease in housed calves. 1) Epidemiological, clinical and microbiological findings. Vet. Rec. 103: 485-489.

Bryson, D.G., McFerran, J.B., Ball, H.J. and Neill, S.D., 1979. Observation on outbreaks of respiratory disease in calves associated with PI 3 virus and RS virus infection. Vet. Rec. 104: 45-49.

Corboz, L., 1982. Haemophilus somnus als Erreger von bronchopneumonie bei Mast kälbern. Proc. XIIth World Congress on Diseases of Cattle, Amsterdam: 35-39.

Duffel, S.J. and Harkness, J.W., 1985. Bovine Virus diarrhoea-mucosal disease infection in cattle. Vet. Rec. 117: 240-245.

Elazhary, M.A.S.Y. and Derbyshire, J.B., 1979a. Effect of medium, temperature and relative humidity on the aerosol stability of infectious bovine rhinotracheitis virus. Can. J. Comp. Med. 43: 158-167.

Elazhary, M.A.S.Y. and Derbyshire, J.B., 1979b. Aerosol stability of Bovine Parainfluenza Type 3 virus. Can. J. Comp. Med. 43: 295-304.

Espinasse, J., 1986. Infectious enzootic bronchopneumonias in young cattle. Proc. 14th World Congress on Diseases of Cattle, Dublin: 423-434.

Filion, L.G., Willson, P.J., Bielefeldt-Ohmann, H., Babiuk, L.A. and Thomson, R.G., 1984. The possible role of stress in the induction of pneumonic pasteurellosis in cattle. Can. J. Comp. Med. 48: 268-274.

Frank, G.H., Briggs, R.E. and Gilette, K.G., 1986. Colonisation of the nasal passages of calves with Pasteurella haemolytica serotype 1 and regeneration of colonisation after experimentally induced viral infection of the respiratory tract. Am. J. Vet. Res. 8: 1704-1707.

Franken, P., Sol, J. and Wentink, G.H., 1986. BVD and IBR: Serological examination on 35 farms. Tijdschr. Diergeneesk. 111: 1205-1207.

Gibbs, H.A., Selman, I.E., Wiseman, A., Allen, E.M., Pirie, H.M. and Watt, N.J., 1982. Pasteurella-Spp. associated respiratory disease in immature cattle. Proc. XIIth World Congress on Diseases of Cattl, Amsterdam: 40-45.

Holzhauer, C., 1978. Bronchopneumonia of yearling cattle (Pinken-griep). Thesis, Utrecht :192 pp.

Holzhauer, C., 1987. Serology in large animal practices. Tijdschr. Diergeneesk. 112: 390-395.

Houghton, S.B. and Gourlay, R.N., 1983. Synergism between Mycoplasma bovis and Pasteurella haemolytica in calf pneumonia. Vet. Rec. 113: 41-42.

Humphrey, J.D. and Stephans, L.R., 1983. Haemophilus somnus a review. Vet. Bulletin 53: 987-1004.

Kimman, T.G., Westerbrink, F. and Straver, P.J., 1986. Studies on the local and systemic immune response to bovine respiratory syncytial virus infection. Proc. 14th World Congress on Diseases of Cattle, Dublin: 482-487.

Lekeux, P., Gustin, P. and Clercx, C., 1986. Shipping fever chez les jeunes bovins: une approche fonctionelle. Proc. 14th World Congress on Diseases of Cattle, Dublin: 452-457.

MacVeau, D.W., Franzen, D.K., Keefe, T.Y. and Bennet, B.W., 1986. Airborne particle concentration and meteorologic conditions associated with pneumonia incidence in feedlot cattle. Am. J. Vet. Res. 47: 2676-2682.

Nieuwstadt, A.P.K.M.I. van, Verhoeff, J., Ingh, T.S.G.A.M. van den and Hartman, E.G., 1982. Epizootiology of PI 3-, bovine RS- and IBR-virus infections in dairy herds with traditional calf rearing. Proc. XIIth World Congress on Diseases of Cattle, Amsterdam:

124-130.

Nieuwstadt, A.P.K.M.I. van and Verhoeff, J., 1983. Serology for diagnosis and epizootiological studies of bovine respiratory syncytial virus infection. Res. Vet. Sci. 35: 135-159.

Pirie, H.M., 1982. Respiratory tract reactions in young bovine animals and their significance. Proc. XIIth World Congress on Diseases of Cattle, Amsterdam: 57-65.

Ploeger, H.W., Boon, J.H., Klaassen, C.H.L. and Florent, G. van, 1986. A Sero-Epidemiological Survey of Infections with the Bovine Respiratory Syncytial Virus in First-Season Grazing Calves. J. Vet. Med. B 33: 311-318.

Postema, H.J., 1985. Veterinary and zootechnical aspects of veal production. Thesis, Utrecht: 121 pp.

Postema, H.J., Franken, P. and Ven, J.B. van der, 1987. Studies in veal calves for a possible correlation between the concentration of immune globulin in the serum, the plane of nutrition and the risk of disease during the first few weeks of the fattening period. Tijdschr. diergeneesk. 112: 665-671.

Potgieter, L.N.D., McCracken, M.D., Hopkins, F.M., Walker, R.D. and Guy, J.S., 1984. Experimental production of bovine respiratory tract disease with bovine viral diarrhoea virus. Am. J. Vet. Res. 45: 1582-1582.

Pritchard, D.G., Carpenter, C.A., Morzaria, S.P., Harkness, J.W., Richards, M.S. and Brewer, J.I., 1981. Effect of air filtration on respiratory diseaese in intensively housed Veal Calves. Vet. Rec. 109: 5-9.

Reed, D.E., Wheeler, J.G. and Lupton, H.W., 1978. Isolation of Bovine Adenovirus type 7 from Calves with Pneumonia and Enteritis. Am. J. Vet. Res. 39: 1965-1970.

Reggiardo, C., 1975. Cell-mediated immune response in cattle. Thesis, Iowa.

Roy, J.H.B., 1980. The Calf, Butterworth, London, ISBN 0-408-70941-3: 397-402.

Slocombe, R.F., Derksen, F.J. and Robinson, N.E., 1984. Interactions of cold stress and Pasteurella haemolytica in the pathogenesis of pneumonic pasteurellosis in calves: Changes in pulmonary function. Am. J. Vet. Res. 45: 1764-1770.

Stalheim, O.H.V., 1983. Mycoplasmal respiratory diseases of ruminants. A review and update. J.A.V.M.A. 182: 403-406.

Stott, E.J., Thomas, L.H., Collins, A.P., Crouch, S., Jebbett, J., Smith, G.S., Luther, P.D. and Caswell, R., 1980. A survey of virus infections of the respiratory tract of cattle and their association with disease. J. Hyg. Comb. 85: 257-270.

Terpstra, C., Eikelenboom, J.L. and Glas, C., 1982. Experiences with early vaccination of fattening calves against infectious bovine rhinotracheitis, bovine virus diarrhoea and parainfluenza type 3. Proc. XIIth World Congress on Diseases of Cattle, Amsterdam: 177-181.

Verhoeff, J., 1983. Some Aspects of Respiratory Viral Infections in Diary Cattle. Thesis, Utrecht: 109 pp.

Wathes, C.M., Jones, C.D.R. and Webster, A.J.F., 1983. Ventilation, air hygiene and animal health. Vet. Rec. 113: 554-559.

Webster, A.J.F., 1981. Weather and infectious disease in cattle. Vet. Rec. 108: 183-187.

Wellemans, G., 1982. Evaluation du programme de vaccination anti-virus respiratoire syncytial bovin en Belgique. Proc. XIIth World Congress on Diseases of Cattle, Amsterdam: 146-152.

Willadsen, C.M., Aalund, O. and Christensen, L.G., 1977. Respiratory diseases in calves. An economic analysis. Nord. Vet. Med. 29: 513-528.

Yates, W.D.G., 1982. A review of infectious bovine rhinotracheitis, Shipping fever, pneumonia and viral-bacterial synergism in respiratory disease of cattle. Can. J. Comp. Med. 46: 225-263.

Yates, W.D.G., Jericho, K.W.F., Doige, C.E., 1983. Effect of bacterial dose on pneumonia induced by aerosol exposure of calves to bovine herpesvirus-I and Pasteurella haemolytica. Am. J. Vet. Res. 44: 238-243.

Zimmer, G.M. and Wierenga, H.K., 1987. Gezondheidsaspecten van de huisvesting en verzorging van vleesstieren. In: Onderzoek welzijn landbouwhuisdieren, Wageningen, Puduc: 97-102.

THE EFFECT OF A SUBCLINICAL HAEMONCHUS INFECTION ON THE METABOLISM OF SHEEP (A PILOT STUDY)

P.W.M. VAN ADRICHEM, M.J.N. LOS, J.E. VOGT AND Y. WETZLAR†

ABSTRACT

In a pilot study with two specific pathogen free sheep, one was infected with a single dose of 50.000 Haemonchus larvae. Nitrogen and carbon balances just as heat production were monitored in respiration chambers for 6 periods of 10 days each. Eight weeks after infection both sheep were treated with an anthelmintic. The appetite of the infected sheep was not disturbed. The digestibility of crude protein decreased with 17.6 units 14 days post infection. Due to an adaptation to housing the heat production in both sheep decreased 20-24 percent over a period of 102 days. The economic result of this acute subclinical parasitic infection was a reduced ME intake of 7 percent and an increased heat production of 6 percent. This together amounts to 3.65 guilders per animal over a period of 74 days. After treatment with an anthelmintic the sheep recovered within three weeks.

INTRODUCTION

Clinical infections with gastro-intestinal parasites are characterized by a decrease in appetite and the occurrence of diarrhoea. It is obvious that parasitic diseases in farm animals are an economic loss to the farmer because the digestion, absorption and utilization of nutrients are diminished, the growth is retarded and the production reduced. In severe cases even death may follow.

M. W. A. Verstegen and A. M. Henken (eds.), Energy Metabolism in Farm Animals.
ISBN 0–89838–974–7. © 1987, Martinus Nijhoff Publishers, Dordrecht.

In most farms however, there are many animals who do not clinically suffer from a parasitic gastro-intestinal disease although adult nematode worms and larvae are present. In that case they maintain a normal appetite and the faeces excreted have a normal appearance. Under experimental conditions this can be observed in young animals at an early stage of a low or moderate artificial infection which does not exceed the resistance of the host. Also in adult animals which have developed an immunity against some nematodes, we will find a low number of eggs in the faeces but usually not a decreased appetite or diarrhoea. Apparently these parasites do no harm to the host, however when a certain degree of immunity has been developed many experiments with adult dairy cows for instance have demonstrated that a treatment with an anthelmintic even in that stage may improve milk production (Bliss and Todd, 1974; Pluimers, 1979). We should therefore be careful in judging the economic significance of a subclinical infection with nematode parasites.

In order to establish whether we can neglect an acute subclinical nematode infection from an economical viewpoint we have carried out a pilot study with two adult Texel sheep kept under strictly controlled conditions.

MATERIALS AND METHODS

Two Texel wethers, age 3 years and 9 months, free of endoparasites were originally kept in loose boxes and fed daily 600 g hay and 350 g mixed concentrates.

Feed was supplied twice a day and drinking water was continuously available.

Experimental design

After an adaptation period of 120 days during which time the sheep were kept in metabolism crates which were placed in separate respiration chambers the following experiment was carried out. During 6 subsequent periods of 10 days each the energy, carbon (C) and nitrogen (N) input and output were analyzed and the balance determined (Figure 1). In each balance period the gas exchange was measured for 3 times 48 h. The temperature of the respiration chambers was kept constant at 15°C and the relative humidity varied between 60-70%. The heat production of the animals was calculated from the respiratory gas exchange (O_2 consumption, CO_2 and CH_4 production) and urinary N-excretion with the equation of Brouwer (1965).

After period 2, one sheep was orally infected once with 50.000 <u>Haemonchus</u> <u>contortus</u> larvae provided by the Central Veterinary Institute at Lelystad (Dr. F.H.M. Borgsteede).

After period 5 both the infected and the control sheep were treated with the anthelmintic Thiabendazole at a dose of 44 mg per kg body weight. Before and after treatment blood samples from a jugular vein were taken and analyzed for haemoglobin, packed cell volume, red and white blood cell counts. In the serum total protein, albumin and pepsinogen analyses were performed. In periods 2, 4, 5 and 6 the number of nematode eggs in the faeces were counted after concentration of the eggs according to the flotation method.

The sheep were weighed each period at the same moment of the day.

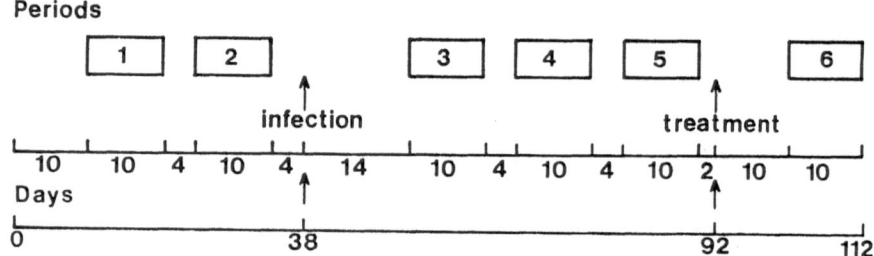

Figure 1.   Experimental outline.

RESULTS

In the whole experimental period both sheep ate their complete ration at any time. Even after the subclinical infection of the infected sheep no feed residues were left. Diarrhoea was never observed.

Blood and serum examination

The results of blood cell counts and serum analysis are given in

Table 1. Blood cells.

Sheep	Period	Hb mmol.1^{-1}	PCV o/o	RBC $\times 10^{-4}$ml^{-1}	WBC $\times 10^{-2}$ml^{-1}	Retic o/oo
79 infection	2	7.3	32	1006	95	0
treatment	5	5.8	27	646	84	12
	6	7.9	35	1008	67	0
102	2	7.1	32	942	75	0
treatment	5	9.3	42	1188	85	0
	6	7.4	33	880	63	0

Hb   = haemoglobin
PCV   = packed cell volume
RBC   = red blood cells
WBC   = white blood cells
Retic = reticulocytes

Tables 1 and 2.

Due to the infection with the blood sucking Haemonchus parasite, the infected sheep became clearly anaemic. The haemoglobin content decreased in the 5th period to 5.8 mmol.1^{-1} while PCV and RBC decreased drastically and newly formed red blood cells (reticulocytes) appeared in the blood. After treatment the blood picture became normal at the end of the 6th period.

A sharp decrease in protein content and a rise in pepsinogen concentration in the 5th period in the infected sheep indicate a damage of the abomasal wall.

Faeces examination

In the infected sheep the egg count per gram faeces (EPG) rose from

Table 2. Serum analysis.

Sheep	Period	Protein g.l^{-1}	Albumin g.l^{-1}	Pepsinogen U.l^{-1}
79 infection	2	69.8	32.4	145
	5	52.6	24.8	300
treatment	6	72.2	26.3	-
102	2	68.7	34.2	110
	5	70.7	40.3	128
treatment	6	66.5	33.9	-

zero before infection to 50.000 and 86.700 respectively 28 and 42 days after infection.

After treatment the EPG returned back to zero again. In the control sheep never any eggs were observed in the faeces.

Digestibility

Each animal consumed daily 821 gram of dry matter. The digestibility of the ration is shown in Table 3.

In the infected sheep the digestibility of the ration decreased sharply in the 3rd period. This is most pronounced for the crude protein digestion which decreased 17.6 units in comparison with the previous period. In all other periods the digestibility appeared quite normal, although in the control animal the digestibility tended to increase slightly during the experiment.

The temporary decrease in digestibility did not affect body weight.

Metabolizable energy, heat production and energy balances

Mainly due to the decrease in digestibility the intake of metabolizable energy (ME) of the infected sheep went down to 320 kJ per kg metabolic

Table 3. Digestibility of dry matter (DM), crude protein (CP), energy (E) and body weight (BW).

Sheep	Period	DM %	CP %	E %	BW kg
79	1	69.1	65.6	68.3	63
	2	69.1	68.3	68.3	64
infection					
	3	61.6	50.7	59.9	64
	4	70.4	67.3	69.6	65
	5	68.1	65.2	67.3	65
treatment					
	6	69.1	68.4	68.1	64
102	1	68.4	66.7	67.8	64
	2	69.5	68.5	68.6	64
	3	68.5	67.3	67.6	63
	4	69.4	70.2	68.5	64
	5	71.2	70.9	70.5	66
treatment					
	6	70.2	72.1	69.1	66

weight ($W^{0.75}$) in the 3rd period (Table 4). The average ME intake in the periods after infection appeared to be 7% less than in the pre-infection periods (354 kJ versus 381 kJ). In the control animal these values were not different (374 kJ versus 371 kJ). The total loss of energy via the urine and as methane gas varied between the periods only from 10.1 to 12.6 percent of GE. The ratio between the ME and the gross energy (GE) of the ration, ME/GE, varied between 54.7 and 58.6 with an exception in the 3rd period where this ratio reached a minimum value of 49.0 in the infected sheep (Figure 2).

The heat production of both sheep declined throughout the experimental periods to 79.4 (sheep 79) and 75.9 percent (sheep 102) of the

Table 4. Metabolizable energy intake (ME), heat production (H) and energy balances (EB) expressed as $kJ/W^{0.75}$.

Sheep	Period	ME	H	EB*
79	1	382	408	-26
	2	380	391	-11
infection				
	3	320	400	-83
	4	369	374	- 7
	5	355	362	- 5
treatment				
	6	373	324	53
102	1	370	427	-53
	2	378	412	-36
	3	387	391	- 6
	4	366	371	- 4
	5	367	345	28
treatment				
	6	364	324	41

* EB is the average value of the energy retention calculated from C- and N-balance and the value ME minus H

original value (period 1). This means for the control sheep a decrease of 1.09 kJ per day and for the infected sheep of 0.90 kJ per day.

Although the ME intake after infection decreased the heat production temporarily increased in the third period from 391 to 400 $kJ/W^{0.75}$. If we assume that at the end of the 6th period the infected animal has recovered from the parasitic infection and the heat production is again close to normal we may calculate that the actual heat production in periods 3 to 6

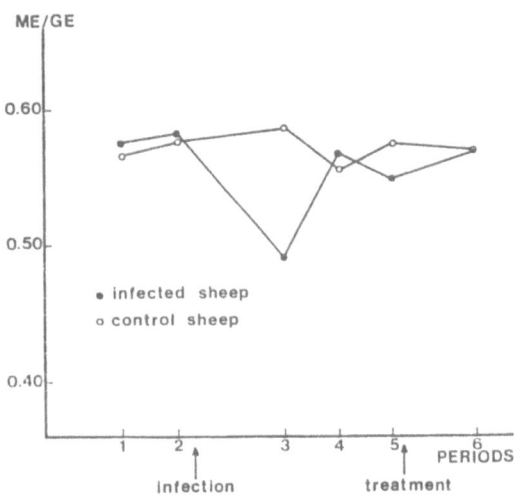

Figure 2.   Ratio between metabolizable energy (ME) and gross energy
(GE).

was 6 percent higher than the estimated value when the animal would not
have been infected (Figure 3). The EB for both sheep was originally negative but without changing diet and due to the decrease in heat production the EB gradually became positive in the 5th and 6th period (sheep
102) and 6th period (sheep 79). In the 3rd period of the infected sheep,
starting 14 days after infection, the EB went down to -83 kJ per kg metabolic weight (Table 4).

DISCUSSION

    The main goal of this pilot study was to control the effect of an acute
Haemonchus infection on the metabolism of parasite free sheep and to determine the economic significance of a subclinical infection. The infection
has well taken as is observed from the EPG and blood picture. The effect on the metabolism is primarily a reduction in the digestibility and by
that a reduction in the metabolizability of the ration. This is also seen in
previous studies (a.o. MacRae et al., 1982). Although the intake of ME
in the infected sheep is temporarily reduced, the heat production increases with 6 percent. The energy expenditure (heat production) is increas-

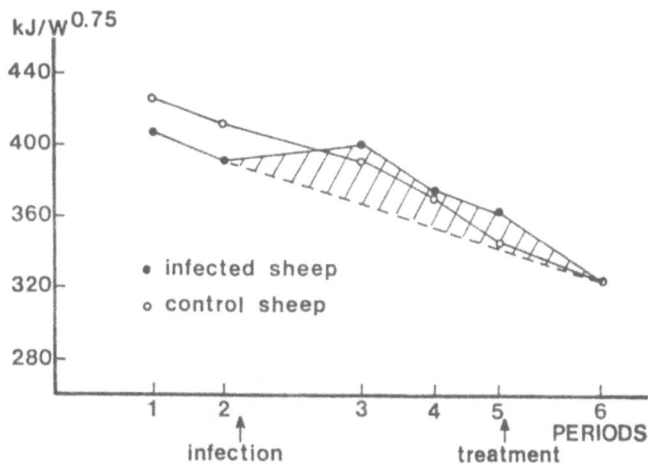

Figure 3. Heat production.

ed when the body performs either physical or chemical work. We have no indication that the infected sheep became more active and we assume that the rise in energy expenditure is the result of an increase in the activity of the defense mechanism and of the recuperation of damaged tissue. Below maintenance level the substrate for this extra energy expenditure derives from own body tissues resulting in an increased negative energy balance.

It is remarkable that during the whole experimental period which lasted for 102 days the heat production decreased with 20 to 24 percent in both sheep. Earlier studies have confirmed that certain sheep are strongly stressed for a long time when they are placed in metabolism crates and kept isolated from their fellow sheep in respiratory chambers. Restless behaviour and an increased activity of the sympatic nervous system account for the initial high heat production.

The economic significance of this acute subclinical parasitic infection may be calculated from the reduced ME intake of 7% and the increased heat production of 6% during a period of 74 days after infection. The price of the daily ration taken was 38 cents. The economic loss amounts to 74 x 0.13 x 38 cents = 365 cents per animal.

In this pilot study a single infection of 50.000 <u>Haemonchus</u> larvae did not disturb the appetite of the animal. However, the efficiency of feed utilization was reduced with 13 percent. After treatment with an anthelmintic the animal recovered within 3 weeks.

REFERENCES

Bliss, D.H. and Todd, A.C., 1974. Milk production by Wisconsin Dairy Cattle after Deworming with Thiabendazole. Vet. Med./Small An. Clin. 69: 638-640.

Brouwer, E., 1965. Report of Sub-committee on constants and factors. In: Energy Metabolism (K.L. Blaxter, ed.) E.A.A.P. Publ. no. 11: 441-443.

MacRae, J.C., Smith, J.S., Sharman, G.A.M., Corrigall, W. and Coop, R.L., 1982. Energy metabolism of lambs infected with <u>Trichostrongylus colubriformis</u>. In: Energy Metabolism of Farm Animals (A. Ekern and F. Sundstøl, eds.), E.A.A.P. Publ. no. 29: 112-115.

Pluimers, E.J., 1979. Milk production increase following treatment of Dutch dairy cattle with thiabendazole. The Veterinary Quarterly vol. 1: 82-89.

COCCIDIOSIS: A PROBLEM IN BROILERS

A.C. VOETEN

ABSTRACT

In broilers coccidiosis occurs in a subclinical form. Normally it is caused by an infection with Eimeria acervulina or Eimeria maxima. The most reliable diagnosis is testing the litter for oocysts. Subclinical coccidiosis increases the feed to gain ratio and decreases the rate of body weight gain. These negative effects last about two to three weeks after infection. Thus, infection two to three weeks before slaughtering gives the greatest loss in production, because no compensation can take place.

More and more the prevention of subclinical coccidiosis by hygienic measures or by the use of anti-coccidial therapeutics in feeds gives satisfactory results. Reduction of the damage by treatments with sulfa-drugs offers little perspective. New developments in the area of therapeutics may be more successful. In the long run vaccines can be expected.

INTRODUCTION

In 1891, i.e. almost one hundred years ago, the disease caecal coccidiosis was described in poultry for the first time.

This disease is caused by protozoa of the genus Eimeria. The name coccidiosis derives from the berry-like appearance (Greek kokkos = berry). Virtually all birds are susceptible to coccidiosis. In poultry the disease causes intestinal damage caused by the various developmental stages of the Eimeria in the intestinal wall. In poultry a total of 8 different species causing coccidiosis are known, of which 5 occur in broilers. Eimeria acervulina and Eimera maxima both cause enteric coccidiosis and are very common. Eimeria tenella, which causes caecal coccidiosis, is less common. Enteric coccidiosis can also be caused by Eimeria necatrix and Eimeria brunetti, which however are very rare in poultry.

The clinical course of coccidiosis is highly variable, varying from a mild subclinical form with no symptoms to a clinical form which in the acute phase can attack a high percentage of birds and has a high mortality and in the chronic form is characterized by emaciated, debilitated birds that finally die. Enteric coccidiosis caused by Eimeria acervulina and Eimeria maxima is usually subclinical in poultry.

Broiler houses with a high density of birds per square meter where the moisture content of the litter, the temperature and the regular turning of the litter by the birds create the optimal sporulation conditions for the oocysts provide almost ideal conditions for development of the disease.

Without effective preventive measures it would be impossible to keep broilers under these conditions. Coccidiosis would have such a negative effect on production results that the business would not be viable.

DIAGNOSIS OF SUBCLINICAL COCCIDIOSIS

Coccidiosis exists as soon as Eimeria starts to multiply in the body of the bird. If this happens without any outward signs of sickness, one speaks of subclinical coccidiosis. The diagnosis can be made by removing a few birds from the flock and examining the intestine under a microscope to detect the oocysts or other intermediate forms. This is known as the indicator-bird method. It is also possible to look for pathological-anatomical alterations. This is the so-called lesion scoring method, which is not very suitable in the industrial situation. Finally one can attempt to detect oocysts in litter and faeces or in the caecal contents. The detection of oocysts in litter is somewhat complicated but is very reliable. In older droppings, which contain a lot of ammonia, the oocysts are no longer present. Any oocysts that are found have therefore been recently excreted by the birds and represent the current situation. If the litter

412

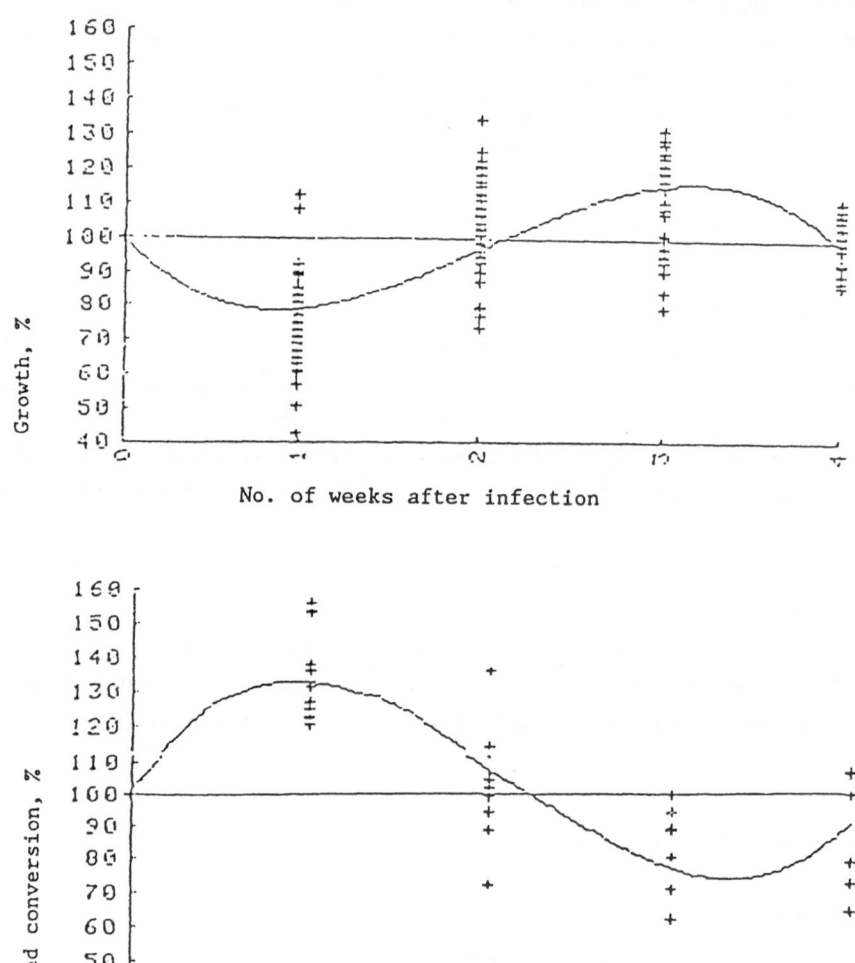

Figure 1. Growth and feed conversion of chickens infected with <u>Eimeria</u> <u>acervulina</u> as percentages of the same traits in uninfected chickens.

sampling sites are uniformly distributed over the broiler house, this test is reproducible and is thus representative. Since the faeces examination is highly dependent on the individual chicken, a mixed faeces sample from a number of birds must be taken. The reproducibility of the results

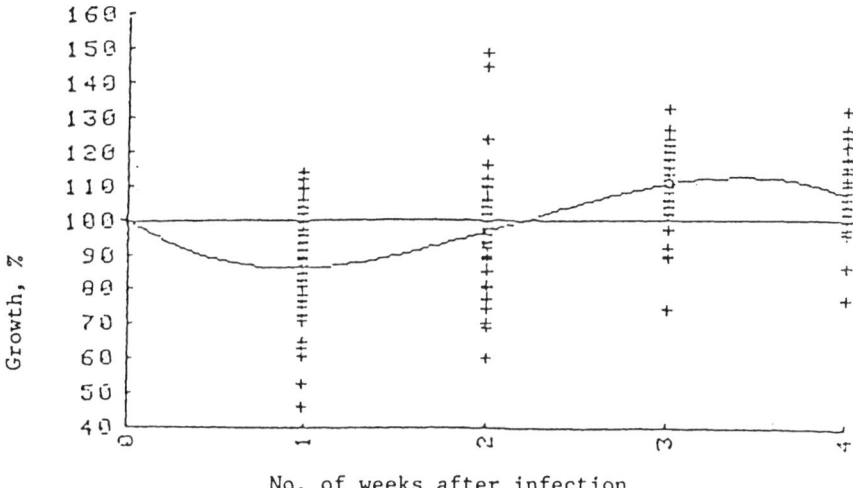

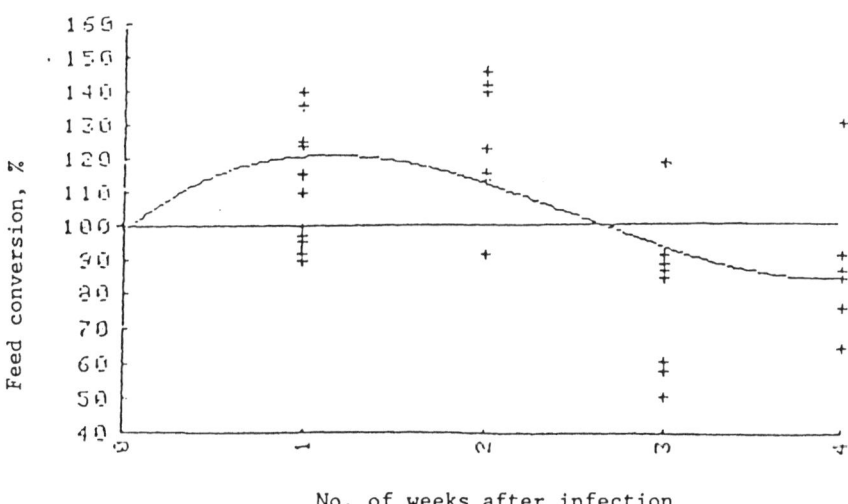

Figure 2. Growth and feed conversion of chickens infected with Eimeria maxima as percentages of the same traits in uninfected chickens.

of the test is not satisfactory. Examination of the caecum is commonly used in the veterinary service but poses the same problems as the faeces test.

414

DAMAGE DUE TO SUBCLINICAL COCCIDIOSIS IN BROILERS

In broilers the damage is manifested in higher feed conversion, inadequate growth rate and somewhat higher spoilage. This damage is due to a complex of causes: damage to the gut wall, pH changes, suboptimal enzyme activity, serum leakages and other changes. A good survey of the background to the damage has been given by Ruff (1985).

It has long been assumed that the only factor determining the extent of damage due to subclinical coccidiosis is the degree of infection. Comprehensive damage analysis has shown otherwise: in addition to the degree of infection the age of the bird when it becomes infected and the time of infection relative to the slaughter date are also important factors. The following study was carried out: Several groups of chickens were kept in batteries in an isolated room. Every week, one group of chickens was transferred to another isolated room and was simultaneously infected with Eimeria acervulina or Eimeria maxima oocysts, thus inducing subclinical coccidiosis in the birds. The growth rate and feed conversion of all the groups were determined every week. A damage model was constructed from the results of the study. If the growth rate and feed conversion are expressed as percentages of the same traits in non-infected chickens it emerges that the age at which the bird is infected plays no role whatever. Irrespective of the age at infection, the infection has an uniform negative effect for about 2 weeks and which is then followed by a compensatory effect (Figures 1 and 2).

The Figures show that the greatest damage occurs if the chickens contract the infection about 2 weeks before slaughter: the damage becomes steadily greater until slaughter and there is no compensation. If however the chickens sustain the infection in their first week, full compensation is possible. This is shown in Table 1, which gives the degree of damage due to coccidiosis at slaughter for various ages at infection. From the table it is clear that the greatest damaged is caused by infections contracted 14 days before slaughter.

These findings prompted the question whether this model is representative of the industrial situation.

To investigate this question a large-scale study was set up. The study comprised 4 consecutive flocks at 20 broiler producers. For these 80 flocks the number of Eimeria acervulina and Eimeria maxima oocysts in

Table 1.   Relation between the age at which infection occurs and the damage to growth and to feed conversion at slaughter ages of 42 and 49 days. Damage is expressed as a percentage of the maximum damage.

Infected with Eimeria acervulina oocysts

Age at which birds are infected (days)	Growth damage to birds slaughtered at		Feed conversion damage to birds slaughtered at	
	42 days	49 days	42 days	49 days
7	-	-	-	-
14	27	27	0	0
21	39	39	22	0
28	100	46	100	28
35	83	100	89	100
42	0	85	0	87
49		0		0

Infected with Eimeria maxima oocysts

Age at which birds are infected (days)	Growth damage to birds slaughtered at		Feed conversion damage to birds slaughtered at	
	42 days	49 days	42 days	49 days
7	-	-	-	-
14	0	0	0	0
21	34	0	66	12
28	100	40	100	71
35	80	100	66	100
42	0	82	0	66
49		0		0

the litter was determined every week. All the production results were collected and for each grow-out a calculation was made to determine the extent to which the production number (Voeten and Brus, 1966) of the flock exceeded or fell short of the mean. The production number (PN) is defined as follows:

$$PN = \frac{\text{growth per bird per day (in g)} \times \text{survival rate (\%)}}{\text{feed conversion} \times 10}$$

This index seems to be very representative of the results of slaughter houses.

A statistical evaluation yielded the following findings. The production results are highly dependent on the interval between the date of infection and the date of slaughter and also on whether there is a single Eimeria acervulina or Eimeria maxima infection or a combination of the two. Under the conditions of this study it was again found that the greatest damage due to coccidiosis occurred in flocks that had been infected 2-3 weeks before slaughter. The damage is particularly great if there is a combined Eimeria acervulina and Eimeria maxima infection. It was confirmed that very early infections caused no damage. The greatest difference in production results as expressed in the production number, between flocks where there was no coccidiosis and flocks with a combined infection occurring 2-3 weeks before slaughter, was about 30 production number points. Table 2 gives the exact number of production number points in relation to the age at which subclinical enteric coccidiosis is contracted.

From the findings an exact calculation can be made of the damage due to subclinical coccidiosis for all the 80 flocks investigated. This damage is as follows:

Feed conversion   per bird = 0.044 too high;

Growth rate   per bird = 1.32 g growth per day too low;

Spoilage   per bird = 1.59% too high.

Under Dutch conditions with a final weight of 1750 g, a feed price of ƒ 63 per 100 kg and a product price of ƒ 0.50 per chicken, this represents damage of ƒ 0.064 per bird. This 6.4 cent damage due to subclinical coccidiosis seems to be representative of the total damage due to coccidiosis in The Netherlands. This means that the Dutch broiler industry suffers damage of about 20 million guilders per year owing to sub-

Table 2. Influence of age (expressed in days prior to slaughter) of _Eimeria_ _acervulina_ or _Eimeria_ _maxima_ infection on production number*.

Days before slaughter at which infection occurs	Eimeria acervulina	Eimeria maxima
0	0	0
7	- 9	- 11
14	- 13	- 16
21	- 13	- 16
28	- 9	- 10
35	- 1	1

* In the event of a combined infection with _Eimeria_ _acervulina_ and _Eimeria_ _maxima_ it appears that the values for the two infections can be added together

clinical enteric coccidiosis (350 million chickens per year x f 0.064).

A second notable outcome is that anticoccidiosis preparations seem to become less and less effective. There are broadly two groups of anticoccidiosis preparations: ionophores and chemical agents. In the above study all the birds received ionophoric anticoccidiosis preparations continuously in their feed. This evidently was not successful in preventing subclinical coccidiosis. However, clinical outbreaks of coccidiosis due to _Eimeria_ _necatrix_, _Eimeria_ _brunetti_ and _Eimeria_ _tenella_ are in practice prevented to virtually 100%.

Chemical anticoccidiosis agents apparently give comparable results in preventing coccidiosis.

PROPAGATION OF THE INFECTION

Oocysts are found outside the body. The oocyst has an exceptionally protective wall, enabling it to survive for long periods under unfavourable conditions. It is resistant to large variations in humidity and temperature and to virtually all disinfectants.

Only ammonia can destroy the wall and annihilate the oocysts. This actually happens in the litter, where the oocysts are exposed to low concentrations of ammonia for a long period and are thus destroyed. Outside the litter, however, they can survive many months.

Broilers suffering from coccidiosis excrete massive numbers of oocysts. These oocysts enter the litter, where many of them perish. Oocysts are apparently also spread with dust particles by air currents. These oocysts may reach places in the broilerhouse where they are not attacked by the ammonia or may escape through ventilation openings and infect the entire site. This infection is persistent: the oocysts can be reintroduced into the broilerhouse by the owner and thus infect a subsequent flock. The oocysts can also be transferred to other farms by shoes or car tyres. Coccidiosis spreads fairly rapidly from farm to farm.

PREVENTION OF COCCIDIOSIS IN BROILERS

A number of measures to prevent coccidiosis in broilers are possible.

Use of anticoccidiosis preparations

It is now 35 years since sulpha preparations in low concentration were first added to feed to prevent coccidiosis. The entire pharmaceutical industry subsequently plunged into the development of products. The first modern product became commercially available 30 years ago.

Anticoccidiosis preparations can be divided into coccidiostats and coccidiocides. Coccidiostats inhibit reproduction of the _Eimeriae_. Coccidiocides kill them. In The Netherlands less and less importance has been attached to this distinction because many substances that were originally coccidiocidal have partly or totally lost their activity via the coccidiostatic stage. In other countries, however, coccidiocides are preferred to coccidiostats. There are now 15 preparations on the market, some of which are combinations. These products are as follows:

Single preparations

Trade name	Chemical name
Amprolium	amprolium
Arpocox	arprinocid
Avatec	lasalocid sodium (ionophore)
Coyden	meticlorpindol
Cycostat	robenidine
Deccox	quinoline
Elancoban	monensin sodium
Nicrazin	nicarbazine
Monteban	narasin (ionophore)
Sacox	salinomycin sodium (ionophore)
Stenorol	halofuginone
Zoalene	3,5-dinitro-o-toluamide (DOT)

Combination preparations

Amprol plus	amprolium and ethopabate
Lerbek	methiclorpindol and methyl benzoquate
Pancoxin plus	amprolium, ethopabate, sulphaquinoxaline and pyrimethamine

If an anticoccidiosis agent is used for some time there is a risk that an Eimeria strain will become resistant to it. The resistance can develop in an one-step manner, as happened e.g. with Coyden.

Resistance can also develop in a step-by-step manner, as happened e.g. with the ionophores and nicarbazine.

In general it can be stated that the pathogens responsible for sub-clinical coccidiosis, Eimeria acervulina and Eimeria maxima, have become partly or entirely resistant to virtually all anticoccidiosis products. Eimeria tenella often seems to have become resistant too. The Eimeria species that induce clinical forms of coccidiosis, Eimeria necatrix and Eimeria brunetti, generally still seem to be sensitive to these products.

Hygiene

Hygiene plays an important part in any attempt to breed broilers that are entirely free of coccidiosis. In practice this means extremely thorough cleansing of the broilerhouse after the birds have been transferred, disinfecting the building with ammonia and changing one's shoes when entering the building. To date these are the only measures that have produced any success. Many disinfectants have been developed, but almost always without success. If one wishes to avoid the damage due to subclinical coccidiosis by allowing the disease to take its course at a very early age, disinfection is inadvisable.

Damage limitation by treating chickens with sulpha preparations

For years it has been customary to examine indicator birds removed from flocks of broilers. If coccidiosis is detected, treatment is initiated. The usual treatment is 60 mg sulphadimidine sodium per kg body weight per day. The drug is administered in the drinking water for 5 h on 3 successive days. An analysis of the results of this treatment yielded the following findings. In 2 groups of producers where coccidiosis was confirmed in all flocks and the first group was treated while the second was not, the production result was examined to see whether it was better in the treated flock. It was found that the treatment produced no improvement whatever, see Table 3. Small-scale studies were then performed to investigate the effect of the sulpha treatment. Infected and uninfected birds were subjected to various treatments. From these studies the following conclusions were drawn.

Treatment with sulphadimidine sodium 2 days after infection has a positive effect on grwoth rate but hardly any effect on feed conversion. If treatment takes place 4 days after infection any positive result vanishes. In practice an infection is present for 2-3 weeks before it has passed through the whole flock. Thus new birds are constantly being infected. Treatment will produce an effect in perhaps a few percent of the birds in a flock. Treatment with a larger quantity of this drug does not improve the anticoccidiosis effect, but inhibits growth.

Treatment with 30 mg sulphachlorpyrazine (EsB3) has the same disappointing result as sulphadimidine sodium as far as elimination of coccidiosis is concerned, but there are indications that treatment with this sulpha preparation has a growth-promoting effect. Treatment with sul-

Table 3.   Subclinical coccidiosis in broilers (diagnosis from indicator birds): results of treatment
with sulphadimidine sodium and non-treatment.

	treated	not-treated
No. of flocks	278	147
result*	- 1.2	- 1.7

*   The result is expressed as the difference between
    the production number of the flock and the mean
    production number of the group (same breed,
    feed and month of hatching). T-test, p > 0.10,
    i.e. statistically there is no difference

pha-chlorpyrazine thus has some advantage over treatment with sulphadimidine sodium. Any treatment should be given in the second half of
the grow-out period. Treatment against subclinical coccidiosis during the
first half of the grow-out period is no longer advised.

Trials with triazinone (Baycox) are currently in progress. Initial results with this drug show that it has a distinctly better effect than the
sulpha preparations because it still protects the birds from coccidiosis
several days after treatment. Treatment with this drug at an age of
about 3 weeks seems to protect the birds from coccidiosis at their most
vulnerable age, i.e. 2-3 weeks before slaughter. This treatment in combination with nicarbazine in the first 3 weeks of life may offer possibilities for the future.

PROSPECTS FOR THE FUTURE

As has become clear in this discussion, subclinical enteric coccidiosis
is a serious source of damage. The number of Eimeria strains that are
resistant to anticoccidiosis preparations is increasing steadily. Thus the

prevention achieved by existing products is steadily diminishing.

The pharmaceutical industry has little interest in developing new products. High research costs, strict safety requirements, ecological problems, anxiety about drugs in poultry feed and poor patent protection are factors that deter producers from going further in this direction.

Damage prevention by hygienic measures is impossible in modern broiler farms. New housing systems must be sought in which e.g. all chickens are kept on mesh floors to prevent contact with droppings.

Damage prevention by treating infected flocks with sulphadimidine sodium is very disappointing. Sulphachlorpyrazine is perhaps somewhat better. The triazinone preparations currently seem quite promising for damage prevention. Many industries have now turned to the development of vaccines. Research is in progress into the value of live vaccine based on less virulent _Eimeria_ strains (precocious strains).

Biotechnology has also been applied to this problem. Already insight is being gained into the antigen structures of _Eimeriae_ which offer prospects for development of immunity. This is the most promising direction in which to search for a solution to the coccidiosis problem. It may be 10 or 20 years until the solution is found, however.

REFERENCES

Ruff, M.D., 1985. Reasons for inadequate nutrient utilization during avian coccidiosis: a review. Proceedings of the Georgia Coccidiosis Conference: 169.
Voeten, A.C. and Brus, D.H.J., 1966. Production number as a measure of production results for broilers: Tijdschrift voor Diergeneeskunde 91: 1233-1240.

CHAPTER V.   VARIATION IN ENERGY METABOLISM CHARACTERISTICS
DUE TO FEEDING LEVEL AND DIFFERENCES BETWEEN BREEDS/STRAINS

EFFECT OF FEEDING LEVEL ON MAINTENANCE REQUIREMENTS OF GROWING PIGS

C.P.C. WENK AND M. KRONAUER

ABSTRACT

In a total of 311 respiration experiments with pigs from 6 to 93 kg body mass (BM) given feed of widely varying composition at strongly differing levels the utilization of metabolizable energy (ME) for growth was studied. The results have been calculated for all data together and for three subsets of data separately (related to body weight, feeding level and ME intake).

If all 311 experiments were included, the energy costs for maintenance were calculated to be about 440 kJ/$BM^{3/4}$. The efficiency of the utilization of ME for growth amounted to k_g = 0.66, k_p = 0.57 and k_f = 0.79. In the heavier animals (BM > 40 kg) the k_p- and k_g-values were statistically slightly higher than in the lighter pigs. This was in combination with an increased maintenance. Energy retention at a medium ME-intake of 1.3 MJ per kg $BM^{3/4}$ was far more constant than at maintenance and amounted to about 565 kJ per kg $BM^{3/4}$. Feeding level had an influence on the composition of retained energy (protein and fat), but the utilization of ME was not changed markedly.

It is suggested that RE at an intake level of 1.3 MJ ME per kg $BM^{3/4}$ will give a better basal value for energy utilization in growing animals than RE at slightly above maintenance feeding. Maintenance requirement determination tend to have a lot of variation due to accumulation of variance in the estimation procedure.

INTRODUCTION

Growing pigs of modern breeds show an enormous growth capacity almost up to the body mass of 100 kg. This is partly due to consequent breeding strategies towards animals with a high protein deposition capacity and therefore to slaughter bodies with high amounts of meat. Adequate supply of nutrients to the pigs with a high growth capacity is an important prerequisite. Under modern production conditions the growth capacity is widely utilized.

In this respect the nutrient content of the rations and the daily feed intake must be considered. During recent years ad libitum feeding is more and more used in practice mainly during the growing period.

The level of feeding (or feeding level; i_F) as a measure for the intensity of feed intake was defined by Brody (1945) as the quotient between the total intake of metabolizable energy (ME) and the ME-needs for maintenance (ME_m). The influence of feeding level on growth performance and feed utilization was discussed by Wenk et al. (1980).

$$i_F = \frac{ME \text{ intake}}{ME \text{ needs for maintenance}} = \frac{ME}{ME_m}$$

The energetic maintenance needs of the growing animal are defined as the amount of ME, which is completely excreted as heat at a energy balance = 0. That means, that no energy is stored as retained protein or fat and that no energy is mobilized from the body reserves (Wenk, 1981; Van Es, 1972 and 1982). Under maintenance conditions the feeding level amounts to 1.

The experimental evaluation of ME_m in fast growing animals is not possible under normal feeding conditions. At low feeding levels they will try to continue to grow with a positive protein deposition, whereas they will cover in the mean time their energy needs by mobilizing energy reserves from the fatty tissues. The feeding condition at maintenance in the fast growing animal is therefore not a normal physiological status and has mainly a statistical meaning (Figure 1).

The description of the quantitative aspects of the utilization of ME in the intermediary metabolism is usually based on a linear model (Fig. 1). Moreover the intake of ME as well as retained energy (RE) are mostly expressed as a function of metabolic body mass (BM, ME* and RE* respectively). Kleiber (1932 and 1967) suggested originally an exponent of

3/4 for adult animals. Breirem (1939) and later other authors found lower values for * in experiments with growing pigs. (For a better comparison of the presented results nevertheless $BM^{3/4}$ is widely used as metabolic body mass).

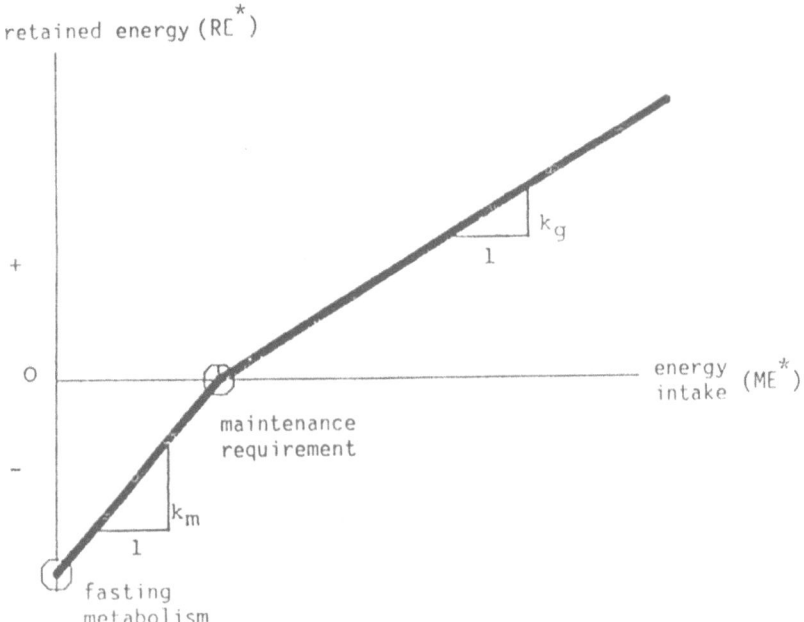

retained energy (RE^*)

maintenance requirement

energy intake (ME^*)

fasting metabolism

k_m = partial efficiency of ME for maintenance
k_g = partial efficiency of ME for growth

Figure 1. Quantitative description of the utilization of metabolizable energy for growth.

In growing animals the separation of ME^* into a part for maintenance and a part for production is usually not based on direct measurements, but corresponds to the extrapolation of a linear regression to $RE^* = 0$. ME_m^* is therefore the part of ME, which statistically cannot be explained by energy retention as deposited in protein or fat. Physiologically ME_m^* comprises the energy needs for the basal metabolism, homeostase and minimal physical activity as well as the calorigenic effect of the intake of the feed for maintenance.

EXPERIMENTAL

At our Institute in Zürich we perform respiration experiments with a simple flat deck system which is transformed into metabolism cages well suited for the balance trials. The results of the balance studies (n = 311) have been collected and used for interpretation of the influence of feeding level on energy metabolism and mainly ME_m* in growing pigs (Wenk et al., 1976; Halter et al., 1980; Hofstetter and Wenk, 1982; Hofstetter et al., 1984; Hofstetter and Wenk, 1987; Kronauer, 1987). In Table 1 the range of the experimental conditions is summarised.

Table 1. Experimental conditions.

		mean	range from - til
body mass of the pigs	kg	31.0	5.9 - 93.0
feed intake per day	kJ ME*	1296	543 - 1965
feed composition:	MJ, g per kg dry matter (DM)		
gross energy (GE)	MJ		17.7 - 21.4
crude protein (CP)	g		112 - 426
crude fat (CL)	g		19 - 126
crude fiber (CF)	g		21 - 90

* Per kg metabolic body mass

The feeding level was varied over a wide range; from 1.2 to 4.5 times ME_m*. The protein content of the rations varied from 112 to 426 g CP/kg DM. Also crude fat and crude fiber were changed markedly, whereby the CL-rich rations contained also a lot of CF.

The respiration experiments were carried out with the equipment described by Wenk et al. (1970) over 4 days every second week until 40 kg BM and every fourth week for pigs heavier than 40 kg BM; always after an adaptation period of at least 3 days.

Water was at free disposal. The rations were offered ad libitum or restricted in the experiments with a limited energy supply. Mainly male castrated piglets of the Swiss landrace, largewhite and crossbreeds were used. In one series of experiments the two lines of landrace pigs of our institute, selected for high and low growth rate as well as thin and thick backfatthickness (Hofstetter and Wenk, 1987) were used. In that series also boars and females were included.

The utilization of ME* has been calculated with the three following regression equations:

$$RE^* = a + b \cdot ME^* \tag{I}$$
$$ME^* = a + b \cdot RE_p^* + c \cdot RE_F^* \tag{II}$$
$$ME^* = a + b \cdot RE^* + c \cdot RE_p/RE \tag{III}$$

(Van der Honing et al., 1981)

ME = metabolizable energy intake
RE = retained energy
RE_p = retained energy protein
RE_F = retained energy fat

Estimation of ME_m^* as well as of the partial efficiencies of the utilization of metabolizable energy for growth was done under the assumption, that the independent parameters are not intercorrelated.

RESULTS AND DISCUSSION

Table 2 contains the mean values of the energy balances, all values are calculated per kg metabolic body mass. They are given for the whole data set as well as for the following subsets:
- body mass below and above 40 kg
- ad libitum or restricted feeding (only for pigs below 40 kg BM)
- feed intake above or below 1.3 MJ ME*

Mean body mass did not vary markedly between the two subsets on intake level (ad libitum and restricted feeding, and ME greater and less than 1.3 MJ/day respectively), although a tendency towards smaller pigs existed at restricted intake levels.

The amount of ME*-intake depended to some extent on age or body

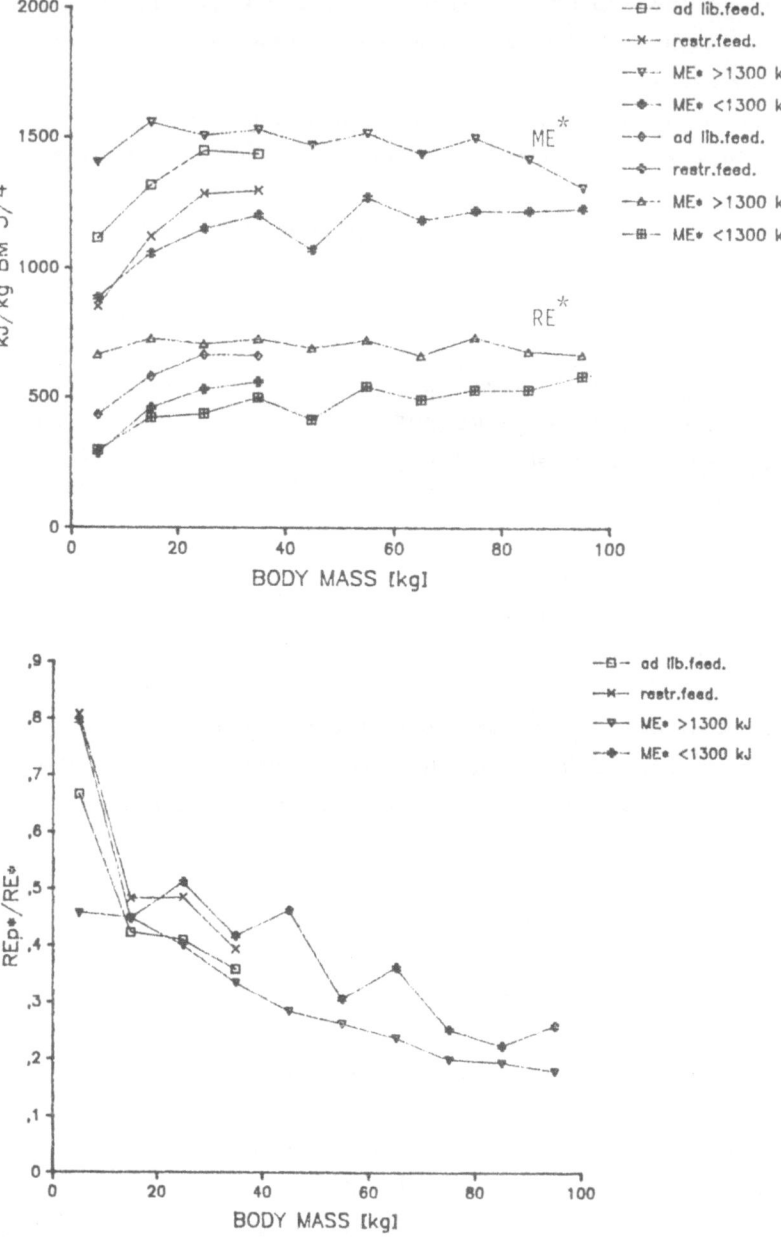

Figure 2. Intake of metabolizable energy, energy retention as well as protein as a part of total energy retention in relation to body mass at ad libitum and restricted feeding.

Table 2. Mean body mass (kg), energy balance data (kJ/kg $BM^{3/4}$ and ratio RE_p/RE.

	n	body mass	ME*	RE_p*	RE_F*	RE_p/RE
all values	311	31.1	1295	212	355	0.420
body mass < 40 kg	233	20.3	1264	229	313	0.473
body mass > 40 kg	78	63.3	1391	162	478	0.262
body mass < 40 kg:						
ad libitum feeding	136	21.7	1367	246	364	0.421
restricted feeding	97	18.2	1119	205	242	0.544
ME* > 1.3 MJ/day	164	37.5	1501	239	464	0.343
ME* < 1.3 MJ/day	147	23.8	1067	182	233	0.505

* per kg metabolic body mass

mass of the pigs. This can be seen in Figure 2, where ME*-intake and RE*-deposition as well as protein as a part of total energy deposition are plotted as a function of body mass for ad libitum and restricted feeding.

A difference in ME*-intake between high feeding level (> 1300 kJ ME*) and low feeding level (< 1300 kJ ME*) could be observed over the whole body mass range. The highest values of ME* were found at a body mass of about 40 kg. Consequently the highest energy retention was observed at the same body mass. The difference in RE_p/RE between these two groups was most pronounced in the youngest pigs. At a high feeding level (> 1300 kJ ME*) the ratio RE_p/RE decreased from about 0.6 to less than 0.2, animals receiving less than 1300 kJ ME* having a higher ratio over the whole range of body mass. The same differences could be observed in the subset on ad libitum and restricted feeding with piglets weighing less than 40 kg.

In Figure 3 all values of RE* are plotted as a function of ME*-intake. Data from the wide range of body mass (6 to 93 kg BM), of feed composition as well as of feeding level are comprised. Linear regression lines (model I) are added for the whole data set as well as for the selections "ad libitum-", "restricted feeding", and body mass above 40 kg respectively. The numerical description of the regressions are included in Table 3.

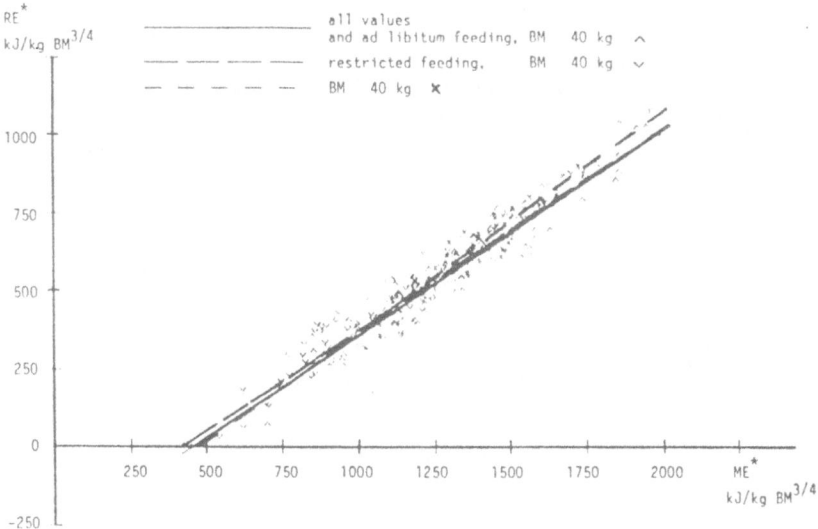

Figure 3.   Retained energy in relation to the intake of metabolizable energy.

Despite the wide variation of experimental conditions over the whole range of ME* from 0.5 to almost 2 $MJ/BM^{3/4}$ a highly significant linear regression is shown. The big number of observations allows the estimation of the maintenance requirements and partial efficiencies with a high accuracy. The overall mean of ME_m* amounts to about 440 $kJ/BM^{3/4}$, the partial efficiency of ME for growth being 0.66. Ad libitum-feeding at less than 40 kg BM gave almost the same results. At restricted feeding below 40 kg BM slightly smaller values for ME_m* and k_g could be observed.

If rations with extreme chemical compositions are used, the estimation of ME_m* is less accurate. In Figure 4 two examples are plotted for a ration with an extremely low protein content of 112 g CP per kg DM (Hof-stetter et al., 1984) as well as for rations with more than 70 g CL and 60 g CF per kg DM for piglets (Wenk and Kronauer, 1986). At a low protein content of the rations ME_m^* was estimated to be only 368 kJ/BM$^{3/4}$

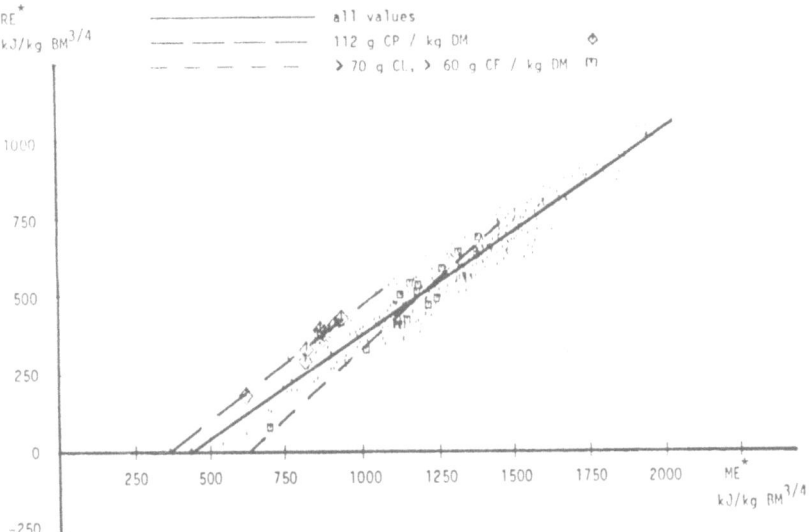

Figure 4. Retained energy as a function of ME-intake in experiments with strongly deviating rations (low protein content as well as high content of crude fat and crude fiber).

In the experiments with the fat- and fiber-rich rations a value of 633 kJ/BM$^{3/4}$ was found.

In the case of low protein content the regression line was almost parallel to the overall curve. Consequently partial efficiencies of ME-utilization were similar. However, ME_m^* was significantly reduced, probably due to a smaller protein turnover and therefore reduced general metabolism. In the case of the rations with a high fat and fiber content the regression curve crossed the general one at a ME*-intake of about 1.2 MJ per kg BM$^{3/4}$. The high value of estimated ME_m^* (633 kJ per kg BM$^{3/4}$) can primarily be explained by the high k_g-value (0.89) due to the extrapolation to RE* = 0. Therefore it has mainly to be interpreted statistically and not physiologically.

Table 3. Maintenance requirements, retained energy and partial efficiency of the utilization of metabolizable energy for growth.

	ME_m^* kJ	k_g	k_p	k_f	Coeff. RE_p^*/RE^*	CV %	RE^{*2} kJ
Model I							
all values	437	0.66				8.1	570
body mass < 40 kg	429	0.65				8.9	565
body mass > 40 kg	482	0.71				5.8	576
ad libitum feeding[1]	437	0.66				8.1	566
restricted feeding[1]	418	0.64				10.5	563
ME* > 1300 kJ/day	453	0.67				6.1	568
ME* < 1300 kJ/day	420	0.64				12.0	565
mean	439	0.66					588
range	418-482						563-576
Model II							
all values	447		0.57	0.75		4.7	562
body mass < 40 kg	433		0.54	0.77		4.9	555
body mass > 40 kg	518		0.60	0.79		3.5	572
ad libitum feeding[1]	468		0.55	0.81		4.5	560
restricted feeding[1]	416		0.54	0.76		5.6	546
ME* > 1300 kJ/day	581		0.64	0.85		3.4	549
ME* < 1300 kJ/day	473		0.58	0.83		5.9	565
mean	477		0.57	0.79			558
range	416-581						546-572
Model III							
all values	435	0.68			65.7	5.1	570
body mass < 40 kg	424	0.67			65.5	5.5	566
body mass > 40 kg	483	0.74			159.8	3.6	573
ad libitum feeding[1]	382	0.68			201.3	4.7	564
restricted feeding[1]	416	0.66			46.4	6.1	566
ME* > 1300 kJ/day	492	0.76			256.8	8.1	549
ME* < 1300 kJ/day	479	0.73			42.2	6.6	586
mean	444	0.70			119.7		568
range	382-492						549-586

[1] body mass < 40 kg

[2] at a ME*-intake = 1.3 MJ ME per kg $BM^{3/4}$

In Table 3 the results of the regression calculations for the different models and selections are summarised.

The mean values for ME_m^* and the partial efficiencies for growth correspond well with other data published e.g. by Verstegen et al. (1973), Thorbek (1975), Hoffmann et al. (1977), Berschauer et al. (1979), Böhme et al. (1980), Close and Fowler (1985). Based on a comprehensive study of the available literature the Agricultural Research Council (1981) estimated ME_m^* at 458 kJ per kg $BM^{3/4}$ and the partial efficiencies of the ME-utilization for energy retention in protein k_p = 0.54 and energy retention in fat k_f = 0.74. Mean ME_m^* did not vary markedly between the three calculation models (about 440 kJ per kg $BM^{3/4}$ in the models I and III as well as 477 kJ per kg $BM^{3/4}$ in model II). The higher k_g-value in model III compared to model I can be explained by the correction coefficient of RE_p/RE. These positive values are in good agreement with the lower efficiency of ME-utilization for protein synthesis in relation to fat synthesis. Van der Honing et al. (1985) found comparable RE_p/RE-values also in experiments with growing pigs.

Table 4.   Estimation of ME_m^* and k-values with fixed k_p- and ME_m^*-values (n = 311).

ME_m^* constant at	400	450	500 kJ per kg $BM^{3/4}$
k_p	0.53	0.57	0.61
k_f	0.72	0.75	0.78
CV%	6.90	7.20	7.80
k_p constant at	0.50	0.55	0.60
ME_m^* kJ	359	401	432
k_f	0.69	0.70	0.70
CV%	12.90	12.40	12.40

A wider variation for ME_m^* was found for the single subsets of data. As an explanation for that variation can be taken the arguments made by Menke (1985). He pointed out, that the parameters used in the different models are usually not independent and often the experimental design does not allow precise estimation of each parameter.

Often a high correlation exists between ME_m^* and k_p. This can be seen in Table 3 for the selections for a body mass above and below 40 kg. ME_m^* and k_p were higher in the heavier compared to the lighter pigs, while k_f was not influenced markedly. The same relationship can be derived from Table 4, where the k_p- and k_f-values have been calculated for fixed ME_m^*-values of 400, 450 and 500 kJ per kg $BM^{3/4}$ as well as ME_m^*- and k_f- with fixed k_p-values of 0.5, 0.55 and 0.6 respectively.

If ME_m^* or foremost k_p was fixed, the k_f-values remained remarkably constant. This can be interpreted as a sign of the close relationship of protein formation and the maintenance requirements of growing animals. Fat formation on the other hand seems to be closely correlated with ME-intake.

Besides the mentioned statistical influences on the height of the maintenance requirements other factors like stress or physical activity of the animals must be taken into consideration. Stress during the experiments has to be avoided as completely as possible. In young growing animals physical activity cannot be reduced whatever. Wenk and Van Es (1976) and developed therefore a method to measure the energy costs for physical activity in respiration experiments.

In experiments with piglets kept in groups of two and three animals per pen in a flatdeck-system (Wenk, 1981) the energy costs for physical activity amounted as an average to about 24% of the maintenance needs. That percentage was distinctly increased at restricted feeding. A higher protein supply caused a reduction of that percentage.

The values of RE* at the fixed energy intake of 1.3 MJ ME per kg $BM^{3/4}$ (see Table 3) showed a far smaller variation than the corresponding values for the maintenance needs. The differences of the parameters of the utilization of ME, estimated in the different subsets, were therefore far smaller at the medium feeding level ($i_F \sim 3$) than what could be expected from the variation of the values of the maintenance requirements.

It is therefore suggested, that RE* at a medium ME-intake of 1.3 MJ ME per kg BM$^{3/4}$ is a better basal value for the description of the utilization of ME (maintenance and production) in respect to growing animals than the maintenance requirements. Because maintenance requirements are always estimated indirectly after determination of RE and partial efficiencies they will vary considerably due to accumulation of variance, which is a result of extrapolation to RE* = 0.

REFERENCES

Agricultural Research council, 1981. The nutrient requirements of farm livestock. No. 3 Pigs. Publ. Commonwealth Agricultural Bureaux, Farnham Royal, Slough, GB.

Berschauer, F., Gaus, G. and Menke, K.H., 1979. Effect of body weight on efficiency of utilization of energy and proteins in pigs. In: L.E. Mount (Ed.), Energy metabolism. Proc. 8th Symp. Energy Metabolism, Cambridge. E.A.A.P. publ. no. 26: 101-105. Butterworths, London, GB.

Böhme, H., Gädecken, D. and Oslage, H.J. 1980. Untersuchungen über den Energieaufwand für den Protein- und Fettansatz bei frühentwöhnten Ferkeln. Landw. Forschung 33: 261-271.

Breirem, K., 1939. Der Energieumsatz bei den Schweinen. Tierernährung 11: 487-528.

Brody, S., 1945. Bioenergetics and growth, with special reference to the efficiency complex in domestic animals. Reinhold, New York, NY.

Close, W.H., Fowler, V.R., 1985. In: D.J.A. Cole, W. Haresign (Eds.), Recent Developments in Pigs Nutrition: 1-17. Butterworths, London, GB.

Es, A.J.H. van, 1972. Maintenance. 1. Kapitel. In: W. Lenkeit and K. Breirem (Eds.), Handbuch der Tierernährung, Bd. 2. P. Parey, Hamburg u. Berlin.

Es, A.J.H. van, 1982. Energy metabolism in pigs: a review. In: A. Ekern and F. Sundstøl (Eds.), Energy metabolism of farm animals. Proc. 9th Symp. Energy Metabolism Lillehammer. E.A.A.P. publ. no. 29: 249-256.

Halter, H.M., Wenk, C. and Schürch, A., 1980. Effect of feeding com-

position on energy utilization, physical activity and growth performance of piglets. Proc. 8th Symp. Energy Metabolism, Cambridge,
E.A.A.P. publ. no. 26: 395-398. Butterworths, London, GB.

Hoffmann, L., Jentsch, W., Klein, M. and Schiemann, R., 1977. Die
Verwertung der Futterenergie durch wachsende Schweine: 1. Energie
und Stickstoffumsatz im frühen Wachstumsstadium. Archiv. f. Tierernährung 27: 421-438.

Hofstetter, P. and Wenk, C., 1982. Influence of protein supply on
energy metabolism of growing pigs. In: A. Ekern and F. Sundstol
(Eds.), Energy metabolism of farm animals. Proc. 9th Symp. Energy
Metabolism, Lillehammer. E.A.A.P. publ. no. 29: 233-236.

Hofstetter, P., Bickel, H. and Wenk, C., 1984. Die Verwertung der
umsetzbaren Energie bei unterschiedlicher Rohprotein-, Rohfaser- und
Fettzufuhr bei wachsenden Schweinen. Z. Tierphysiol. Tierern. Futtermittelkde 52: 58-62.

Hofstetter, P. and Wenk, C., 1987. Energy metabolism of growing pigs
selected for growth performance, thin and thick backfat. In: P.W.
Moe, H.F. Tyrrell and P.J. Reynolds (Eds.), Energy Metabolism of
Farm Animals. Proc. 10th Symp. Energy Metabolism, Airlie. E.A.A.P.
publ. no. 32: 122-125.

Honing, Y. van der, Jongbloed, A.W., Wiemann, B.J. and Es, A.J.H.
van, 1985. Report 1964, IVVO, Lelystad, The Netherlands.

Kleiber, M., 1932. Body size and metabolism. Hilgardia 6, nr. 11: 315.

Kleiber, M., 1967. Der Energiehaushalt von Mensch und Haustier. Verlag P. Parey, Hamburg u. Berlin.

Kronauer, M., 1987. Stoffliche und energetische Ausnützung von Fut-
terprotein mit unterschiedlicher Aminosäuren - Zusammensetzung beim
wachsenden schwein. Diss. ETH Zürich, in Vorbereitung (in preparation).

Menke, K.H., 1985. Faktoren, die die Verteilung der Energie und damit
den Fettansatz beim Tier massgeblich beeinflussen. In: Wenk, C. et
al., Proc. Dreiländertagung DGE, OeGE, SGE; St. Gallen: 43-56. Wis-
senschaftl. Verlagsges. Stuttgart.

Thorbek, G., 1975. Studies on energy metabolism in growing pigs, 424.
Beretning fra Statens Husdyrbrugforsog. Kopenhagen: 153 pp.

Verstegen, M.W.A., Close, W.H., Start, I.B. and Mount, L.E., 1973.
The effects of environmental temperature and plane of nutrition on

heat loss, energy retention and deposition of protein and fat in groups of growing pigs. Brit. Journal of Nutrition 30: 21-35.

Wenk, C., Prabucki, A.L. and Schürch, A., 1970. Beschreibung einer Apparatur zur automatischen Durchführung von Respirationsversuchen an Schafen und Schweinen. Mitt. Lebensm. Unters. Hygiene 61: 378-387.

Wenk, C. and Es, A.J.H. van, 1976. Eine Methode zur Bestimmung des Energieaufwandes für die körperliche Aktivität von Küken. Schweiz. Landw. Monatsh. 54: 232-236.

Wenk, C., Pfirter, H.P. and Schürch, A., 1976. Zu Energie- und Proteinversorgung des Mastschweines. Z. Tierphysiol. Tierern. Futtermittelkde. 36: 249-266.

Wenk, C., Pfirter, H.P. and Bickel, H., 1980. Energetic aspects of feed conversion in growing pigs. Livestock Production Science 7: 483-495.

Wenk, C., 1981. Zur Verwertung der Energie beim wachsenden, monogastrischen, landwirtschaftlichen Nutztier. Habilitationsschrift ETH Zürich: 107 pp.

Wenk, C and Kronauer, M., 1986. Energieumsatz beim jungen wachsenden Schwein. J. Anim. Physiol. Anim. Nutr. 56: 13-23.

GENETIC VARIATION OF ENERGY METABOLISM IN POULTRY

P. LUITING

ABSTRACT

A literature survey was performed to quantify the variation between poultry strains and between individual animals within strains with respect to some energy metabolism parameters: metabolizability of dietary gross energy (ME %), net energy for maintenance (NEm), heat increment of maintenance (HIm), and heat increment of production (HIprod). Both laying hens and growing chicks were studied.

Genetic differences in ME% are found to be of limited magnitude; the coefficient of variation is around 3%, both in laying hens and in growing chicks.

When multiple linear regression of ME intake of strains of laying hens on metabolic body weight (MBW), body weight gain (BWG), and egg production (EP) is calculated, a standard deviation of 24-97 kJ ME/bird/day remains unexplained by these effects; between individuals within strains, 70-80 kJ ME/bird/day. The unexplained fraction, called "residual feed consumption", has a heritability of 20-60%.

From calorimetric experiments on laying hens, the between strains standard deviation of heat production (HP) is found to be 24-88 $kJ\cdot kg^{3/4}$/day (6-19% of the mean level); the individual standard deviation is of the same magnitude (15-90 $kJ/kg^{3/4}$/day; 5-23%).

The standard deviation between strains of HP at maintenance is reported to be 46-69 $kJ/kg^{3/4}$/day (10-19% of the mean); variation of HIprod (constituting the difference between total HP and maintenance HP) and, therefore, variation of partial efficiency of production, seems to be

441

small in laying hens.

The standard deviation between strains of fasting HP of laying hens appears to be 22-60 kJ/kg$^{3/4}$/day (6-15%), whereas the individual variation is of the same order again (sd = 19-127 kJ/kg$^{3/4}$/day; 6-22%).

Comparing this to the maintenance HP variation leads to the conclusion that variation of HIm (the difference between maintenance HP and fasting HP) and, therefore, variation of partial efficiency of maintenance, is small.

In conclusion, the main component of HP variation in laying hens seems to be the variation of maintenance net energy.

For growing chicks, literature points to the same direction, although evidence is more scarce; the coefficient of variation between strains of total HP is found to be around 12%, whereas those of maintenance and fasting HP are of about the same magnitude (8-31%).

The variation in maintenance net energy can be partially explained by differences in egg production, physical activity, feathering density, area of bare skin (comb, wattles, legs), body temperature, and body composition.

INTRODUCTION

Quantitative research of metabolic processes usually aims at estimation of average levels of the parameters under study. In this paper the fact will be considered that these parameters also express variation, not only as a result of external factors (e.g., experimental treatments), but also as a result of genetic differences between animals.

The quantitative distribution of feed energy over the various energy demanding processes is different for growing versus egg laying animals. Therefore, these groups will be treated separately. The average levels are given in Table 1; the genetic variation that occurs with respect to the various metabolic processes will be considered following the scheme given in Figure 1. The emphasis will be on the maintenance related fraction of energy expenditure in laying hens. The energy contents per kg of egg production and body weight gain of these animals show very little variation, which means that the major part of metabolizable energy intake differences (after correction for egg mass production and body weight

442

Table 1.   Average distribution (%) of gross energy over the various energy demanding processes in laying hens and in broilers.

	Laying hens	Broilers
Gross energy (GE)	100	100
Faecal energy (FE)	23	23
Digestible energy (DE)	77	77
Urine energy (UE)	4	4
Metabolizable energy (ME)	73	73
Heat increment of maintenance (HIm)	6	6
Net energy for maintenance (NEm)	38	29
Heat increment of production of eggs and/or body weight growth (HIprod)	10	17
Net energy for production of eggs and/or body weight growth (NE prod)	19	25

gain) can be explained from the variation of maintenance requirements and heat increment of production. In growing chicks, energy costs (both NE and heat increment) of production are known to show much variation over time, mainly as a result of the rapidly changing body weight gain composition (and, hence, growth energy content and efficiency) of these juvenile animals. The variation of their energy partitioning pattern is a much more dynamic phenomenon than in laying hens; it falls beyond the scope of this paper. The maintenance requirements of growing chicks will be compared to those of laying hens.

Here, the reference quantity will be the true feed energy consumption, assuming absence of any feed wastage.

Heil and Hartmann (1980) could not detect any differences with respect to feed wastage between two White Leghorn (WL) crosses, but found much variation between animals within crosses (σ = 8 g/bird/ day). The heritability was estimated at 13%. Hurnik et al. (1973) found significant feed wastage differences between two strains of commercial type layers, both from 8 to 12 weeks of age (16 vs 23 g/bird/ day; group housing) and from 18 to 21 weeks of age (0.9 vs 1.2 g/bird/day; individual housing). They also noted large variability within strains.

Morrison and Leeson (1978) measured feed wastage in a WL population. They could not detect any influence of wastage on the individual variation of feed efficiency. Finally, Damme (1984) stated: "Feed wastage may largely be excluded by using the right type of trough and feeding technique".

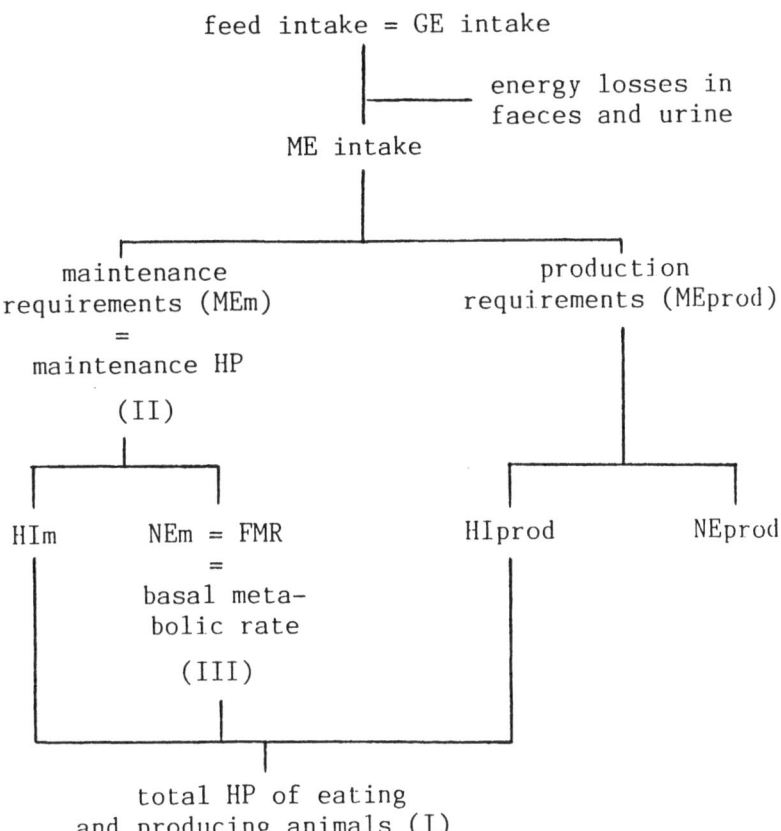

Figure 1.   Partitioning of consumed energy.

LAYING HENS

Genetic differences of metabolizability of gross energy

Foster (1968) studied metabolizability of GE (ME %) in hens of Rhode Island Red (RIR), Light Sussex (LS), Brown Leghorn (BL), the LS*BL and BL*LS crosses, and four WL strains. Three experiments were performed; the difference between the lowest and highest ME% found was 3.4% of the mean metabolizability. The individual coefficient of variation was 1-2% (CV, standard deviation as fraction of the mean), and the heritability was estimated to be 17%.

Hoffmann and Schiemann (1973) found a significant individual variance component for ME% in laying hens (σ = 2-6%), and stated that this value is about twice as large as that for Cornish cocks. The major part of this variance is reported to be caused by variation of digestibility.

Kirchgessner and Vorreck (1980) found a linear decrease in ME% of 0.055 per gram/day increase of feed consumption in laying hens consuming 72-110 gram feed per day.

Leeson and Morrison (1978) compared two groups of laying hens of about equal BW and EP, but extremely different with respect to feed consumption. They could not detect ME/GE differences between these groups. Bentsen (1983b) and Luiting (unpublished results) could not find significant phenotypic relations between ME/GE and feed consumption corrected for BW and EP. However, the genetic correlation coefficient between ME/GE and feed consumption corrected for BW and EP found by Bentsen (1983b) was -0.3; although non-significant, this estimate has the same sign as the phenotypic (not corrected for BW and EP) results of Kirchgessner and Vorreck (1980).

Summarizing, the metabolizability of GE in laying hens shows a coefficient of variation of around 3% of the mean level, mainly caused by variation in feed intake. However, when metabolizability is corrected for body weight (BW) and egg mass production (EP), a large part of this variation seems to be removed.

The possibilities of genetic improvement of metabolizability of GE seem to be small as a result of limited variation and low heritability. High ME% levels seem to be genetically related to efficient feed conversion levels, which means that selection on feed conversion is likely to result, among others, in improvement of metabolizability.

Genetic differences of ME partitioning: regression studies

ME is used for maintenance and for production. In laying hens, ME for maintenance represents roughly 65% of the total ME intake. It seems to be linearly related to metabolic body weight (MBW, in $kg^{3/4}$), whereas production ME requirements seem to be linearly related to EP and, when relevant, to body weight gain (BWG in grams). Thus, the major part of differences of ME requirements between strains and between individual animals is caused by differences in MBW and production. Still, some variation of ME requirements remain unexplained.

Variation between strains

Multiple linear regression of Waring and Brown's (1967) feed consumption data on MBW and EP of hens of three strains strongly differing with respect to MBW (WL, Thornber 404, and BL*LS) resulted in only a small unexplained fraction of variation of feed consumption (3.2 g/bird/day); the residual standard deviation (RSD) was 40 kJ ME/bird/day.

McDonald (1978) performed multiple regression of feed consumption on MBW , BWG and EP of laying hens of many strains, situated on many worldwide locations, with many husbandry systems, diets, and years; 19% of the variance of feed consumption could not be explained by variation of the independent variables (RSD= 24 kJ ME/ bird/day).

Byerly et al. (1980) compared a WL line and two WL hybrids of different BW, a broiler dam line and an adult broiler hybrid line; when performing multiple regression techniques, a standard deviation of 8 g feed consumption per day (RSD=97 kJ ME/day) remained unexplained.

Bentsen (1983b) found 10-20% of total variance of feed efficiency between various layer strains unexplained.

Leeson et al. (1973) fitted two linear regression equations for a light hybrid and for two heavy ones. When these equations were used to predict the feed consumption of other strains of about equal body weight, systematic deviations from realized feed consumption of 9.3-11.4% were observed.

Bentsen (1983a) found significantly different regression equations between WL and RIR hens: the RIR regression coefficient of feed consumption on MBW was significantly smaller (41.3 vs 49.9 g/kg$^{3/4}$/day, 451 vs 545 kJ ME/kg$^{3/4}$/day), and the coefficients on BWG (18 vs 11 kJ/g BW/day) and EP (9 vs 8 kJ ME/g EP) were significantly larger. No significant differences were found between three strains within the WL.

Damme (1984) also found different regression coefficients of feed consumption on MBW between two medium heavy strains and their reciprocal crosses (21.7-30.0 g/kg$^{3/4}$/day, 234-324 kJ ME/kg$^{3/4}$/day).

Summarizing, multiple regression of ME consumption on MBW, EP, and BWG leaves an unexplained standard deviation of 24-97 kJ ME/day; the residual standard deviation of maintenance ME requirements is about 90-94 kJ ME/kg$^{3/4}$/day.

Variation between individuals

Differences between individual hens in MBW, EP and BWG explain the major part of differences of ME requirements, too. But again, some variation remains unexplained: laying hens with equal MBW, EP and BWG still may have a different ME consumption.

When ME consumption, EP and BW are measured individually on large numbers of hens, only 30-90% of the ME consumption variance is explained by variation of MBW, EP, and BWG variation (Arboleda et al., 1976; Hagger and Abplanalp, 1978; Bordas and Merat, 1981; Bentsen, 1983a; Schild, 1983; Damme, 1984; Luiting, unpublished results).

This means that a large fraction of genetic ME intake differences between hens corresponds to the well-known genetic differences of MBW, EP, and BWG; still, 10-70% of the ME intake variance appears to be unrelated to these traits. The individual residual term seems to show systematic variation; Bentsen (1983a) and Luiting (unpublished results) find standard deviations of 70-80 kJ ME/bird/day/ over the entire laying period. It is referred to by various authors as "residual feed consumption" (RFC), being the difference between the true ME intake of a hen and the intake predicted from MBW, EP, and BWG.

This residual ME intake may be caused by true deviations from the linear model used, by unexplained sources of variation, by individual variation of the linear coefficients, and by measuring errors. The inclusion into the model of interaction and quadratic terms, and of other powers than 3/4 for estimation of MBW (McDonald, 1978) do not result in a decrease of the RFC variance (which is also caused by the fact that BW is close to 1 kg). The same is true for the inclusion into the model of age at first egg, for replacing EP by egg number and mean egg weight, and for separating BWG into its positive and negative components (Bentsen, 1983a; Luiting, unpublished results). Hagger and Abplanalp (1978) found a significant, but small, effect of age at first egg.

RFC variance not only consists of measuring and random error; this is shown by the heritability estimates for this trait, ranging from 20 to 60% (Hagger and Abplanalp, 1978; Wing and Nordskog, 1982; Bentsen, 1983b; Pauw et al., 1986; Luiting, unpublished results).

Summarizing, multiple regression of ME consumption on MBW, EP, and BWG leaves an unexplained individual standard deviation of 70-80 kJ ME/day.

Genetic differences of ME partitioning: physiological experiments

The causes of differences between strains in the regression coefficients of feed consumption on MBW, EP, and BWG, and the individual RFC differences (for which it remains unclear to what extent they are caused by different ME intake either per MBW unit, or per unit of production) have been studied in several physiological experiments.

The regression coefficient of feed consumption on MBW quantifies the maintenance requirements per $kg^{3/4}$; it may be estimated experimentally by determining the total maintenance requirements (II in Figure 1), e.g., by applying various feeding levels, and extrapolating to zero the linear regression of energy retention on ME intake. Many experiments reported in literature, however, measure the total heat production (HP) of ad libitum fed producing animals (I in Figure 1); this quantity consists of the maintenance HP and the heat increment of production (EP and/or BWG). Another quantity that has been measured in many experiments as being part of the maintenance requirements is fasting HP (III in Figure 1), defined as basal metabolic rate. The relation between these parameters is visualized in Figure 1.

To investigate the sources of variation between strains, and between individual animals, the literature will be reviewed in the order: total HP of producing and eating animals (I), maintenance HP (II), and fasting

HP (III).

Variation between strains

(I) Variation of HP of producing and eating hens: Berman and Snapir (1965) used 10 WL hens and 10 White Plymouth Rock (WPR) hens (about 40 weeks of age; ambient temperature 23.4°C) to estimate total HP by 3-4 hours mask respiration measurements. They obtained values of 425 and 379 kJ/kg$^{3/4}$/ day, respectively. Total HP of 20 New Hampshire (NH)*WL hens (temperature 22.6°C) was 416 kJ/kg$^{3/4}$/day. The standard deviation between strains is 24.4 kJ/kg$^{3/4}$/ day (cv=6.1%).

Farrell (1975) estimated total HP of 4 WL, 4 Australorp (ALP), and 6 WL*ALP hens (temperature 22°C) by 3-4 days respiration measurements to be 515, 453, and 478 kJ/kg$^{3/4}$/day, respectively. Standard deviation is 31.2 kJ/kg$^{3/4}$/day (cv=6.5%).

Lundy et al. (1978) performed four 6-hours respiration measurements on 10 Warren-SSL hens (mean BW 2.268 kg) and 10 Babcock-W300 hens (1.527 kg) at 28 weeks of age (temperature 20°C). They found total HP values of 393 and 453 kJ/kg$^{3/4}$/day, respectively, corresponding to a standard deviation of 42.4 kJ/kg$^{3/4}$/day (cv = 10.0%). The same animals were measured at 35 weeks of age for 8*22 hours (Figure 2).

MacLeod et al. (1982) determined total HP in 5 WL and 5 RIR*Sussex hens, and found 480 and 379 kJ/kg$^{3/4}$/day, respectively. The standard deviation is 31.4 kJ/kg$^{3/4}$/ day (cv = 16.6%).

Bentsen (1983 b) measured two 24-hours respiration of 48 WL and 16 RIR hens at 66 weeks of age, and found total HP values of 527 and 403 kJ/kg$^{3/4}$/day, respectively. The standard deviation is 87.7 kJ/kg$^{3/4}$/day (cv = 18.9%).

Summarizing, the standard deviation between strains of total HP of eating and egg producing hens is found to be 24-88 kJ/kg$^{3/4}$/day, or 6-19% of the mean level.

(II) Variation of maintenance HP: Balnave et al. (1978) reviewed the literature on maintenance HP of various strains with different BW, as measured by calorimetry (see Table 2). The difference between the light WL strains and the heavier ALP and Thornber-404 strains was about 19% of the mean. The level of the WL*ALP cross was intermediate between the WL and ALP values; however, the heavy WL strain had a much lower maintenance HP than expected from its strain and BW. The standard de-

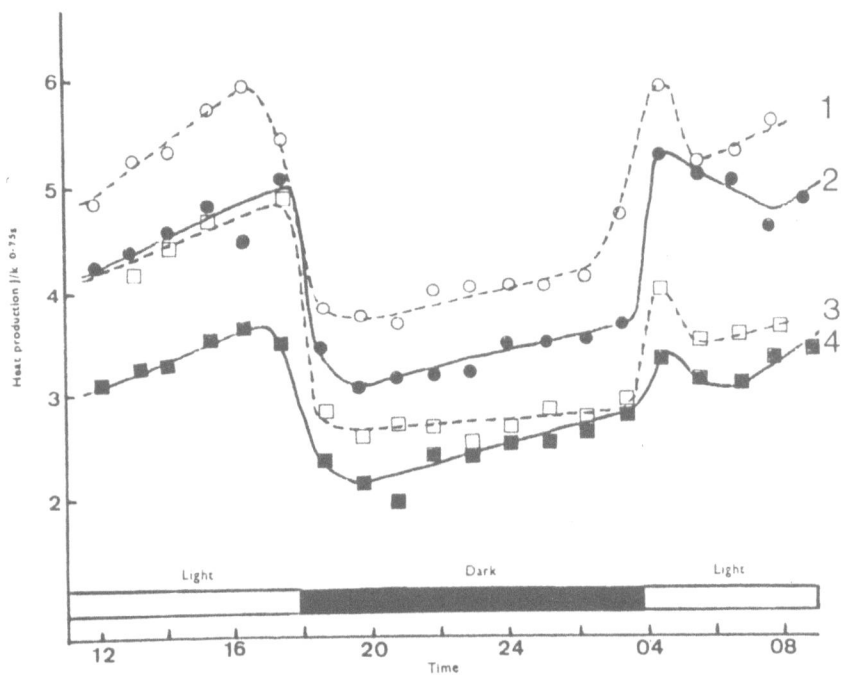

Figure 2.   Diurnal variations in starving and fed heat production (Lundy et al., 1978).

1: Babcock B300, fed HP

2: Warren SSL, fed HP

3: Babcock B300, starving HP

4: Warren SSL, starving HP

viation between the different strains is 45.7 $kJ/kg^{3/4}$/day (cv = 9.9%).

When comparing this variation of maintenance HP (II) to that for total HP (I), it may be concluded that the heat increment of production contributes little to the total variation between strains; it also constitutes only a small fraction of total HP level (Table 1). This means that partial efficiency of production is not expected to vary much between strains.

Table 2. Maintenance requirement estimates $(ME/kg^{3/4}/day)$ of laying hens of different body weights, adjusted to 22°C (Balnave et al., 1978, modified).

Strain	BW	Maintenance requirement
	(kg)	(kJ)
WL	1.70	536
WL	2.00	415
Thornber 404	2.00	445
WL*ALP	2.38	461
ALP	2.62	442

When assuming no variation between strains of the heat loss during egg formation per unit of energy content of EP, and when correcting for EP level, Bentsen (1983b) found a difference between WL and RIR hens with respect to maintenance HP of 443 vs 345 $kJ/kg^{3/4}/day$. The standard deviation between these strains is 69.3 $kJ/kg^{3/4}/day$, 17.6% of the mean.

Thus, even after correction for the EP energy content, the variation of maintenance HP remains of a similar magnitude as the variation of total HP of eating and egg producing hens. The major fraction of variation appears to be associated with the maintenance ME requirements.

(III) Variation of fasting HP: Berman and Snapir (1965) found fasting metabolic rates (FMR) of 10 WL (mean BW 1.66 kg) and 10 WPR hens (3.50 kg), starved for 24 hours (temperature 23.4°C), to be 367 and 324 $kJ/kg^{3/4}/day$, respectively; the value for 20 NH*WL hens (1.94 kg; temperature 22.6°C) was 357 $kJ/kg^{3/4}/day$. The standard deviation between the strains is 22.5 $kJ/kg^{3/4}$day (cv = 6.4%).

FMR was measured in 24 hours respiration by Waring and Brown (1965) in 4 WL hens (mean BW 1.692 kg, 24-48 hours of starvation), and by Waring and Brown (1967) in 5 Thornber-404 hens (1.983 kg, 48-72 hours of starvation) to be 445 and 360 $kJ/kg^{3/4}/day$, respectively. The standard deviation between strains is 60.1 $kJ/kg^{3/4}/day$ (cv = 14.9%).

Farrell (1975) found in 3 WL (mean BW 1.563 kg), 4 ALP (2.525 kg), and 5 WL*ALP hens (2.322 kg), when measuring respiration for 24 hours after 30 hours of starvation, FMR values of 387, 312, and 398 kJ/kg$^{3/4}$/day; the standard deviation is 46.8 kJ/kg$^{3/4}$/day (cv = 12.8%).

Lundy et al. (1978) measured FMR in their Warren-SSL and Babcock-B300 hens in 8 respiration periods after 48 hours of starvation; fasting HP was 29% and 27% lower than HP when fed ad libitum, respectively. The difference between these strains as found when fed (I) becomes slightly larger when starved (see Figure 2)

Damme (1984) compared FMR of two WL strains (22 and 70 weeks of age, mean BW 1.346 and 1.315 kg), two different broiler parent strains (26 and 56 weeks, 2.922 and 3.614 kg), Sussex (58 weeks, 2.078 kg) and RIR hens (58 weeks, 1.985 kg), measured during 10 min respiration periods after 24 hours of starvation (measured at daylight, temperature 22°C). The largest difference, observed between the Sussex and the WL hens, was 129 kJ/kg$^{3/4}$ per 24 hours. The standard deviation between strains was 56 kJ/kg$^{3/4}$ per 24 hours (cv = 13%). For measurements during nighttime, cv reduces to 10% of the mean value; hence, over the whole day it would range from 10 to 13%.

Summarizing, the standard deviation between strains of fasting HP appears to be about 22-60 kJ/kg$^{3/4}$/day (6-15% of the mean level). The strain differences do not totally reflect the various mean BW values of the strains: correction of FMR for BW does not remove the strain differences.

This standard deviation is of the same magnitude as the one of maintenance HP, which indicates that variation between strains of heat increment at maintenance feeding is of limited magnitude (the absolute level also constitutes only a small fraction of total HP, see Table 1). Therefore, the variation of the partial efficiency of maintenance seems to be small in laying hens.

Variation between individuals

(I) Variation of HP of producing and eating hens: Leeson and Morrison (1978), Katle et al. (1984), and Luiting (1986) found a difference of 13-25% in HP between groups of ad libitum fed WL hens with about equal BW and EP, but an extremely different feed consumption (Table 3). When assuming a Normal HP distribution in the populations, the individual coefficients of variation may be (under)estimated as 2.9-6.0% ($\sigma \geq$ 15.3, 33.6 and 37.8 kJ/kg$^{3/4}$/day, respectively.

Correction for the difference in EP between these two groups by using equal HP factors per unit of production energy increases the group difference found by Katle et al. (1984) to 78 kJ/kg$^{3/4}$/day, resulting in a cv of 3.2%.

Table 3.   Heat production (kJ/kg$^{3/4}$/day) of two groups of WL hens with equal BW and EP and an extremely different feed consumption

Reference	Efficient hens	Inefficient hens	Measurements
Morrison and	(n=4)	(n=5)	3*24 hours
Leeson (1978)	730	897	
Katle et al.	(n=16)	(n=16)	2*24 hours
(1984)	491	561	
Luiting (1986)	(n=6)	(n=6)	10*48 hours
	494	637	

Berman and Snapir (1965), Burlacu and Baltac (1971), Farrell (1975), and Lundy et al. (1978) estimated individual variation coefficients of total HP of eating and egg producing hens at 5-23% (σ = 23-90 kJ/kg$^{3/4}$/day).

Relevant information on individual variation of maintenance HP (II) has not been found in the literature.

(III) Variation of fasting HP: Berman and Snapir (1965), Burlacu and Baltac (1971), Waring and Brown (1965 and 1967) and Damme (1984) found individual coefficients of variation for FMR of 6-22% (σ = 19-85 kJ/kg$^{3/4}$/ day). Morrison and Leeson (1978) also measured HP of the above mentioned extreme groups after starvation periods of 24-48 hours, 48-72 hours, and 72-96 hours; the differences ranged from 67 to 127 kJ/kg$^{3/4}$/day, the inefficient hens showing the higher levels. An (under-)estimate of the individual coefficient of variation would be 3.2% ($\sigma \geq$ 20.1 kJ/kg$^{3/4}$/day).

When measuring FMR at 22°C during ten minutes in daylight after 24 hours of starvation in 1131 hens of two medium heavy strains and two reciprocal crosses, Damme (1984) found a heritability of 7-35% using data not corrected for MBW; after correction, heritability decreased to 0-29%.

Individual variation between laying hens for total HP appears to be between 3 and 23% of the mean; the variation of fasting HP is of the same magnitude, indicating again little individual variation for heat increment values.

GROWING CHICKS

Genetic differences of metabolizability of gross energy

When growing chicks and laying hens are fed the same diet, the former seem to metabolize GE less efficiently; Burlacu and Baltac (1971) found values of ME% of 72.9-75.5 for chicks, and 80.1 ± 1.7% for laying hens on the same feed.

Begin (1967) did not find any significant differences in ME per gram feed when comparing growing chicks of WL, RIR, Silver Cornish (SC), NH, WPR, and the SC*WPR cross. There was some indication, however, that the WL and RIR chicks metabolized less energy per gram feed than the other strains; the difference is 1.4% of the mean. Begin (1969) did not find any significant ME% differences in growing WL, NH and WPR chicks each fed three different diets; the range is 1.4-4.7% of the mean. Sibbald and Slinger (1963) found WL chicks to metabolize 2-3% more GE per gram feed than WPR chicks.

The presence of genetic variation of metabolizability was shown in a selection experiment carried out by Pym et al. (1984), with four different lines (selected on increased 5-9 week BWG (W), on increased 5-9 week feed consumption (F), on decreased 5-9 week feed/gain ration (E), and an unselected control line (C)). After twelve generations of selection, they found significant ME% differences between the lines (Table 4).

The W line did not deviate from the control line, which indicates absence of genetic relationships between ME% and BWG. The F and E lines did deviate from the control as would be expected; the relationships of ME/GE with feed intake and feed conversion follow the same trend as was observed in laying hens.

Generally, the variation of metabolizability in growing chicks seems to be of the same magnitude as in laying hens.

Genetic differences of ME partitioning

Growing chicks are always housed in groups; hence, no data on individual variation of ME partitioning are available.

(I) variation of HP of growing and eating chicks: Guillaume (1969) found different oxygen consumption values from 1 to 11 weeks of age be-

Table 4. ME/GE of unselected broilers at 6 weeks of age and of broilers selected for daily gain, feed consumption, and feed conversion (Pym et al., 1984).

Selection line	ME/GE (%)
C	72.7^{a} [1]
W	73.0^{a}
F	62.7^{b}
E	76.0^{c}

[1] different superscripts indicate significant differences ($P < 0.05$)

tween full sibs from seven families that differed with respect to the sex-linked Dwarf gene (dw). Dwarfs consume more O_2 than normal chicks when compared at equal age, but less when compared at equal BW (Figure 3); a clear difference between the genotypes is shown to exist.

After ten generations of selection, Pym et al. (1984) found significant line differences for HP at equal age when fed ad libitum (Table 5); the standard deviation between the lines is 100 $kJ/kg^{3/4}/day$ (cv = 12.2%).

These line differences do not fully correspond with the BW differences between the lines.

(II) Variation of maintenance HP: Pym et al. (1984) used their selection lines to estimate maintenance requirements as well (Table 5). The standard deviation is 85.8 $kJ/kg^{3/4}/day$ (cv = 11.5%).

From these data, a significant variation between lines of the heat increment of BWG becomes apparent; because the absolute value of the

Table 5. Heat production (HP) of different broiler selection lines when fed ad libitum, maintenance requirements (MEm), heat increment (HI) of BWG, and fasting metabolic rate (FMR) (Pym et al., 1984).

Selection line	Mean BW (g)	HP	MEm	HI of BWG	6 wk BW (g)	FMR
		$(kJ/kg^{3/4}/day)$				$(kJ/kg^{3/4}/day)$
C	694	816	742	74	653	508
W	891	767	671	96	787	481
F	794	960	866	94	730	569
E	783	733	701	32	707	485

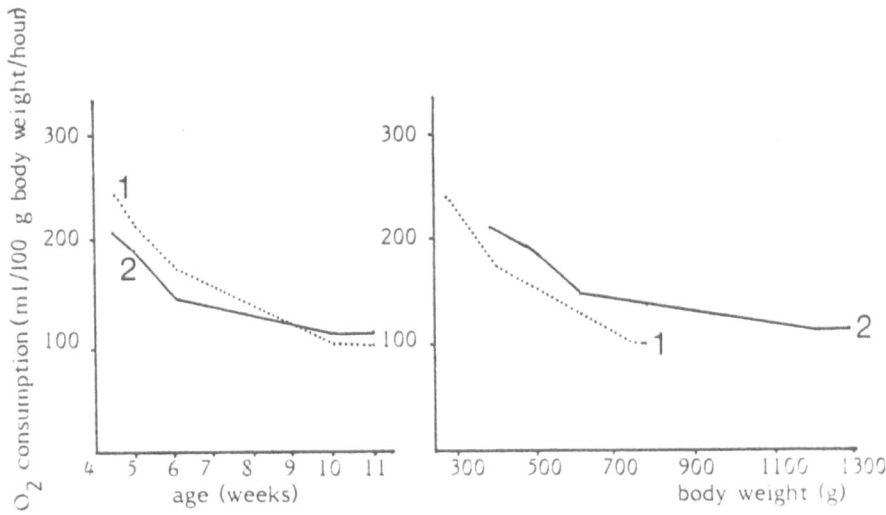

Figure 3. Oxygen consumption of dwarf (1) and normal (2) growing chicks (Guillaume, 1969).

456

heat increment is limited, its effect on the variation of maintenance HP is small. The heaviest line appears to have the lowest maintenance HP level.

(III) Variation of fasting HP: Pym et al. (1984) measured fasting HP after 24 hours of starvation in their selection lines, (Table 5). The standard deviation between lines is 40.6 (kJ/kg$^{3/4}$/day) (cv = 7.9%). Again, the heaviest line showed the lowest FMR level.

In the 12th generation of a divergent selection experiment in WPR broilers, with two lines selected for a high and a low 8-week BW, Owens et al. (1971) found a significant difference for oxygen consumption (during daytime, at 4 weeks of age after 12 hours of starvation) between the lines. The male chicks from the heavy line were 130 g heavier than those from the light line, and consumed 790 ml O_2/kg/hour less. The standard deviation between the lines was 559 ml O_2/kg/hour (cv= 31%).

The same birds were 532 g heavier at 8 weeks of age than the light line mean, and consumed 210 ml O_2/kg/hour less (standard deviation 148 ml O_2/kg/hour, cv = 22%).

Direct divergent selection on oxygen consumption after 14-32 hours of starvation at 3 weeks of age in broilers was performed by MacLaury and Johnson (1972). After 11 generations of selection, the realized heritability of O_2 consumption was estimated to be 8%. In the 11th generation, the birds of the low O_2 consumption line consumed (at 3 weeks of age) 840 ml O_2/kg/hour less than those of the high line (standard deviation 508 ml/kg/hour, cv = 20.4%), and were (at 8 weeks of age) 62 g heavier.

Künzel and Künzel (1977) studied growing male chicks of a WL layer type hybrid and of a WPR broiler hybrid (0-8 weeks of age) after 24-50 hours of starvation. On average, FMR values were 140 kJ/kg/day lower for the broiler chicks (standard deviation is 99.0 kJ/kg/day, cv = 15.5%) when compared at equal age. When compared at equal body weight (corrected by means of regression of FMR on BW) the difference reduced to 117 kJ/kg/day (standard deviation 82.7 kJ/kg/day, cv = 13.0%). The difference disappeared beyond 500 g BW (Figure 4).

Summarizing, the variation between strains of total HP of growing chicks is of the same order as found for laying hens: about 12% of the mean. Variation of maintenance HP and fasting metabolic rate is of the same magnitude (coefficients of variation are 8-31%), which leaves little

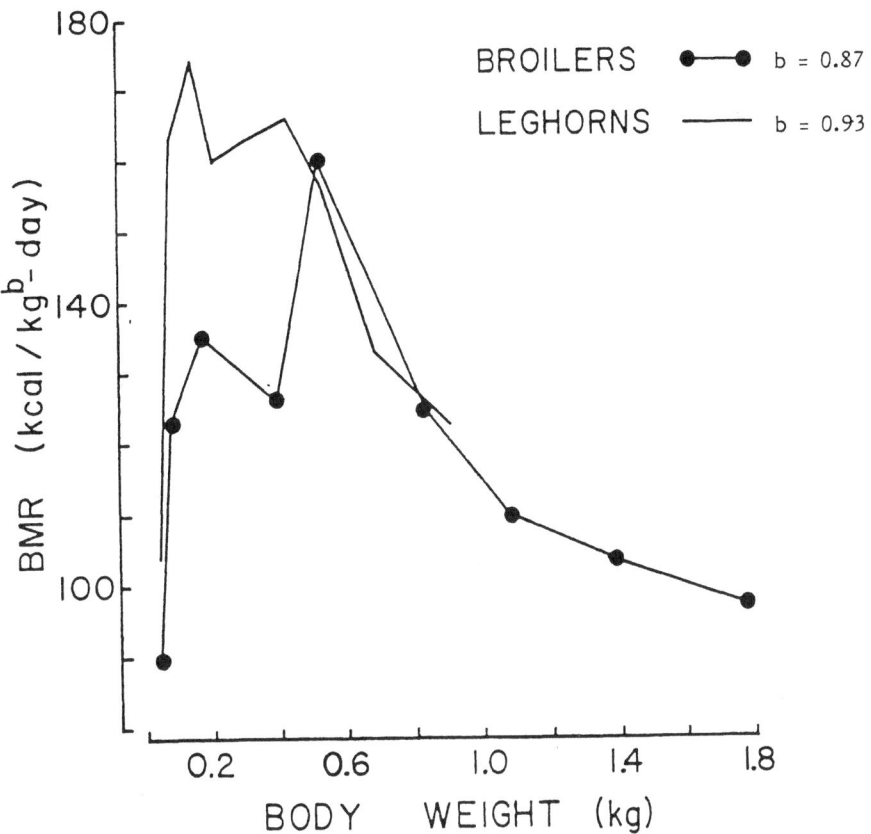

Figure 4.   Fasting metabolic rate expressed in terms of metabolic body weight (Künzel and Künzel, 1977).

variation for heat increment values.

Again, the main source of variation of HP seems to be the NE requirements for maintenance.

DISCUSSION

The genetic variation of maintenance HP per MBW unit may be associated with differences in egg mass production, physical activity, feathering quality, area of bare skin (the latter two being related to thermal insulation of the body), body temperature, body composition, etcetera.

Laying hens appear to have higher basal metabolic levels than non-laying hens. Waring and Brown (1965) found 19% higher FMR values for laying hens than for non-laying hens of equal BW; Balnave et al. (1978) found 42% larger maintenance requirements for laying hens than for ova-riectomized hens of equal BW. This phenomenon could be involved in the differences between total HP and maintenance HP of Bentsen's (1983b) WL (EP 43.1 g/day) and RIR hens (EP 33.6 g/day), or FMR of Waring and Brown's (1967) WL (EP 43.6 g/day) and Thornber-404 (EP 38.2 g/day) hens. The FMR differences between the strains of Damme et al. (1984) correspond only partly with their egg mass differences, which means that production differences cannot be the only factor explaining maintenance HP differences. Of course, the same becomes apparent from the comparisons of groups of hens with equal HP and BW but an ex-tremely different feed consumption, showing large HP differences (both total HP and FMR: Morrison and Leeson, 1978; for total HP see also Katle et al., 1984, and Luiting, 1986).

Physical activity (or "movement") is reported to cause 10-25% of total HP levels in growing chicks (Wenk and Van Es, 1976) and in laying hens (MacLeod et al., 1982, and Luiting, 1986). MacLeod et al. (1982) found around 24% of the HP differences between 5 WL and 5 RIR*Sussex laying hens to be caused by differences in activity levels. Especially the move-ments made during feed consumption, pre-ovipositional behaviour, stand-

Table 6. Estimated energy costs of various types of physical activity of laying hens (Kemp, 1985; modified).

Type of activity	duration (% of daylight phase)	energy costs ($kJ/kg^{3/4}/hr$)	total extra costs ($kJ/kg^{3/4}/day$)
feed consumption	13-50%	2-14	4-112
pre-oviposn. behaviour	0-19%	5-14	0-43
standing	10-55%	1-5	2-44
preening	8-14%	6	8-13

ing, and preening seem to cause high absolute levels and large variation of HP between animals (Table 6). Heil et al. (1982) estimated the heritability of the daily duration of pre-ovipositional behaviour (the mean was 51 min per day, with an individual standard deviation of 44 min/day) to be < 64%. Subjective activity assessment of the three strains in Farrell's (1975) research showed primarily the docile nature of the ALP hens compared to the WL and the WL*ALP cross; this is in good agreement with the differences of both total HP and FMR. Bentsen (1983b) mentioned the generally higher activity level of WL hens compared to RIR hens, which corresponds to the total HP difference. MacLeod et al.(1982) showed the activity level of WL hens to be around six times as high as the level of RIR and RIR*Sussex hens.

Damme (1984) also measured the FMR of hens of six strains during 10 minutes at nighttime (i.e., without physical activity). The coefficient of variation between strains decreased from 13% during daylight to 10% at night (standard deviation 56 and 35 $kJ/kg^{3/4}/day$, resp.), but the relative order of the strains did not change. This indicates that physical activity differences do explain a part of the FMR variation.

The same is indicated by the results of three experiments on hens with equal EP and BW, but an extremely different feed consumption. Morrison and Leeson (1978) recorded activity on video during 3 days, and noticed that the efficient animals spent more time resting (58% vs 50% of the day) and less time standing (42% vs 50%), including less time for feed consumption (12% vs 13%). All differences were of importance for total HP, but the last one does not play a role in FMR; thus, the FMR difference between the groups may be lower.

Katle et al. (1984) observed their groups by video during one day, and found the efficient hens to be less active and less sensitive to disturbances than the inefficient hens (Table 7). Luiting (1986) detected (by means of ultrasonic waves) a 77 $kJ/kg^{3/4}/day$ difference of activity related HP between extreme groups, constituting 54% of the total HP difference.

Damme's (1984) heritability estimates for FMR of 0-29% and 7-35% did not change after correction for activity levels.

The density of the feather cover of the body (partly related to physical activity) is another explanatory factor for maintenance HP differences. Thermal insulation depends on changes in the spatial arrangement of

460

the feathers and, thus, on the depth of the air layer around the body. This density may be appraised subjectively by means of visual scoring, usually in five classes, with "1" denoting a fully intact feather cover, and "5" an almost totally nude bird.

Minor feather damage does not affect total HP of ad libitum fed hens;

Table 7. Behaviour frequencies (mean % of 24 hours) in low RFC and high RFC test groups (Katle et al, 1984).

	low RFC	high RFC
Number of hens	12	10
Active	45.6	49.0[1]
Sleeping	41.7	37.2
Standing while sleeping	5.5	7.9
Threatening neighbour	2.9	1.4
Low activity while standing	15.5	17.4
Facing back wall while standing inactive	3.4	6.3[2]
Resting after eggs were collected (% of first hour)	30.2	11.1

[1] Groups differed significantly, P < 0.05
[2] Groups differed significantly, P < 0.01

the difference between more extreme scores may be as large as 206 kJ/kg$^{3/4}$/day (Damme, 1984). Lee et al. (1983) found 42 kJ/kg$^{3/4}$/day more total HP in hens with clipped back and breast feathers. Damme (1984) and Damme et al. (1982) found a 142 kJ/kg$^{3/4}$/day larger FMR for hens with feather score "4" versus score "1" during daylight, and 103 kJ/kg$^{3/4}$/day at night (i.e., independent from activity). Lee et al (1983) found FMR of two hens with feather score "4" to be 223 kJ/ kg$^{3/4}$/day larger than of two hens with score "1"; of course, both subjective scoring methods will not be fully comparable.

Clear differences between strains for feathering density have been found (Balnave, 1974); correction for feathering score did not reduce, however, the FMR variation between the six different strains of Damme et al. (1984), neither during daytime nor nighttime.

Leeson and Morrison (1978) determined the total feather weight in addition to feathering scores; they found a 13.9 g difference between groups of hens with equal BW and EP but extremely different feed consumption (efficient hens: 81.5 g, mean score: 1.3; inefficient hens: 67.6 g, mean score: 2.0).

Luiting (1986) also found a feathering score difference between her extreme groups, the efficient group scoring almost one point less (i.e., better) than the inefficient one. Damme (1984) calculated a positive phenotypic correlation (0.22-0.53) between feathering score and FMR, during daytime and at nighttime, and both independent from and including activity. Hence, correction for feathering score reduced the individual genetic variance (heritability) of FMR to some extent.

As shown in Table 5, the F line in the selection experiment of Pym et al. (1984) had the largest total HP, maintenance HP, and FMR; it was found to have become homozygous for the "slow" allele (K) of the gene coding for feathering rate. The W and E lines were homozygous for rapid feathering (k), whereas both alleles were found in the C line. It might be argued that part of the HP differences (total, maintenance, and fasting) was caused by this difference in feathering rate.

The effects on HP of the nude body area (comb, wattles, and legs) resemble those of feathering. The heat dissipation of the comb was estimated by Van Kampen (1974) to be about 10% of total HP at 22°C.

Leeson and Morrison (1978) found no significant differences with respect to the area scores of comb and wattles between their extreme groups of hens. Damme (1984) estimated positive phenotypic correlations between FMR and wattle length (0.06-0.16), and between FMR and shank length (0.17-0.69). When FMR was measured at night these values were somewhat larger than during daytime (0.17-031 vs 0.29-0.48). The corresponding genetic correlations were somewhat larger. Both types of correlations were reduced to some extent by correction for BW, activity, and feathering.

Information on variation of body temperature is scarce. Freeman (1971) reviewed the literature, and reported the existence of significant strain differences.

The same holds for information on the variation of body composition of laying hens. Egg production poses a large claim on consumed energy, which leaves little opportunity for substantial differences in body protein and fat accretion. However, quantitative information on this topic is scarce.

Vogt and Hamisch (1983) quote two investigations in this field. WL*NH hens were found to contain 23% of crude protein and 13% of fat in the body at the beginning of the laying period (20 weeks of age, BW 1715 g); they lost 0.05 g body protein and gained 0.51 g body fat per day up to 64 weeks of age. Warren-Studler hens contained 17% of crude protein and 15% of fat in the body at the beginning of the laying period (26 weeks of age, BW 1938 g); up to 62 weeks of age, they gained 0.04 g protein and 0.65 g fat per day.

Bentsen (1983b) reports significant differences between RIR and WL hens in the body weight percentage of depot fat in the chest hollow (1.24 vs 0.95%) and the abdominal cavity (6.28 vs 3.72%), and in total body fat content (60.5 vs 49.1% of total dry matter). The maintenance requirement differences between these strains (quoted earlier) are in agreement with these data.

REFERENCES

Arboleda, C.R, Harris, D.L. and Nordskog, A.W., 1976. Efficiency of selection in layer type chickens by using supplementary information on feed consumption II. Application to net income. Theor. Appl. Genet. 48: 75-83.

Balnave, D., 1974. Biological factors affecting energy expenditure. In: T.R. Morris and B.M. Freeman (editors) Energy requirements of poultry. Br. Poult. Sci., Ltd. Edinburgh: 25-46.

Balnave, D., Farrell, D.J. and Cumming, R.B., 1978. The minimum metabolizable energy requirement of laying hens. Wrld. Poult. Sci. J. 34: 149-154.

Begin, J.J., 1967. The relation of breed and sex of chickens to the utilization of energy. Poult. Sci. 46: 379-383.

Begin, J.J., 1969. The effect of diet and breed of chicken on the metabolic efficiency of nitrogen and energy utilization. Poult. Sci. 48: 48-54.

Bentsen, H.B., 1983 a. Genetic variation in feed efficiency of laying hens at constant body weight and egg production. I. Efficiency measured as a deviation between observed and expected feed consumption. Acta Agr. Scand. 33: 289-304.

Bentsen, H.B., 1983 b. Genetic variation in feed efficiency of laying hens at constant body weight and egg production. II. Sources of variation in feed consumption. Acta Agr. Scand. 33: 305-320.

Berman, A. and Snapir, N., 1965. The relation of fasting and resting metabolic rates to heat tolerance in the domestic fowl. Br. Poult. Sci. 6: 207-216.

Bordas, A. and Merat, P., 1981. Genetic variation and phenotypic correlations of food consumption of laying hens corrected for body weight and production Br. Poult. Sci. 22: 25-33.

Burlacu, G. and Baltac, M., 1971. Efficiency of the utilization of energy of food in laying hens. J. Agric. Sci., Camb. 77: 405-411.

Byerly, T.C., Kessler, J.W., Gous, R.M. and Thomas, O.P., 1980. Feed requirements for egg production. Poult. Sci. 59: 2500-2507.

Damme, K., Eichinger, H., Willeke, H., and Pirchner, F., 1982. Heat production of two different lines and their reciprocal crosses. 2nd Wrld. Cgrs. Gen. Appl. Livest. Prodn., Madrid; Vol 8: 835-840.

Damme, K., 1984. Genetische un phänotypische Beziehungen zwischen Produktionsmerkmalen und dem Energiestoffwechsel von Legehennen. PhD Thesis, TU München-Weihenstephan; 166 pp.

Damme, K., El-Sayed, T. and Pirchner, F., 1984. The fasting metabolic rate of white and brown egg layers and broiler dams of different age. Arch. Geflügelkunde 48: 77-81.

Farrell, D.J., 1975. A comparison of the energy metabolism of two breeds of hens and their cross using respiration calorimetry. Br. Poult. Sci. 16: 103-113.

Foster, W.H., 1968. Variation between and within birds in the estimation of the metabolizable energy content of diets for laying hens. J. agric. Sci., Camb. 71: 153-159.

Freeman, B.M., 1971. Body temperature and thermoregulation. In: D.J. Bell and B.M. Freeman (editors). Physiology and biochemistry of the domestic fowl. Academic Press, London/New York. Vol. 2: 1115-1151.

Guillaume, J., 1969. Conséquences de l'introduction de gène de nanisme dw sur l'utilisation alimentaire chez le poussin femelle. Ann. Biol. anim. Bioch. Biophys. 9: 369-378.

Hagger, C. and Abplanalp, H., 1978. Food consumption records for the genetic improvement of income over feed costs in laying flocks of White Leghorn. Br. Poult. Sci. 19: 651-667.

Heil, G., Otto, C. and Sodeicat, G., 1982. Zur Unruhe von Legehennen vor der Eiablage bei Haltung in Einzelkäfigen. Arch. Geflügelk. 46: 62-69.

Heil, G. and Hartmann, W., 1980. Feed wastage in strains of crossbred hens from Leghorn lines selected for egg production and feed efficiency. 6th European Poultry Conference, Hamburg; Vol. 2: 147-155.

Hoffmann, L. and Schiemann, R., 1973. Die Verwertung der Futterenergie durch die legende Henne. Arch. Tierernährung 23: 105-132.

Hurnik, J.F., Summers, J.D. and Morrison, W.D., 1973. Some factors influencing feed wastage. Poult. Sci. 52: 1665-1667.

Kampen, M. van., 1974. Physical factors affecting energy expenditure. In: T.R. Morris and B.M. Freeman (editors). Energy requirements of poultry. Br. Poult. Sci., Ltd. Edinburgh: 47-59.

Katle, J., Bentsen, H.B. and Braastad, B.O., 1984. Correlated traits with residual feed consumption. 17th World Poultry Conference, Helsinki: 136-138.

Kemp, B., 1985. Beschrijving, meetmethoden en energiekosten van activiteit bij legpluimvee. Een literatuuroverzicht. MSc. Thesis, Dept. Anim. Breeding, Agric. Univ. Wageningen: 28 pp.

Kirchgessner, M. and Vorreck, O., 1980. Zur Umsetzbarkeit der Futterenergie bei der Legehenne in Abhängigkeit von der Energie- und Proteinversorgung. Arch. Geflügelk. 44: 61-66.

Kuenzel, W.J. and Kuenzel, N.T., 1977. Basal metabolic rate in growing chicks Gallus domesticus. Poult. Sci. 56: 619-627.

Lee, B.D., Morrison, W.D., Leeson, S. and Bailey, H.S., 1983. Effects of feather cover and insulative jackets on metabolic rate of laying hens. Poult. Sci. 62: 1129-1132.

Leeson, S. and Morrison, W.D., 1978. Effect of feather cover on feed efficiency in laying birds. Poult. Sci. 57: 1094-1096.

Leeson, S., Lewis, D. and Shrimpton, D.H., 1973. Multiple linear regression equations for the prediction of food intake in the laying fowl. Br. Poult. Sci. 14: 595-608.

Luiting, P., 1986. Metabolic differences between White Leghorns selected for high and low residual feed consumption. 7th European Poultry

Conference, Paris; Vol.1: 118-121.

Lundy, H., MacLeod, M.G. and Jewitt, T.R., 1978. An automated multi-calorimeter system: preliminary expriments on laying hens. Br. Poult. Sci. 19: 173-186.

McDonald, M.W., 1978. Feed intake in laying hens. Wrld. Poult. Sci. J. 34: 209-221.

MacLaury, D.W. and Johnson, T.H., 1972. Selection for high and low oxygen consumption in chickens. Poult. Sci. 51: 591-597.

MacLeod, M.G., Jewitt, T.R., White, J., Verbrugge, M. and Mitchell, M.A., 1982. The contribution of locomotor activity to energy expenditure in the domestic fowl. 9th Symp. Energy Metabolism Farm Anim.: 197-300.

Morrison, W.D. and Leeson, S., 1978. Relationship of feed efficiency to carcass composition and metabolic rate in laying birds. Poult. Sci. 57: 735-739.

Owens, C.A., Siegel, P.B. and Van Krey, H.P., 1971. Selection for body weight at 8 weeks of age. VIII. Growth and metabolism in two light environments. Poult. Sci. 50: 548-553.

Pauw, R., Petersen, J. and Horst, P., 1986. Comparison of several indicators of food conversion efficiency in laying hens from genetic point of view. 7th European Poultry Conference, Paris; Vol. 1: 104-108.

Pym, R.A.E., Nichols, P.J., Thomson, E., Choice, A. and Farrell, D.J., 1984. Energy and nitrogen metabolism of broilers selected over ten generations for increased growth rate, food consumption and conversion of food to gain. Br. Poult. Sci. 25: 529-539.

Schild, H.J., 1983. Genetische Parameter von Leistungs- und Eiqualitätsmerkmalen von Legehennen in der zweiten Legeperiode. PhD Thesis TU München-Weihenstephan.

Sibbald, I.R. and Slinger, S.J., 1963. The effects of breed, sex and arsenical and nutrient density on the utilization of dietary energy. Poult. Sci. 42: 1325-1332.

Vogt, H. and Hamisch, S., 1983. Veränderung der Zusammensetzung der Legehennenkörper während des Legejahres. Arch. Geflügelk. 47: 142-147.

Waring, J.J. and Brown, W.O., 1965. A respiration chamber for the study of energy utilization for maintenance and production in the laying hen. J. Agric. Sci., Camb. 65: 139-146.

Waring, J.J. and Brown, W.O., 1967. Calorimetric studies on the utilization of dietary energy by the laying White Leghorn hen in relation to plane of nutrition and environmental temperature. J. Agric. Sci., Camb. 68: 149-155.

Wenk, C. and Van Es, A.J.H., 1976. Eine Methode zur Bestimmung für die körperliche Aktivität von wachsenden Küken. Schweiz. Landw. Monatshefte 54: 232-236.

Wing, T.L. and Nordskog, A.W., 1982. Use of individual records in a selection programme for egg production efficiency. I. Heritability of the residual component of feed efficiency. Poult. Sci. 61: 226-230.

GENETIC VARIATION OF ENERGY METABOLISM IN MICE

E.J. VAN STEENBERGEN

ABSTRACT

A review and own results are presented of genetical differences in energy metabolism in mice. The parameters considered are: within and between-line genetic variation in growth rate and body weight; correlated response to selection for weight or weight gain on body composition. Furthermore effects of selection on feed efficiency and maintenance energy are described in detail.

INTRODUCTION

Laboratory mice have been extensively used as a mammalian model to study various aspects of growth. Growth is the end product of many different physiological processes and is controlled by many genes and therefore easy to manipulate by selective breeding. Growth rate is determined by intake and utilization of the feed. During digestion and metabolism of the dietary nutrients direct and indirect energy losses occur. Metabolizable energy intake (ME), i.e. the gross energy of the feed minus the energy lost in faeces and urine, can be divided into energy required for maintenance (ME_m) and production (ME_p). Partitioning of the total feed intake of growing animals into maintenance and growth components can not easily be done because of measurement problems. Both heat production associated with maintenance and synthesis of protein and fat are measured together and can only be distinguished by calculation.

In addition few information is available about the genetic control of this partitioning. Results of most long term selection experiments with mice have indicated that selection for growth rate has little effect on the shape of the growth curve but that it can alter the chemical composition of the body particularly carcass fat (ct. Malik, 1984). This indicates the existence of genetic variation for body composition.

GROWTH RATE AND BODY WEIGHT

Over the past 30 years many long term selection experiments with growth rate or body weight have been carried out. Selection usually resulted in marked changes in weight and the limits to selection response are not reached for about 20 generations (Roberts, 1966). Direct additive genetic effects are demonstrated by realized within line heritability estimates which range from 0.08 to 0.17 for preweaning body weights (Eisen et al., 1970; Eisen, 1972; Robinson et al., 1974; Frahm and Brown, 1975), 0.2 to 0.5 for postweaning body weights (Falconer, 1953; Wilson et al., 1971; Bakker, 1974) and 0.1 to 0.4 for postweaning gains (Rahnefeld et al., 1963; Sutherland et al., 1970; Bradford, 1971; Bakker, 1974; Frahm and Brown, 1975). In general heritability estimates of body weights and gains increase with the age at which weights and gains are measured.

Malik (1984) gave a detailed review of non-additive genetic effects on body weight. He concluded that, in general, the maternal effects account for an increasing proportion of the total variance in body weight from birth to 3 or 4 weeks of age and steadily decrease thereafter. Heterosis has been reported in a number of studies, but it is not an invariable feature in the mice data.

BODY COMPOSITION

Studies on direct selection for body composition traits in mice have so far not been reported. Nevertheless several studies deal with the effects on body composition as a correlated response to selection for weight or weight gain, on full or restricted feeding. The majority of reports

Table 1. Percentual body composition of two divergent selected lines of mice at two feeding levels and four ages.

Parameter	Age*	W56H		W56L	
		Ad lib	Restricted	Ad lib	Restricted
Fat %	20	12.1	12.1	9.9	9.9
	30	13.3	12.7	9.1	7.2
	40	19.4	19.0	8.2	6.9
	50	24.9	24.8	7.6	6.9
Protein %	20	17.0	17.0	16.6	16.6
	30	16.5	16.2	17.9	18.7
	40	17.0	17.1	19.7	19.9
	50	16.3	16.1	19.9	20.6
Water %	20	68.2	68.2	70.8	70.8
	30	67.5	68.4	70.1	70.8
	40	61.1	61.1	68.7	69.5
	50	56.3	56.4	69.0	68.5

* Age in days. Weaning was at 20 days and different feeding levels were applied thereafter

(Fowler, 1958; Timon and Eisen, 1970; Bakker, 1974; Hayes and McCarthy, 1976; Allen and McCarthy, 1980) has indicated that at equal ages lines selected for high body weight tend to be fatter as compared with unselected lines or low body weight selection lines. Lang and Legates (1969) and Brown et al. (1977) did not find significant differences in fat percentage between their high, low and control lines. On a fat-free basis, selection for body weight in mice has been unsuccesfully in changing the percentage composition of protein, water and ash in mice carcasses (Robinson and Bradford, 1969; Timon et al., 1970; Sutherland et al., 1974). The effect of restricted feeding on body composition, after many generations of selection on body weight or weight gain with full access to feed, has been reported by Timon et al. (1970), Stanier and Mount (1972) and Meyer and Bradford (1974). Timon et al. (1970) compared the carcass composition of mice selected for nine generations for rapid postweaning gain with a control line at 57 days of age. In both lines mice were fed either ad libitum or restricted from weaning to 57 days of age. Determination of body composition was based on skinned and eviscerated carcasses. No differences in percentual body composition

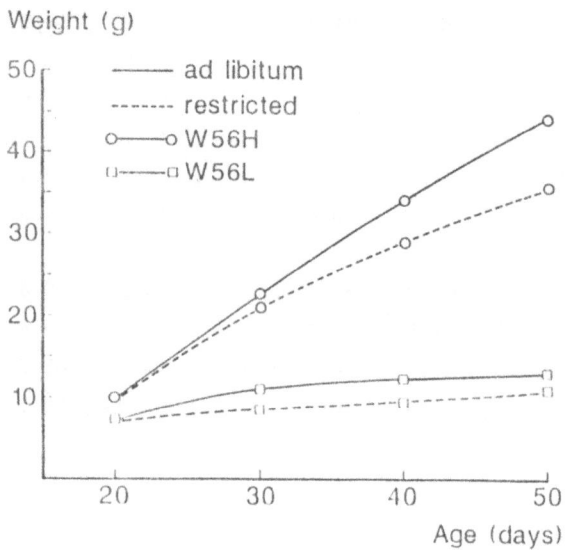

Figure 1.   Growth curves of two lines of mice at two feeding levels.

between ad libitum and restricted fed mice were found. This is in agreement with data found by Stanier and Mount (1972). Meyer and Bradford (1974) however, applied feed restriction from 6 to 10 weeks of age and found a significant lower fat % in restricted fed mice.

Results of an experiment at the Agricultural University in Wageningen, on body composition with two feeding levels (ad lib and 80% ad lib) with two lines of mice which have been selected for high (W56H) and low (W56L) 8 wk body weight for more then 20 generations are presented in Table 1.

The data represent results of whole carcass analyses, after removal of gut fill, of 710 mice (all males). Number of mice per treatment group varied from 29 to 82. Males of the base population weighed 31.1 g at day 56 (Bakker, 1974). After more then 20 generations of selection the high and low line weighed resp. 44.6 and 12.3 g at day 50. Growth curves of both lines are presented in Figure 1. It is clear that in both lines a big selection response has been obtained.

Restricted feeding after weaning decreased growth rate in both lines. At day 50 the difference in wieght due to feeding regime was 19 and 15% for the resp. high and low line. Line differences in percentual body composition are obvious at all ages. In W56L the peculiar phenomenon of a steadily decrease of fat % irrespective of feeding level can be observed. When fed ad libitum the amount of fat in the carcass did not increase from day 30 to 50. A possible explanation for this phenomenon can be a correlated selection response for feed intake capacity and activity. Activity measurements of W56H and W56L showed the W56L 1.5 times more active. The combination of limited feed intake capacity and increased activity will strongly reduce the available production energy. The remaining energy for production will be mainly used for protein and ash accretion because of their priority above fat deposition. On a fat-free basis protein and water % did not differ between lines, which is in good agreement with literature.

Feeding had a significant effect on percentual body composition for fat % at day 30. This may possibly be because in the W56L line the restriction was most severe. Moreover this may be the combined effect of weaning and feed restriction. Absence of an effect of feed restriction on body composition at older ages is in agreement with results of Timon et al. (1970) and Stanier and Mount (1972). Evaluation of the effect of feed restriction in W56H mice on a weight instead of an age scale shows the restricted W56H mice turn faster to fat. Apparently, age is also a determining factor for fat deposition.

FEED EFFICIENCY

Feed efficiency (gain: feed ratio) is difficult to interpret biologically because it is the end product of complex metabolic processes and it will vary with feeding level, age, sexe, stage of reproduction, activity, season, temperature, humidity and possibly many other factors. Differences in feed efficiency may arise from differences in the amount of energy required for:

1. maintenance, basal metabolism and physical activity,

2. lean growth and fat accretion.

In the literature only a few studies deal with direct selection on feed efficiency. Heritability estimates of feed efficiency in mice range from 0.1 to 0.4 (Sutherland et al., 1970; Yuksel et al., 1981).

Malik (1984) reviewed the relationship of feed efficiency with body weight and feed intake. He pointed at the considerable evidence showing genetic and phenotypic relationships between feed efficiency, feed consumption and postweaning growth. McCarthy (1980) stated that "... there is no case for straightforward selection for efficiency, since selection for weight achieves similar results without the expense of food recording". From the review of Malik (1984) it appears that increased gross efficiency of the lines selected for increased growth rate may be due to their increased capacity for feed consumption. After meeting energy requirements for body maintenance and the complex processes of protein synthesis and degradation, the surplus ingested energy is then stored as fat. Alteration of amount of energy and/or protein required for each g of protein deposition would require biochemical changes. So evidence for efficiency of protein deposition to be altered is not very likely. Positive correlated responses in feed intake invariably accompany selection for body weight or growth rate. Many reports indicate that selection for increased growth per se may not be the most desirable. Although an increase in feed efficiency is observed, the correlated increase in feed consumption above that which is needed for maintenance requirements and protein synthesis results in a greater fat deposition in most lines. The problem of increased fatness in carcasses of slaughter animals can be avoided to some extent by restricting feed energy intake.

There is a disagreement between reports available over the effect of selection for body weight on energetic efficiency. Partly this deals with the expression of energetic efficiency. There is no evidence of increased digestibility due to selection for body weight. However changes in energy requirements for body maintenance and growth due to selection have sometimes been observed (Malik, 1984). In his review Malik concluded that there is little doubt about the corresponding increase in feed consumption as a result of selection for body weight or growth rate. Increased feed efficiencies were also observed, but the reported interpretations are somewhat conflicting and ambigious.

MAINTENANCE

In mice energy not required for depositing protein and fat is mostly used for basal metabolism but also for physical activity. McCarthy (1980) did not find differences in basal metabolic rate per unit body weight, based on oxygen consumption, between his small and large lines at thermoneutrality except possibly prior to weaning. He suggested that differences in energy cost of maintenance per unit time per unit body weight between large and small mice at fixed ages arise mainly through scaling differences in the ratio of surface area to weight which affects heat loss. This does not rule out the possibility that large lines may reduce their maintenance cost through other means.

At our own laboratory oxygen consumption and carbon dioxide production of three lines of mice have been measured from 20 to 50 days of age. The lines were selected over more than 20 generations for:

1. high 56 day weight (W56H),
2. low 56 day weight (W56L),
3. large littersize (L).

Each line was divided into two groups which were either fed ad libitum or restricted (80% ad libitum). Heat production (H) has been estimated from oxygen consumption and carbon dioxide production per 10 day period between 20 to 50 days of age at a temperature of 24°C. Within each 10 day period gas metabolism of about 8 mice per cell has been measured over 3 to 5 48-hour periods. Metabolizable energy used for maintenance (ME_m) has been calculated as: $ME_m = (H-(1-b)ME)/b$; where b = partial efficiency coefficient (0.7) which is the energy deposited in gain per kJ of metabolizable energy ingested above maintenance, ME = amount of con-

Table 2. Number of mice per subclass of which gas metabolism has been measured.

Line	W56H		W56L		L	
Feeding level	ad lib.	restr.	ad lib.	restr.	ad lib.	restr.
Age period 20-30	16	24	8	32	14	15
30-40	8	24	8	40	16	16
40-50	16	24	24	31	16	16

Table 3. Mean and standard error of ME_m ($kJ \cdot kg^{-0.75} \cdot day^{-1}$) in three lines of mice on two feeding levels in three age periods (days).

Line	W56H				W56L				L			
Feeding level	ad lib.		restr.		ad lib.		restr.		ad lib.		restr.	
Age period												
20-30	553	±25	439	±26	502	±39	610	±14	499	±27	522	±14
30-40	600	±19	505	±17	584	±19	581	±22	493	±23	531	±16
40-50	486	±25	446	±24	487	±17	511	±16	511	±19	561	±22

sumed metabolizable energy. Distribution of number of mice over line, feeding level and age period is given in Table 2.

Mean ME_m is presented in Table 3. When fed ad libitum the three lines do not differ much in ME_m (approx. 500 $kJ \cdot kg^{0.75} \cdot day^{-1}$) except for 30-40 days of age. In that period the lines selected for body weight produced ± 600 $kJ \cdot kg^{-0.75} \cdot day^{-1}$. When compared to ad libitum feeding, the restricted W56H mice showed a lower ME_m while the restricted L mice showed a slightly higher ME_m. There is considerable variation in estimated ME_m of various subclasses. This may be due to the method of determination.

It can be concluded that selection on weight and littersize did not clearly change ME_m. In some line * age subclasses big differences in ME_m occurred between ad libitum and restricted feeding. An explanation for these differences could not be found. Further study is needed to verify the observed differences and to understand the underlying processes.

REFERENCES

Allen, P. and McCarthy, J.C., 1980. The effects of selection for high and low body weight on the proportion and distribution of fat in mice. Anim. Prod. 31:1-12.

Bakker, H., 1974. Effect of selection for relative growth rate and body weight of mice on rate, composition and efficiency of growth. Meded. Landbouwhogeschool 74-8, Wageningen: 94 pp.

Bradford, G.E., 1971. Growth and reproduction in mice selected for rapid body weight gain. Genetics 69: 499-512.

Brown, M.A., Frahm, R.R. and Johnson, R.R., 1977. Body composition of mice selected for pre-weaning and post-weaning growth. J. Anim. Sci. 45: 18-23.

Eisen, E.J., 1972. Long-term selection response for 12-day litter weight in mice. Genetics 72: 129-142.

Eisen, E.J., Legates, J.E. and Robison, O.W., 1970. Selection for 12-day litter weight in mice. Genetics 64: 511-532.

Falconer, D.S., 1953. Selection for large and small size in mice. J. Genet. 51: 470-501.

Fowler, R.E., 1958. The growth and carcass composition of strains of mice selected for large and small body size. J. Agr. Sci. (Camb.) 51: 137-148.

Frahm, R.R. and Brown, M.A., 1975. Selection for increased preweaning and postweaning weight gain in mice. J. Anim. Sci. 41: 33-41.

Hayes, J.F. and McCarthy, J.C., 1976. The effects of selection at different ages for high and low body weight on the pattern of fat deposition in mice. Genet. Res. (Camb.) 27: 389-403.

Lang, B.J. and Legates, J.E., 1969. Rate, composition and efficiency of growth in mice selected for large and small body weight. Theor.

Appl. Genet. 39: 306-314.

Malik, R.C., 1984.   Genetic and physiological aspects of growth, body composition and feed efficiency in mice: a review. J. Anim. Sci. 58: 577-590.

McCarthy, J.C., 1980.   Morphological and physiological aspects of selection for growth rate in mice In: A. Robertson (ed.) Proc. Symp. Selection Experiments in Laboratory and Domestic Animals: 100-109. C.A.B. Slough, U.K.

Meyer, H.H. and Bradford, G.E., 1974.   Estrus, ovulation rate and body composition of selected strains of mice on ad libitum and restricted feed intake. J. Anim. Sci. 38: 271-278.

Rahnefeld, G.W., Boylan, W.J., Comstock, R.E. and Madho Sing, 1963. Mass selection for post weaning growth in mice. Genetics 48: 1567-1583.

Roberts, R.C., 1966.   Limits to artificial selection for body weight in the mouse.   I. The limits attained in earlier experiments. Genet. Res. (Camb.), 8: 347-360.

Robinson, D.W. and Bradford, G.E., 1969.   Cellular response to selection for rapid growth in mice. Growth 33: 221-229.

Robinson Jr., W.A., White J.M. and Vinson, W.E., 1974.   Selection for increased 12-day litter weight in mice. Theor. Appl. Genet. 44: 337-344.

Stanier, M.W. and Mount, L.E., 1972.   Growth rate, food intake and body composition before and after weaning in strains of mice selected for mature body weight. Brit. J. Nutr. 28: 307-325.

Sutherland, T.M., Biondini, P.E., Haverland, L.H., Pettus, D. and Owen, W.B., 1970.   Selection for rate of gain appetite and efficiency of feed utilization in mice. J. Anim. Sci. 31: 1049-1057.

Sutherland, T.M., Biondini, P.E. and Ward, G.M., 1974.   Selection for growth rate, feed efficiency and body composition in mice. Genetics 78: 525-540.

Timon, V.M. and Eisen, E.J., 1970.   Comparisons of ad libitum and restricted feeding of mice selected for post weaning gain. I. Growth, feed consumption and feed efficiency. Genetics 64: 41-57.

Timon, V.M., Eisen, E.J. and Leatherwood, J.M., 1970.   Comparisons of ad libitum and restricted feeding of mice selected for post weaning gain.   II. Carcass composition and energetic efficiency. Genetics 65:

145-155.

Wilson, S.P., Goodyale, H.D., Kyle, W.H. and Godfrey, E.F., 1971. Long term selection for body weight in mice. J. Hered. 62: 228-234.

Yuksel, E., Hill, W.G. and Roberts, R.C., 1981. Selection for efficiency of feed utilization in mice. Theor. Appl. Genet. 59: 129-137.

EFFECTS OF BODY WEIGHT, FEEDING LEVEL AND TEMPERATURE ON ENERGY METABOLISM AND GROWTH IN FISH

L.T.N. Heinsbroek

ABSTRACT

The effects of body weight, feeding level and temperature on energy metabolism and growth in fish are discussed, with special reference to research done at the Department of Fish Culture and Fisheries of the Agricultural University Wageningen on the African catfish, Clarias gariepinus.

Metabolizability of the ration varies from 50 - 85% in fish and decreases at increasing feeding levels. Temperature seems to have little effect on ME.

The efficiency of the conversion of ME_p for growth (k_g) is 0.8 - 0.9 in fish and independent of body weight, feeding level and temperature. These factors affect growth therefore mainly through the maintenance requirements and the maximum feed intake/metabolism.

Due to an interactive effect of feeding level and temperature on the values of the weight exponents in the allometric relations of feed intake/ metabolism with body weight the ratio ME_p/ME_m varies with body weight and temperature. Values of this ratio range from 9.5 to 2.6 and decrease with body weight. Temperatures at which this ratio is maximal, i.e. optimal temperatures for growth, also decrease with fish size.

It is concluded that research in the field of energy metabolism and growth of fish is hampered by the problems the aqueous environment poses on determination of the energy balance components.

INTRODUCTION

Many studies on energy metabolism and growth of fish have been primarily concerned with aspects related to fisheries ecology and management. Emphasis has been on the energy costs of swimming and food acquisition. In addition to field studies, laboratory studies were conducted mainly with natural foods and were aimed at predicting growth efficiency of fish in their natural surroundings (Winberg, 1956; Ivlev, 1961; Paloheimo and Dickie, 1965 and 1966; Ursin, 1967; Beamish et al., 1975; Elliot, 1976a and b and 1982).

Research along those lines is still developing (Tytler and Calow, 1985), but the demands of a rapidly growing aquaculture industry has also invoked the expansion of research into energy metabolism and growth under intensive controlled conditions with formulated feeds. Owing much to the conceptual lines layed down by the earlier work, notably that of Winberg (1956) and Ursin (1967), recently some extensive studies have been published on sockeye salmon, Onchorynchus nerka (Brett, 1979), rainbow trout, Salmo gairdneri (Staples and Nomura, 1976; From and Rasmussen, 1984; Cho et al. 1982), carp, Cyprinus carpio (Huisman, 1974; Huisman et al., 1979) and African catfish, Clarias gariepinus (Hogendoorn, 1983; Hogendoorn et al., 1983; Machiels and Henken, 1986).

The laws of thermodynamics and therefore the same principles of energy flow diagram as in homeothermic animals are applicable to studies on fish (Figure 1). Due to the fact that fish are poikilothermic and that they are living in an aqueous environment, the conditions of this environment profoundly affect the relationships between feed intake, energy metabolism and growth.

The effects of environmental factors on energy metabolism and growth in fish have recently been reviewed by Webb (1978) and Brett (1979), following the classification of Fry (1971). It is generally accepted that for fish temperature is the most important controlling factor regulating energy metabolism and growth and also that any effects of temperature, due to the complex interactions with other factors, should not be studied without at least taking the effects of fish size and feeding level into account. The aim of this study is to discuss the effects of these three factors on energy metabolism and growth of fish, with special reference to the work done in our laboratory on the African catfish, Clarias gariepinus

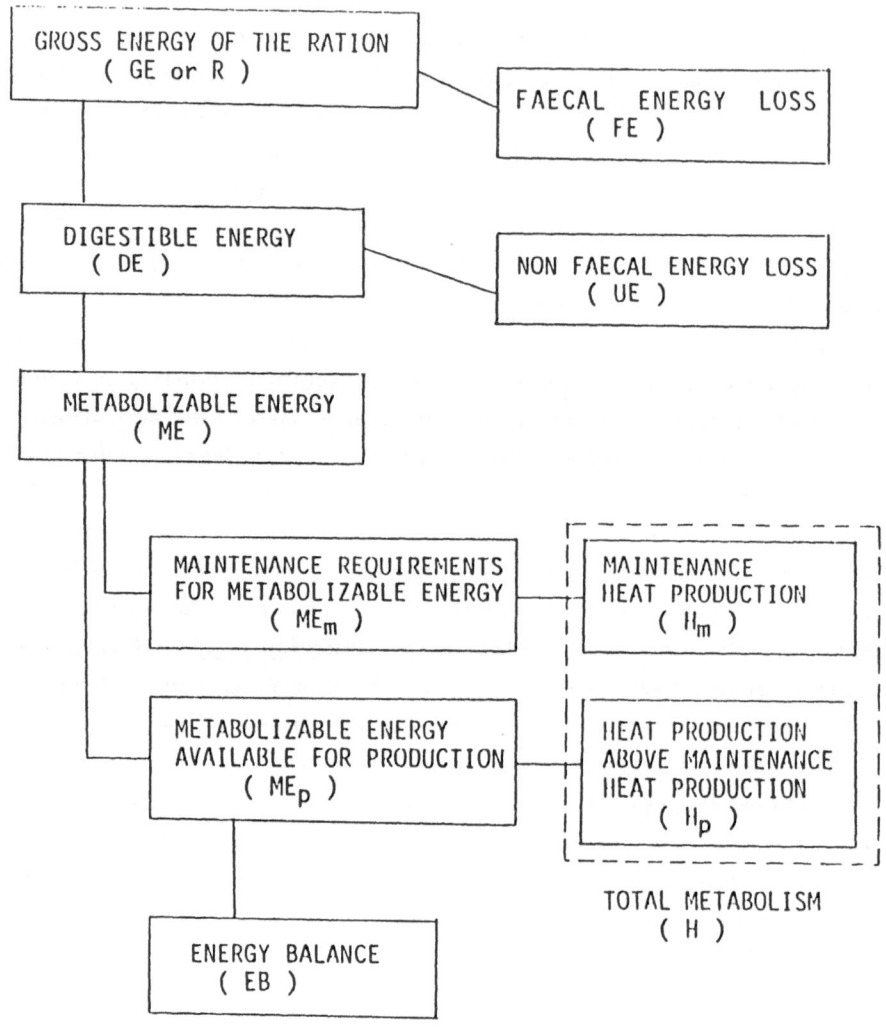

Figure 1.   Energy flow diagram used in bioenergetic studies on fish.

(Hogendoorn,   1983;   Hogendoorn   et   al.,   1983;   Machiels   and   Henken,   1986).

Before doing this, a short review on methodology in fish bioenergetics will be given because the aqueous environment poses special problems on determination of the various energy balance components.

METHODOLOGY IN FISH BIOENERGETICS

A number of excellent reviews have been published recently on methodology in fish bioenergetics (Braaten, 1979; Cho et al., 1982; Jobling, 1983; Brafield, 1985; Talbot, 1985). The reader is referred to those for more detailed information. Some aspects on methodology will be discussed here, in view of how these affect our understanding of energy metabolism and growth.

The starting point in erecting an energy budget, the determination of energy intake, already poses some problems to the experimenter, especially at the higher feeding levels. The energy content of the food can be accurately determined by bomb calorimetry. The use of values calculated from COD or from the chemical composition is not recommended due to uncertainty of the various conversion and/or correction factors (Jobling, 1983; Henken et al., 1986). However the amount of food actually eaten by the fish is less easily determined, depending on food type (natural vs pelleted) and feeding method (one to several feedings per day vs continuous feeding).

The use of pelleted feed which breaks up in the water or in the mouth of the fish with subsequent leaching of materials makes collection of uneaten feed very difficult. Brafield (1985) recommends the use of an alginate or a similar binding agent to reduce the danger of desintegration of the pellets.

Carefull observation when feeding fish one to several times per day can give a fairly reliable estimate of maximum consumption (Elliot, 1976b and 1982; From and Rasmussen, 1984). With continuous feeding maximum consumption can be estimated from the growth-ration curve (Figure 2a) as the ration giving maximum growth (Brett, 1979; Hogendoorn, 1983; Hogendoorn et al., 1983; Machiels and Henken, 1986).

Determination of the apparent digestibility, usually by an indirect method, is also difficult due to the dissolving of materials into the water. Moreover there can be contamination of the collected faeces with uneaten food. To minimize leaching of materials, faeces should be collected as soon as possible after they are voided. Several methods developed to collect faeces include netting by fine dip net, syphoning, mechanical filtration (filtration columns, glass fiber filter discs, mechanically rotating filter screens) both intermittent and continuously, and settling

482

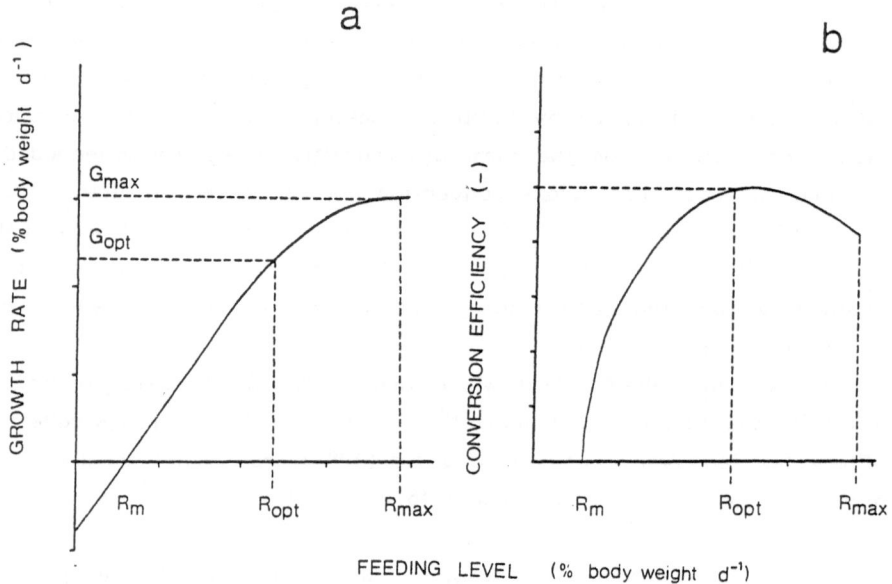

Figure 2.   The "classical" relation between growth rate and ration size.
a) growth-ration curve,   b) conversion efficiency curve (conversion
efficiency = growth/ration).

(Windell et al., 1978b; Hogendoorn et al., 1981; Choubert et al.,
1982; Cho et al., 1982; Talbot, 1985).

Other methods are developed to collect faeces from the intestine before they are expelled from the fish. These include dissection of the intestinal contents, manual stripping and anal suction (Austreng, 1978; Windell et al., 1978b; Henken et al., 1985).

Objective judgement of the different methods is not possible, because there is no true reference. The methods of collecting faeces before they are expelled seem to underestimate the digestibility, possibly due to the collection of incompletely digested food and/or contamination of the faeces with body fluids or intestinal epithelium. Netting seems to over-estimate the digestibility suggesting breaking up and leaching. Mechanical filtration and settling have been reported to give the most accurate estimates of digestibility (Choubert et al., 1982; Cho et al., 1982) provided that they are designed as to effectively trap the faeces as soon as possible (cf. Henken et al., 1985).

Ammonia, which constitutes the main part of the non faecal losses in fish (Elliot, 1976a; Brafield, 1985), should be kept at low concentrations, as the unionized form of ammonia is extremely toxic to fish. This makes accurate measurement of the ammonia excretion difficult. The situation is further complicated by the possible elimination of ammonia by nitrifying bacteria in the experimental system.

Due to the fact that fish use a considerable amount of protein to meet their energy demands, these non faecal losses do however contribute significantly to the energy budget and their ommission in the estimation of the metabolizable energy leads to erroneous results as has been shown by Huisman (1974). This author therefore suggests that metabolizable energy is more reliably determined as EB + H instead of GE - FE - UE. Calculation of the metabolizable energy content of the faeces from the chemical composition is again not recommended due to uncertainty about conversion and/or correction factors. This and the fact that the metabolizable energy content on theoretical grounds is dependent on feeding level and feed composition have led some authors to disencourage the use of the term metabolizable energy (Cho et al., 1982; Jobling, 1983).

The determination of the energy retained as growth (RE) or the energy balance (EB) is most accurately determined by weighing of the fish, together with determination of the energy content of a sample, at the beginning and the end of an experiment. The experimental period should be long enough to create a significant difference between beginning and end however. The difficulties in determining the other components of the energy budget makes calculation of EB from N- and C-balance less usefull. Calculation of the energy content of the fish from their chemical composition is also not recommended. Although in the African catfish, <u>Clarias gariepinus</u>, the common energy conversion factors of 23.64 kJ. g^{-1} for protein and 39.54 kJ. g^{-1} for fat give a good estimate for the energy content (Henken et al., 1986), in other species they do not, e.g. carp, <u>Cyprinus carpio</u> (Nijkamp et al., 1974), perch, <u>Perca fluviatilis</u> (Craig et al., 1978) and rainbow trout, <u>Salmo gairdneri</u> (From and Rasmussen, 1984).

The assessment of the energy lost as heat in fish is usually done by indirect calorimetry due to the fact that the large heat capacity of water makes the use of direct calorimetry very complicated. Because in fish the main nitrogenous end product is ammonia, Brafield (1985) suggested the following equation for fish:

$$H = 11.18 * O_2 + 2.61 * CO_2 - 9.55 * NH_3 \tag{1}$$

where H is the total heat production (in joules) and O_2 consumed and CO_2 and NH_3 produced are expressed in milligrammes. Again, because of the difficulties of accurately measuring the ammonia in the low concentrations that have to be maintained in working with fish, the factor correcting for the N-excretion is usually omitted. This could however lead to overestimation of the heat production especially at and below the maintenance feeding level where progressively more of the protein in the diet (and from the body) is catabolized. This in turn could lead to an overestimation of the metabolizable energy because at and below maintenance all metabolizable energy is lost as heat. Because there is no simple and reliable method for the determination of CO_2 in water the estimation of the heat production is in most cases further simplified by taking an appropriate value for the respiratory quotient (RQ), thereby relating the heat production directly to the oxygen consumption. Accurate measurements of the respiratory quotient have to my knowledge not been reported to date. Commonly used values vary from 0.8 - 0,95, giving oxycaloric equivalents (Q_{ox}) of 13.3 kJ.g^{-1} to 13.7 kJ.g^{-1} (Brafield, 1985).

Summarizing the above, it can be stated that the aqueous environment poses some serious problems to the determination of most of the components of the energy budget. Energy budgets often do not add up to a hundred percent, thereby limiting our understanding of the process of energy metabolism and growth in fish. For example variation in the digestible or metabolizable fraction of the gross energy which is suggested to be caused by variation in feeding level and/or temperature is of an order of magnitude of 5 - 15 percentage points which is of the same order or even less than the deviation of most energy budgets from a hundred percent (Hogendoorn, 1983; Musissi, 1984, cited by Brafield, 1985).

Due to these difficulties and the specific equipment required, many bioenergetic studies on fish are restricted to determination of feed intake, faecal loss and growth. Examples of balance respirometers, which enable the determination of gas and matter balances in fish during prolonged experimental periods are given by Hogendoorn et al. (1981) and Cho et al. (1982).

EFFECTS OF FEEDING LEVEL, BODY WEIGHT AND TEMPERATURE ON ENERGY METABOLISM AND GROWTH IN FISH

Effects of ration and fish size

It is generally accepted that the relationship between growth rate and feed ration, at a given temperature and fish size, follows the classical curve depicted in Figure 2a (Huisman, 1974; Brett, 1979; Elliot, 1982; Hogendoorn, 1983; Hogendoorn et al., 1983; From and Rasmussen, 1984). Growth increases from a negative value at feed deprivation towards a maximum as the ration increases. Huisman (1974) also has shown that the shape of the curve is the same regardless whether growth and/or ration are expressed in terms of wet weight, dry weight or energy. There is some discussion whether the increasing part of the curve should in fact be a straight line (Brett, 1979; Corey et al., 1983; Hogendoorn et al., 1983). Hogendoorn et al. (1983) state that the apparent deflection at near satiation feeding, which causes a distinct optimum in conversion efficiency below the maximum feed intake (Figure 2b), might be an artifact rather than a true depression in feed utilization. For the African catfish, Clarias gariepinus, Hogendoorn et al. (1983) and Machiels and Henken (1986) could model growth assuming a constant feed utilization up to satiation feeding.

For a better understanding of the nature of the relationship between growth and ration and of the effects of the environment on this relation-ship, many workers have adopted a bioenergetic approach based on the Pütter-Von Bertalanffy anabolism-catabolism model and/or Winberg's (1956) balanced energy equation (Ursin, 1967; Elliot, 1976b and 1982; Kitchell et al., 1977; Ricker, 1979; Corey et al, 1983; Hogendoorn, 1983; Hogendoorn et al., 1983; From and Rasmussen, 1984). Both models state that growth can be regarded as the diffference between what enters the body and what leaves it: Growth = In - Out. Formalized according to the energy flow diagram presented in the introduction (Figure 1) this gives:

$$EB = GE - FE - UE - H \qquad (2)$$

or

$$EB = ME - H \qquad (3)$$

Hogendoorn (1983) and Hogendoorn et al. (1983) following Winberg (1956), assumed the metabolizable fraction of the ration independent of the feeding level: $ME = pR$. It seems however that this assumption is not correct. Based on the results of Huisman (1974 and 1976), Elliot (1976a), Windell et al. (1978a), Hogendoorn (1983), From and Rasmussen (1984) and Henken et al. (1985) it has to be concluded that both the digestible and the metabolizable fraction of the ration are decreasing with increas-ing rations, although calculating fractions of a sometimes not well known feed intake is somewhat questionable.

Mean values for the metabolizability of the ration range from 50 to 85%. Whether the magnitude of the decrease in metabolizability is of an order as reported by Huisman (1974 and 1976) and Hogendoorn (1983) remains unclear, because of their use of possibly erroneous Q_{ox} values and because their highest feeding levels seemed to be well in excess of satiation. On the other hand Elliot (1976a) and From and Rasmussen (1984), both calculating ME as GE - FE - UE , also found a significant decrease in the fraction of ME with increasing rations.

The heat production of fish can be considered to originate from a number of different processes, including fasting or routine metabolism (H_O) and feeding metabolism (H_f). The latter, also called the heat in-crement of feeding or the specific dynamic action (SDA), can be divided in a component below or at the maintenance ration ($H_m - H_O$) and a component above the maintenance ration ($H_p = H - H_m$).

At feed deprivation body constituents are used for the fasting metabolism. Propably due to some non metabolic losses, i.e. mucus secretion, the negative energy balance at feed deprivation is usually somewhat larger than this fasting metabolism but the differences are usually small, hence at feed deprivation:

$$H_O \approx - EB \tag{4}$$

Up to the maintenance ration (R_m; ME_m; $EB = O$) all ME is used for maintenance and becomes heat:

$$ME_m = H_m = H_O + (1-k_m) * ME_m \tag{5}$$

where K_m is the efficiency of the conversion of the ME_m for metabolism.

Above the maintenance ration ($R_m \leq R \leq R_{max}$) additional ME comes available for production (ME_p):

$$ME = ME_m + ME_p = ME_m + (1-k_g) * ME_p + k_g * ME_p \tag{6}$$

or

$$ME \qquad = \qquad H \qquad + \qquad EB \tag{3}$$

where k_g is the efficiency of the conversion of the ME_p for growth.

The k_m can be determined as H_O/ME_m and values thus obtained range from 0.6 to 0.7 for the African catfish, Clarias gariepinus (calculated from Hogendoorn, 1983 and Machiels and Henken, 1986) and for carp, Cyprinus carpio (Huisman, 1974). These values are rather low when compared to those of homeotherms (Ekern and Sundstøl, 1982) and are even lower than their respective k_g values. Huisman (1976) states that these low values might be caused by the estimation of the maintenance requirements of fish. If activity of fishes is much lower at fasting than at maintenance, as has been observed for the above mentioned species, than k_m will be underestimated. In this regard it is interesting to note that Huisman (1976) found that for the rainbow trout, Salmo gairdneri, which is a much more active fish than the above mentioned species, k_m had a value of 0.83.

The maintenance requirements for ME (ME_m) of the above mentioned fish species range from 20 to 70 kJ.$kg^{-0.80}$.d^{-1}, which is very low when compared to homeotherms (Ekern and Sundstøl, 1982). This is largely due to the lower body temperatures of fish and the fact that the energy costs of locomotion and maintaining position are lower in water than in air.

The conversion efficiency of ME_p for growth, kg, can be determined from the slope of the line relating EB to ME (7) or better, as ME is usually determined as EB + H, from the slope of the line relating H to EB (8):

$$EB = K_g * (ME - ME_m) \tag{7}$$
$$H = H_m + ((1-k_g)/k_g) * EB \tag{8}$$

Huisman (1976) found values for k_g of 0.78 and 0.89 for rainbow trout, Salmo gairdneri, at 15°C and for carp, Cyprinus carpio, at 23°C respectively, while Hogendoorn (1983) found a value of 0.80 for the African catfish, Clarias gariepinus, at 25°C, independent of feeding level or body weight. These values are somewhat higher than in homeotherms (Ekern and Sundstøl, 1982), which is ascribed by Huisman (1976) to a lower protein turn-over in fish due to their lower body temperature.

From the above mentioned relations between R, ME, H and EB it follows that the increasing part of the growth - ration curve is in fact not linear. The slope of this curve, above maintenance, equals p x k_g and although k_g seems to be independent of feeding level, this seems not to be the case for p. The decrease in p with increasing ration causes the growth - ration curve to flex downwards with increasing ration.

These relations further seem to validate the use of the concepts "metabolic scope for growth" (H_{max} - H_m) and "scope for growth" (R_{max} - R_m) as correlates for the growth potential of a given fish species at a given set of circumstances (Warren, 1971; Brett, 1979; Elliot, 1982).

To account for the effect of fish size on the different aspects of feeding, energy metabolism and growth, the use of allometric relations of the form:

$$y = a * w^b \tag{9}$$

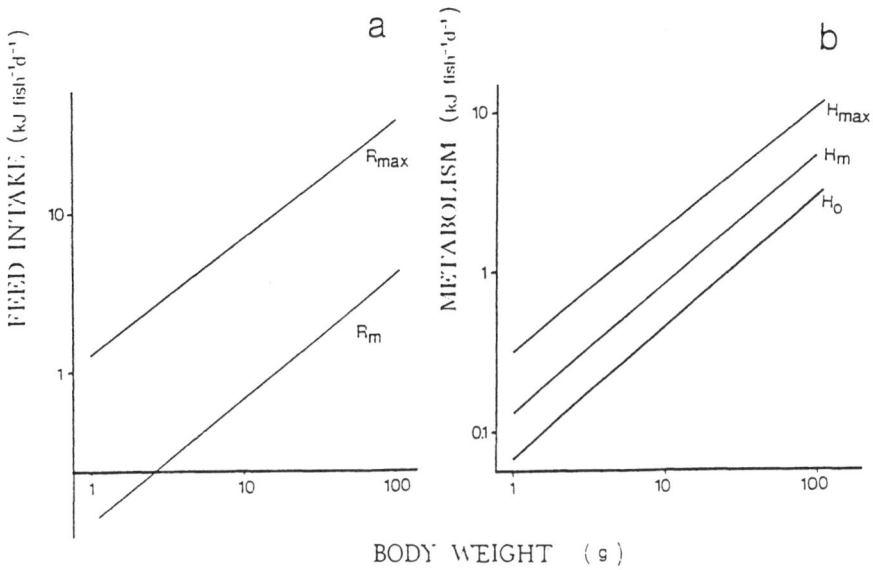

Figure 3. Relations of feed intake (a) and metabolism (b) with body
weight for the African catfish, <u>Clarias gariepinus</u>, at 25°C (R_{max} =
1.32 $W^{0.79}$, R_m = 0.16 $W^{0.83}$, H_{max} = 0.325 $W^{0.77}$, H_m = 0.13 $W^{0.82}$,
H_0 = 0.07 $W^{0.83}$).

where y is some measure of feeding, metabolism or growth, W is the body
weight and a and b are constants, is generally accepted (Winberg, 1956;
Paloheimo and Dickie, 1965 and 1966; Ursin, 1967; Huisman, 1974; Hogendoorn et al., 1983). Figure 3 shows some examples of the relations between feed intake/metabolism and body weight for the African catfish,
<u>Clarias gariepinus</u>, at 25°C (calculated from Hogendoorn, 1983; Hogen-
doorn et al., 1983; Machiels and Henken, 1986). From this figure it can

be seen that the values of the weight coefficient (a) vary strongly with the feeding level as has already been observed by Winberg (1956) and Paloheimo and Dickie (1966) for a number of fish species. These authors also state that the value of the weight exponent (b) is independent of feeding level. However, the differences observed by Hogendoorn (1983) for the weight exponents for fasting/ maintenance metabolism compared to (maximum) feeding metabolism (0.82-0.83 vs 0.77-0.79) are, although statistically not significant, in accordance with a general trend that can be observed in various other studies (Ursin, 1967; Huisman, 1974; Kitchell et al., 1977; Corey et al., 1983; From and Rasmussen, 1984). Although there are also studies reporting no differences or even differences in the opposite direction (Elliot, 1976b; Staples and Nomura, 1976; Brett, 1979), the observation of the weight exponent for fasting/maintenance metabolism being greater than the weight exponent at maximum feeding is believed to be biologically significant, as this could theoretically account for fish not growing indefinitely. The difference of the weight exponents ultimately results in the absence of a metabolic scope for growth at high body weights (Ursin, 1967; Kitchell et al., 1977; Hogendoorn, 1983). This phenomenon will be elaborated upon further in the next paragraph in conjunction with temperature effects.

Effects of temperature

The effect of temperature on the growth-ration curve for a given fish size is shown in Figure 4 for the African catfish, Clarias gariepinus (Hogendoorn et al., 1983). The same pattern is observed for a number of salmonids (Elliot, 1976b and 1982; Brett, 1979; From and Rasmussen, 1984) and for carp, Cyprinus carpio (Huisman et al., 1979; Goolish and Adelman, 1984). Both the negative energy balance at feed deprivation and the maintenance ration normally are larger at higher temperatures. The maximum feed intake also increases with temperature, initially causing an increasing growth rate. Due to the higher maintenance requirements the maximum growth rate is, after this initial increase, decreasing at higher temperatures, despite the fact that the maximum feed intake still shows an increase. At even higher temperatures the maximum feed intake will also decrease.

Considering the balanced energy equation (3), the effect of temperature on the digestible and metabolizable fraction of the feed is not very

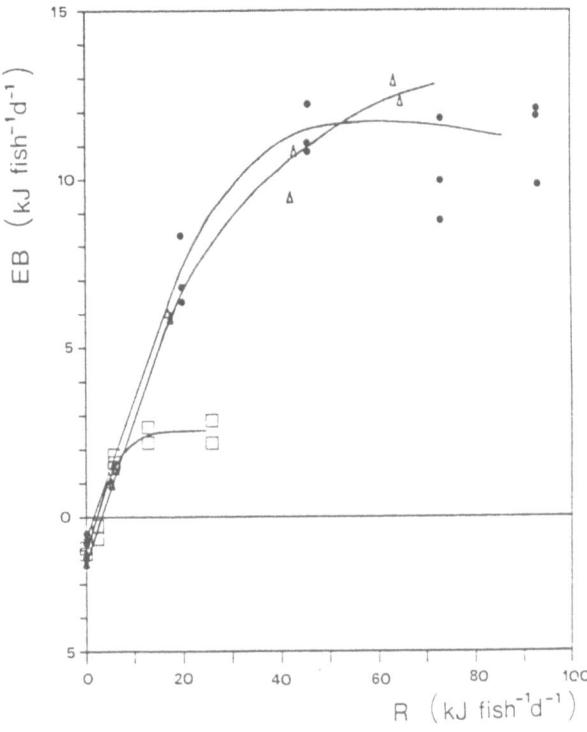

Figure 4. The effect of temperature on the growth-ration curve for African catfish, Clarias gariepinus, with a body weight of 20-25 gram.

clear. The rate of gastric/ intestinal evacuation commonly shows a sharp increase with increasing temperatures as observed for rainbow trout, Salmo gairdneri (From and Rasmussen, 1984) and carp, Cyprinus carpio (Garcia and Adelman, 1985). However the apparent digestibility usually shows only a minor increase with an increase in temperature, with a possible exception for very low temperatures, i.e. physiologically very low regarding to the species investigated (Elliot, 1976a and 1982; From and Rasmussen, 1984; cf. Machiels and Henken, 1986 for ME). Because the fraction of non faecal losses also shows an increase with increasing temperature, the effect of temperature on the metabolizable fraction of the ration also seems to be small (Elliot, 1976a and 1982; From and Rasmussen, 1984).

Temperature appears to have no effect on the conversion efficiencies of ME_m and ME_p as is shown for k_g in Figure 5 for the African catfish,

<u>Clarias gariepinus</u>, based on results presented by Machiels and Henken (1986). Although the value of k_g seems to be slightly higher at 20°C, this effect was not statistically significant and a common value of 0.804 could be adopted for k_g independent of temperature (F_{35}^2 = 0.1475). Furthermore, the good agreement of the output of a dynamic simulation model for growth of the African catfish, based on commonly accepted bio-chemical pathways in the intermediate metabolism, with the above men-tioned results at the different temperatures even indicates that the fish are operating at maximum biochemical efficiency independent of body weight, feeding level and temperature (Machiels and Henken, 1986). Be-cause of this relative constancy of feed utilization efficiency, the main-tenance requirements and the maximum feed intake and/or metabolism seem to be the most important factors through which environmental fac-

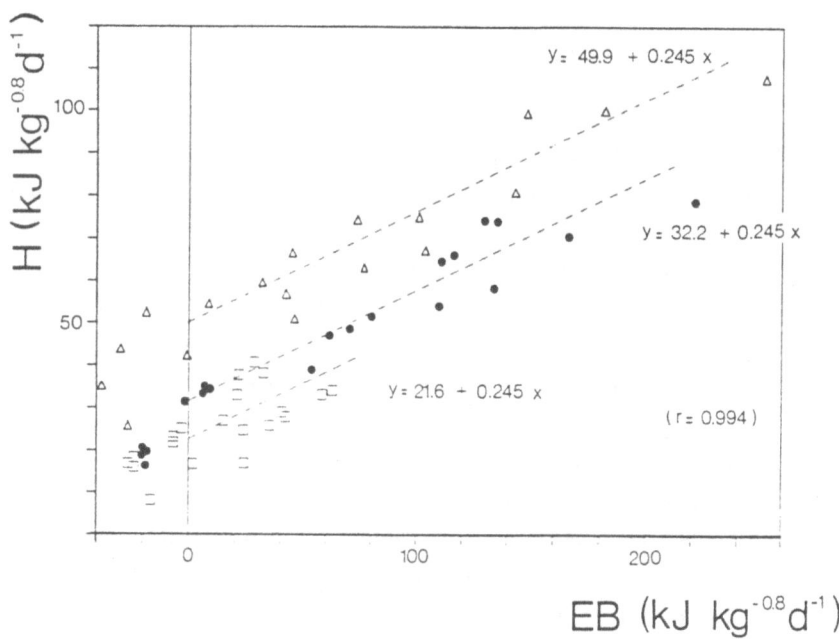

Figure 5.   Metabolic expenditure in relation to energy gain by the Afri-can catfish, <u>Clarias gariepinus</u>, at different temperatures.

Table 1. Allometric relations of feed intake, energy metabolism and growth with body weight at different temperatures[a].

Balance component ($kJ.fish^{-1}.d^{-1}$)	Temperature (°C) 20	25	27.5	30
H_0^b	$0.09\ W^{0.70}$	$0.08\ W^{0.79}$	$0.11\ W^{072}$	$0.16\ W^{0.78}$
H_0^c	$0.08\ W^{0.64}$	$0.07\ W^{0.83}$		$0.12\ W^{0.82}$
H_m	$0.10\ W^{0.73}$	$0.13\ W^{0.82}$		$0.16\ W^{0.85}$
H_{max}	$0.15\ W^{0.76}$	$0.325\ W^{0.77}$		$0.59\ W^{0.68}$
R_m	$0.19\ W^{0.77}$	$0.16\ W^{0.83}$	$0.21\ W^{0.82}$	$0.24\ W^{0.85}$
R_{max}	$0.70\ W^{0.72}$	$1.32\ W^{0.79}$	$1.82\ W^{0.74}$	$2.21\ W^{0.69}$
EB_{max}	$0.23\ W^{0.65}$	$0.56\ W^{0.77}$	$0.73\ W^{0.73}$	$0.90\ W^{0.65}$

[a] W is expressed in grams, Q_{ox} is taken as 13.6 kJ g^{-1}

[b] determined as -EB

[c] determined from oxygen consumption

tors influence growth.

Again based on the results of Hogendoorn (1983), Hogendoorn et al. (1983) and Machiels and Henken (1986), the effect of temperature on the various allometric relations of aspects of feed intake/metabolism with body weight is shown in Table 1 for the African catfish. In accordance with common beliefs the value of the weight coefficient (a) varies with both feeding level and temperature. As already seen in Figure 3 the weight coefficients for feed intake/metabolism increase with the feeding level. The weight coefficients for fasting metabolism and maintainance feed intake/metabolism increase with increasing temperature in a manner comparable to the classical curve of Ege and Krogh (1914, in Winberg, 1956) for the standard metabolism of the goldfish and in accordance with studies on coho salmon, Onchorynchus, (Corey et al., 1983) and rainbow trout, Salmo gairdneri (From and Rasmussen, 1984). The weight coefficients for maximum feed intake/metabolism initially also show an increase with temperature, but are believed to reach a maximum at about 30°C

494

followed by a decrease at higher temperatures (Hogendoorn et al., 1983; cf. Elliot, 1976b and 1982; Corey et al., 1983; From and Rasmussen, 1984).

The observed variation of the weight exponents (b) with temperature is, although again statistically not significant, believed to be biologically significant, at least for maximum feed intake/metabolism. Indirect evidence for this is given in Table 2 and Figure 6 from which can be seen that these differences in weight exponent account for the decrease in optimum temperature for growth with increasing fish size which has also been observed in other fish species (Ricker, 1979; Huisman et al., 1979; Hogendoorn et al., 1983). Assuming a common weight exponent for starvation/maintenance metabolism independent of temperature would not alter the above mentioned relationships whereas assuming a common weight exponent for maximum feed intake/metabolism would.

From these results it is concluded that feed intake, energy metabolism and growth in fish are subject to a complex interaction of body weight and temperature. The growth potential of fish (ME_p/ME_m) shows an optimum temperature which decreases with fish size. This ratio of ME_p/ME_m, which largely explains the highly efficient growth of fish (Hogendoorn, 1983), is also decreasing with fish size.

Table 2. The ratio of ME_p to ME_m in relation to temperature and fish size.

| | Temperature (°C) | | |
	20	25	30
ME_p/ME_m a	$7.5\ W^{0.03} - 5.1$	$13.1\ W^{-0.05} - 5.1$	$19.0\ W^{-017} - 5.1$
5 gram	2.8	7.0	9.4
200 gram	3.7	4.9	2.6

a these equations follow from $ME_p/ME_m = (H_{max}/H_m)/(1-k_g)-1/(1-k_g)$

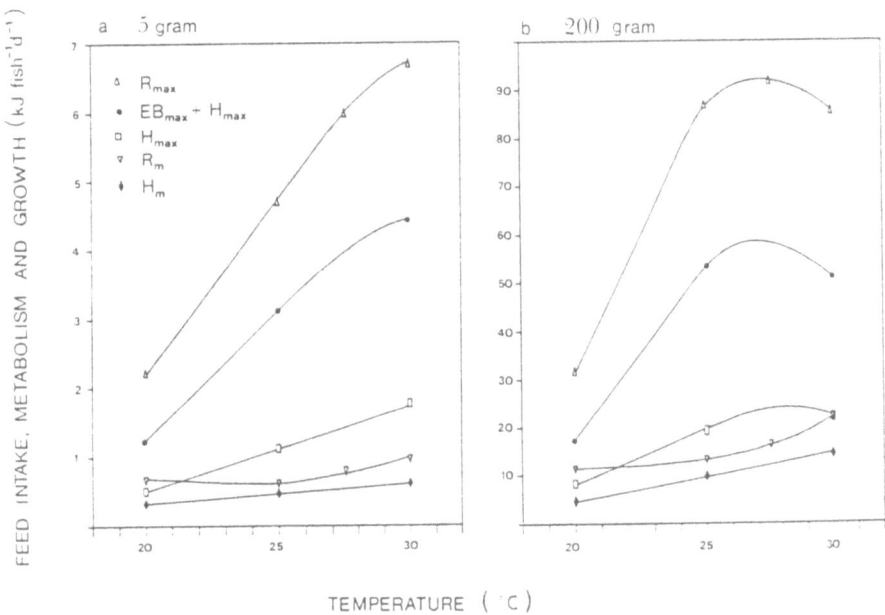

Figure 6. The effect of temperature on feed intake and metabolism of African catfish, Clarias gariepinus, of different body weight (a: 5 gram, b: 200 gram).

CONCLUSION

In conclusion it can be stated that much remains uncertain about the complex effects of temperature, ration and body weight upon energy metabolism and growth in fish. Developments in methodology in fish bioenergetics are proceeding fast and it is hoped that future research will throw some light on for instance the effects of these factors on the digestible and metabolizable fraction of the ration.

Although the observed differences in the weight exponents suggest some differential interplay of surface - and volume dependent processes with regard to feed intake, energy metabolism and growth, attempts at explaining these differences directly from ratio's of various surfaces (i.e. gills, intestine) to body weight seems to be a too simple and not very fruitfull approach. Future research in this area should be aimed at sub-stantiating the evidence for these phenomena and at elucidating the pos-sible roles of uptake, transport and metabolism of oxygen and nutrients and metabolism, transport and excretion of metabolites as limiting factors for growth. In doing this the effects of other environmental influences, notably oxygen and metabolites, should also be taken into account as should effects of feed and body composition as suggested by Machiels and Henken (1987) and Machiels and van Dam (1987).

REFERENCES

Austreng, E., 1978. Digestibility determination in fish using chromic oxide marking and analysis of contents from different segments of the gastro-intestinal tract. Aquaculture 13: 265-272.

Beamish, F.W.H., Niimi, A.J. and Lett, P.F.K.P., 1975. Bioenergetics of teleost fishes: Environmental influences. In: L. Bolis, H.P. Maddrell and K. Schmidt-Nielsen (Eds.), Comparative Physiology - Functional Aspects of Structural Materials: 187-209. North Holland Publishing Compagny, Amsterdam.

Braaten, B.R., 1979. Bioenergetics - a review on methodology. In: J.E. Halver and K. Tiews (Eds.), Proc. World Symp. om Finfish Nu-trition and Fishfeed Technology, vol.: 461-504. Heenemann, Berlin.

Brafield, A.E., 1985. Laboratory studies of energy budgets. In: P. Tytler and P. Calow (Eds.), Fish Energetics: New Perspectives: 257-281. Croom Helm, Beckenham (Kent).

Brett, J.R., 1979. Environmental factors and growth. In: W.S. Hoar, D.J. Randall and J.R. Brett (Eds.), Fish Physiology, Vol. VIII, Bio-energetics and Growth: 599-675. Academic Press, New York.

Cho, C.Y., Slinger, S.J. and Bayley, H.S., 1982. Bioenergetics of sal-monid fishes: energy intake, expenditure and productivity. Comp. Bio-chem. Physiol. 73B(1): 25-41.

Choubert, G., De La Noue, J. and Luquet, P., 1982. Digestibility in fish: improved device for the automatic collection of faeces. Aqua-culture 29: 185-189.

Corey, P.D., Leith, D.A. and English, M.J., 1983. A growth model for Coho salmon including effects of varying ration allotments and temperature. Aquaculture 30: 125-143.

Craig, J.F., Kenley, M.J. and Talling, J.F., 1978. Comparative estimations of the energy content of fish tissue from bomb calorimetry, wet oxidation and proximate analysis. Freshwater Biol. 8: 585-590.

Ekern, A. and Sundstøl, F. (Eds.), 1982. Proc. IXth Symp. on Energy Metabolism of Farm Animals. The Agricultural University of Norway, Aas.

Elliot, J.M., 1976a. Energy losses in the waste products of brown trout (Salmo trutta L.). J. Anim. Ecol. 45:561-580.

Elliot, J.M., 1976b. The energetics of feeding, metabolism and growth of brown trout (Salmo trutta L.) in relation to body weight, water temperature and ration size. J. Anim. Ecol. 45: 923-948.

Elliot, J.M., 1982. The effects of temperature and ration size on the growth and energetics of salmonids in captivity. Comp. Biochem. Physiol. 73B(1): 81-91.

From, J. and Rasmussen, G., 1984. A growth model, gastric evacuation and body composition in rainbow trout, Salmo gairdneri Richardson, 1836. Dana 3: 61-139.

Fry, F.E.J., 1971. The effect of environmental factors on the physiology of fish. In: W.S. Hoar and D.J. Randall (Eds.), Fish Physiology, Vol. VI, Environmental Relations and Behavior: 1-98. Academic Press, New York.

Garcia, L.M. and Adelman, I.R., 1985. An in situ estimate of daily food consumption and alimentary canal evacuation rates of common carp, Cyprinus carpio L. J. Fish Biol. 27: 487-493.

Goolish, E.M. and Adelman, I.R., 1984. Effects of ration size and temperature on the growth of juvenile common carp, (Cyprinus carpio L.). Aquaculture 36: 27-35.

Henken, A.M., Kleingeld, D.W. and Tijssen, P.A.T., 1985. The effect of feeding level on apparent digestibility of dietary dry matter, crude protein and gross energy in the African catfish Clarias gariepinus (Burchell, 1822). Aquaculture 51: 1-11.

Henken, A.M., Lucas, H., Tijssen, P.A.T. and Machiels, M.A.M., 1986. A comparison between methods used to determine the energy content of feed, fish and faeces samples. Aquaculture 58: 195-201.

Hogendoorn, H., 1983. Growth and production of the African catfish, Clarias lazera (C. & V.). III. Bioenergetic relations of body weight and

498

feeding level. Aquaculture 35: 1-17.

Hogendoorn, H., Korlaar, F. van and Bosch, H., 1981. An open circuit balance respirometer for bioenergetic studies of fish growth. Aquaculture 26: 183-187.

Hogendoorn, H., Janssen, J.A.J., Koops, W.J., Machiels, M.A.M., Ewijk, P.H. van and Hees, J.P. van, 1983. Growth and production of the African catfish, Clarias lazera (C. & V.). II. Effects of body weight, temperature and feeding level in intensive tank culture. Aquaculture 34: 265-285.

Huisman, E.A., 1974. Optimalisering van de groei bij de karper Cyprinus carpio L. Ph.D.-thesis. Wageningen: 95 pp.

Huisman, E.A., 1976. Food conversion efficiencies at maintenance and production levels for carp, Cyprinus carpio L., and rainbow trout, Salmo gairdneri R. Aquaculture 9: 259-273.

Huisman, E.A., Klein Breteler, J.G.P., Vismans, M.M. and Kanis, E., 1979. Retention of energy, protein, fat and ash in growing carp (Cyprinus carpio L.) under different feeding and temperature conditions. In: J.E. Halver and K. Tiews (Eds.), Proc. World Symp. on Finfish Nutrition and Fishfeed Technology, vol. I: 175-188. Heenemann, Berlin.

Ivlev, V.S., 1961. Experimental ecology of the feeding of fishes. Yale University Press, New Haven: 302 pp.

Jobling, M., 1983. A short review and critique of methodology used in fish growth and nutrition studies. J. Fish Biol. 23: 685-703.

Kitchell, J.F., Stewart, D.J. and Weininger, D., 1977. Applications of a bioenergetics model to yellow perch (Perca flavescens) and walleye (Stizostedion vitreum vitreum). J. Fish. Res. Board Can. 34: 1922-1935.

Machiels, M.A.M. and Dam, A.A. van, 1987. A dynamic simulation model for growth of the African catfish, Clarias gariepinus (Burchell, 1822). III. The effect of body composition on growth and feed intake. Aquaculture 60: 55-71.

Machiels, M.A.M. and Henken, A.M., 1986. A dynamic simulation model for growth of the African catfish, Clarias gariepinus (Burchell, 1822). I. The effect of feeding level on growth and energy metabolism. Aquaculture 56: 29-52.

Machiels, M.A.M. and Henken, A.M., 1987. A dynamic simulation model for growth of the African catfish, <u>Clarias gariepinus</u> (Burchell, 1822). II. Effect of feed composition on growth and energy metabolism. Aquaculture 60: 33-53.

Nijkamp, H.J., Es, A.J.H. van and Huisman, E.A., 1974. Retention of nitrogen, fat, ash, carbon and energy in growing chickens and carps. In: K.H. Menke, H.J. Lantzsch and J.R. Reichl (Eds.), Proc. VIth. Symp. Energy Metab. of Farm animals: 277-280. Univ. of Stuttgart-Hohenheim.

Paloheimo, J.E. and Dickie, L.M., 1965. Food and growth of fishes. I. A growth curve derived from experimental data. J. Fish. Res. Board Can. 22: 521-542.

Paloheimo, J.E. and Dickie, L.M., 1966. Food and growth of fishes. II. Effects of food and temperature on the relation between metabolism and body weight. J. Res. Board Can. 23: 869-908.

Ricker, W.E., 1979. Growth rates and models. In: W.S. Hoar, D.J. Randall and J.R. Brett (Eds.), Fish Physiology, Vol. VIII, Bioenergetics and Growth: 599-675. Academic Press, New York.

Staples, D.J. and Nomura, M., 1976. Influence of body size and food ration on the energy budget of rainbow trout <u>Salmo gairdneri</u> Richardson. J. Fish Biol. 9: 29-43.

Talbot, C., 1985. Laboratory methods in fish feeding and nutritional studies. In: P. Tytler and P. Calow (Eds.), Fish Energetics: New Perspectives: 125-154. Croom Helm, Beckenham (Kent).

Tytler, P. and Calow, P. (Eds.), 1985 Fish Energetics: New Perspectives. Croom Helm, Beckenham (Kent): 349 pp.

Ursin, E., 1967. A mathematical model of some aspects of fish growth, respiration and mortality. J. Fish. Res. Board Can. 24: 2355-2453.

Warren, C.E. (Ed.), 1971. Biology and water pollution control. W.B. Saunders Company, Philadelphia.

Webb, P.W., 1978. Partitioning of energy into metabolism and growth. In: S.D. Gerking (Ed.), Ecology of Freshwater Fish Production, Blackwell Sci. Publ., Oxford.

Winberg, G.G., 1956. Rate of metabolism and food requirements of fish. Res. Board Can. Transl. Ser. 194 (1960): 253 pp.

Windell, J.T., Foltz, J.W. and Sarokan, J.A., 1978a. Effect of fish size, temperature, and amount fed on nutrient digestibility of a pelleted diet by rainbow trout, <u>Salmo gairdneri</u>. Trans. Am. Fish. Soc. 107(4):

613-616.

Windell, J.T., Foltz, J.W. and Sarokan, J.A., 1978b. Methods of faecal collection and nutrient leaching in digestibility studies. Prog. Fish. Cult. 40: 51-55.

Publication of this book was made possible by support from, among others, the 'Stichting Fonds Landbouw Export Bureau 1916/1918' - LEB fonds - Wageningen. We gratefully acknowledge the centre for word processing of the Agricultural University for carefully typing the manuscript.